Rolf Isermann

Digital
Control Systems

Volume 1:
Fundamentals, Deterministic Control

Second, Revised Edition

With 88 Illustrations

Springer-Verlag Berlin Heidelberg GmbH

Professor Dr.-Ing. Rolf Isermann

Institut für Regelungstechnik
Technische Hochschule Darmstadt
Schloßgraben 1
D-6100 Darmstadt, West Germany

ISBN 978-3-642-86419-3

Library of Congress Cataloging-in-Publication Data
Isermann, Rolf. [Digitale Regelsysteme. English] Digital control systems/Rolf Isermann.
Rev. and enl. translation of: Digitale Regelsysteme. Includes bibliographical
references and index. Contents: v. 1. Fundamentals, deterministic control.
ISBN 978-3-642-86419-3   ISBN 978-3-642-86417-9 (eBook)
DOI 10.1007/978-3-642-86417-9
1. Digital control systems. I. Title.
TJ213.I64713   1989

© Springer-Verlag Berlin Heidelberg 1989
Originally published by Springer-Verlag Berlin Heidelberg New York in 1989
Softcover reprint of the hardcover 2nd edition 1989

Typesetting: Macmillan India Ltd., Bangalore 25.

2161/3020 543210 – Printed on acid-free paper.

Preface

The great advances made in large-scale integration of semiconductors and the resulting cost-effective digital processors and data storage devices determine the present development of automation.

The application of digital techniques to process automation started in about 1960, when the first *process computer* was installed. From about 1970 process computers with cathodic ray tube display have become standard equipment for larger automation systems. Until about 1980 the annual increase of process computers was about 20 to 30%. The cost of hardware has already then shown a tendency to decrease, whereas the relative cost of user software has tended to increase. Because of the high total cost the first phase of digital process automation is characterized by the centralization of many functions in a single (though sometimes in several) process computer. Application was mainly restricted to medium and large processes. Because of the far-reaching consequences of a breakdown in the central computer parallel standby computers or parallel back-up systems had to be provided. This meant a substantial increase in cost. The tendency to overload the capacity and software problems caused further difficulties.

In 1971 the first *microprocessors* were marketed which, together with large-scale integrated semiconductor memory units and input output modules, can be assembled into cost-effective microcomputers. These microcomputers differ from process computers in fewer but higher integrated modules and in the adaptability of their hardware and software to specialized, less comprehensive tasks. Originally, microprocessors had a shorter word length, slower operating speed and smaller operational software systems with fewer instructions. From the beginning, however, they could be used in a manifold way resulting in larger piecenumbers and leading to lower hardware costs, thus permitting the operation with small-scale processes.

By means of these process-microcomputers which exceed the efficiency of former process computers decentralized automatic systems can be applied. To do so, the tasks up to now been centrally processed in a process computer are delegated to various process microcomputers. Together with digital buses and possibly placed over computers many different hierarchically organized automatization structures can be build up. They can be adapted to the corresponding process. Doing so the high computer load of a central computer is avoided, as well as a comprehensive and complex user-software and a lower reliability. In addition decentralized systems can be easier commissioned, can be provided with mutual redundancy (lower susceptibility to malfunctions) and can lead to savings in wiring. The second phase of process automation is thus characterized by decentralization.

Besides their use as substations in decentralized automation systems process computers have found increasing application in *individual elements* of automation systems. Digital controllers and user-programmable sequence control systems, based on microprocessors, have been on the market since 1975.

Digital controllers can replace several analog controllers. They usually require an analog-digital converter at the input because of the wide use of analog sensors, transducers and signal transmission, and a digital-analog converter at the output to drive actuators designed for analog techniques. It is to be expected that, in the long run, digitalization will extend to sensors and actuators. This would not only save a-d and d-a converters, but would also circumvent certain noise problems, permit the use of sensors with digital output and the reprocession of signals in digital measuring transducers (for example choice of measurement range, correction of nonlinear characteristics, computation of characteristics not measurable in a direct way, automatic failure detection, etc.). Actuators with digital control will be developed as well. Digital controllers not only are able to replace one or several analog controllers they also succeed in performing additional functions, previously exercised by other devices or new functions. These additional functions are such as programmed sequence control of setpoints, automatic switching to various controlled and manipulated variables, feedforward adjusted controller parameters as functions of the operating point, additional monitoring of limit values, etc. Examples of new functions are: communication with other digital controllers, mutual redundancy, automatic failure detection and failure diagnosis, various additional functions, the possibility of choice between different control algorithms and, in particular, selftuning or adaptive control algorithms. Entire control systems such as cascade-control systems, multivariable control systems with coupling controllers, control systems with feedforward control which can be easily changed by configuration of the software at commissioning time or later, can be realized by use of one digital controller. Finally, very large ranges of the controller parameters and the sample time can be realized. It is because of these many advantages that, presently various digital devices of process automation are being developed, either completing or replacing the process analog control technique.

As compared to analog control systems, here are some of the characteristics of digital control systems using process computers or process microcomputers:

— Feedforward and feedback control are realized in the form of software.
— Discrete-time signals are generated.
— The signals are quantized in amplitude through the finite word length in a-d converters, the central processor unit, and d-a converters.
— The computer can automatically perform the analysis of the process and the synthesis of the control.

Because of the great flexibility of control algorithms stored in software, one is not limited, as with analog control systems, to standardized modules with P-, I- and D-behaviour, but one can further use more sophisticated algorithms based on mathematical process models. Many further functions can be added. It is especially significant that on-line digital process computers permit the use of process identification-, controller design-, and simulation methods, thus providing the engineer with new tools.

Since 1958 several books have been published dealing with the theoretical treatment and synthesis of linear sampled-data control, based on difference equations, vector difference equations and the z-transform. Up to 1977, when the first German edition of this book appeared, books were not available in which the various methods of designing sampled-data control have been surveyed, compared and presented so that they can be used immediately to design control algorithms for various types of processes. Among other things one must consider the form and accuracy of mathematical process models obtainable in practice, the computational effort in the design and the properties of the resulting control algorithms, such as the relationship between control performance and the manipulation effort, the behaviour for various processes and various disturbance signals, and the sensitivity to changes in process behaviour. Finally, the changes being effected in control behaviour through sampling and amplitude quantization as compared with analog control had also be studied.

Apart from deterministic control systems the first edition of this book dealt also with stochastic control, multivariable control and the first results of digital adaptive control. In 1983 this book was translated into Chinese. In 1981 the enlarged English version entitled "Digital Control Systems" was edited, followed by the Russian translation in 1984, and, again a Chinese translation in 1986. In 1987 the 2nd edition appeared in German, now existing in two volumes. This book is now the 2nd English edition.

As expected, the field of digital control has been further developed. While new results have been worked out in research projects, the increased application provided a richer experience, thus allowing a more profound evaluation of the various possibilities. Further stimulation of how to didactically treat the material has been provided by several years of teaching experience and delivering courses in industry. This makes the *second edition* a complete revision of the first book, containing many supplements, especially in chapters 1, 3, 5, 6, 10, 20, 21, 23, 26, 30, 31. Since, compared with the first edition, the size of the material has been significantly increased, it was necessary to divide the book in two volumes.

Both volumes are directed to students and engineers in industry desiring to be introduced to theory and application of digital control systems. Required is only a basic familiarity of continuous-time (analog) control theory and control technique characterized by keywords such as: differential equation, Laplace-Transform, transfer function, frequency response, poles, zeroes, stability criterions, and basic matrix calculations. The *first volume* deals with the theoretical basics of linear sampled-data control and with deterministic control. Compared with the first edition the introduction to the basics of sampled-data control (part A) has been considerably extended. Offering various examples and exercises the introduction concentrates on the basic relationships required by the up-coming chapters and necessary for the engineer. This is realized by using the input/output-, as well as the state-design. Part B considers control algorithms designed for deterministic noise signals. Parameter-optimized algorithms, especially with PID-behaviour are investigated in detail being still the ones most frequently used in industry. The sequel presents general linear controllers (higher order), cancellation controllers, and deadbeat controllers characteristic for sampled-data control. Also state controllers including observers due to different design principles and the required supplements are considered. Finally, various control methods for deadbeat processes, insensitive and robust controllers are described and different control algorithms are compared by simulation methods. Part C of the *second volume* is dedicated to the control design for stochastic noise signals such as minimum variance controllers. The design of interconnected control systems (cascade control, feedforward control) are described in Part D while part E treats different multivariable control systems including multivariable state estimation. Digital adaptive control systems which have made remarkable progress during the last ten years are thoroughly investigated in Part F. Following a general survey, on-line identification methods, including closed loop and various parameter-adaptive control methods are presented. Part G considers more practical aspects, such as the influence of amplitude quantization, analog and digital noise filtering and actuator control. Finally the computer-aided design of control with special program systems is described, including various applications and examples of adaptive and selftuning control methods for different technical processes. The last chapters show, that the control systems and corresponding design methods, in combination with process modeling methods described in the two volumes were compiled in program systems. Most of them were tried out on our own pilot processes and in industry. Further specification of the contents is given in chapter 1.

A course "Digital Control Systems" treats the following chapters: 1, 2, 3.1–3.5, 3.7, 4, 5, 6, 7, 3.6, 8, 9, 11. The weekly three hours lecture and one hour exercises is given at the Technische Hochschule Darmstadt for students starting the sixth semester. For a more rapid apprehension of the essentials for applications the following succession is recommended: 2, 3.1 to 3.5 (perhaps excluding 3.2.4, 3.5.3, 3.5.4) 4, 5.1, 5.2.1, 5.6, 5.7, 6.2, 7.1, 11.2, 11.3 with the corresponding exercises.

Many of the described methods, development and results have been worked out in a research project funded by the Bundesminister für Forschung und Technologie (DV 5.505) within the project "Prozeßlenkung mit DV-Anlagen (PDV)" from 1973–1981 and in research projects funded by the Deutsche Forschungsgemeinschaft in the Federal Republic of Germany. The author is very grateful for this support.

His thanks also go to his coworkers,—who had a significant share in the generation of the results through several years of joint effort—for developing methods, calculating examples, assembling program packages, performing simulations on digital and on-line computers, doing practical trials with various processes and, finally, for proofreading.

The book was translated by my wife, Helge Isermann. We thank Dr. Ron Patton, University of York, U.K., for screening the translation.

Darmstadt, June 1989   Rolf Isermann

Contents

Graphic Outline of Contents (Volume I)

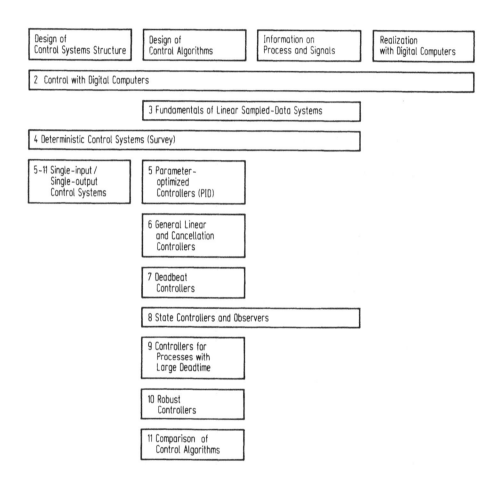

Design of Control Systems Structure

Design of Control Algorithms

Information on Process and Signals

Realization with Digital Computers

2 Control with Digital Computers

3 Fundamentals of Linear Sampled-Data Systems

4 Deterministic Control Systems (Survey)

5~11 Single-input / Single-output Control Systems

5 Parameter-optimized Controllers (PID)

6 General Linear and Cancellation Controllers

7 Deadbeat Controllers

8 State Controllers and Observers

9 Controllers for Processes with Large Deadtime

10 Robust Controllers

11 Comparison of Control Algorithms

Summary of Contents Volume II

Graphic Outline of Contents (Volume II)

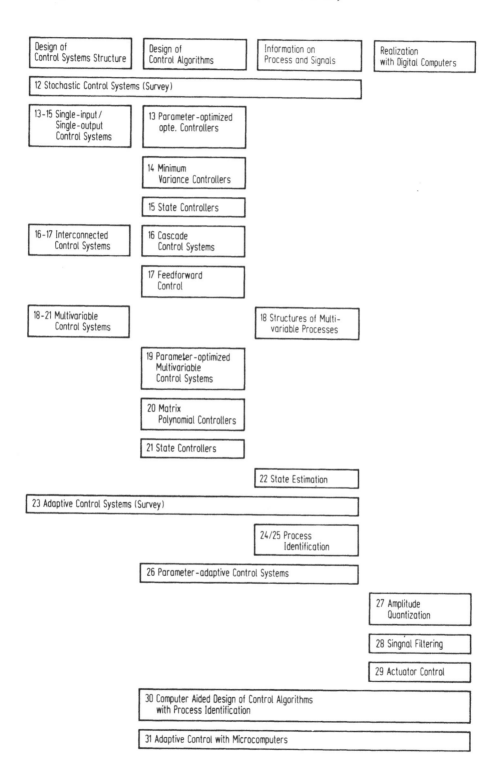

| Design of Control Systems Structure | Design of Control Algorithms | Information on Process and Signals | Realization with Digital Computers |

12 Stochastic Control Systems (Survey)

13–15 Single-input / Single-output Control Systems — 13 Parameter-optimized opte. Controllers

14 Minimum Variance Controllers

15 State Controllers

16–17 Interconnected Control Systems — 16 Cascade Control Systems

17 Feedforward Control

18–21 Multivariable Control Systems — 18 Structures of Multi-variable Processes

19 Parameter-optimized Multivariable Control Systems

20 Matrix Polynomial Controllers

21 State Controllers

22 State Estimation

23 Adaptive Control Systems (Survey)

24/25 Process Identification

26 Parameter-adaptive Control Systems

27 Amplitude Quantization

28 Singnal Filtering

29 Actuator Control

30 Computer Aided Design of Control Algorithms with Process Identification

31 Adaptive Control with Microcomputers

List of Abbreviations and Symbols

This list defines commonly occurring abbreviations and symbols:

$\left.\begin{array}{c} a \\ b \end{array}\right\}$ parameters of the difference equations of the *process*

$\left.\begin{array}{c} c \\ d \end{array}\right\}$ parameters of the difference equations of *stochastic signals*

d deadtime $d = T_t/T_0 = 1, 2, \ldots$

e control deviation $e = w - y$ (also $e_w = w - y$); or equation error for parameter estimation; or the number $e = 2.71828 \ldots$

f frequency, $f = 1/T_p$ (T_p period), or parameter

g impulse response (weighting function)

h parameter

i integer; or index; or $i^2 = -1$

k discrete time unit $k = t/T_0 = 0, 1, 2, \ldots$

l integer; or parameter

m order of the polynomials $A(\ )$, $B(\ )$, $C(\ )$, $D(\ )$

n disturbance signal (noise)

p parameters of the difference equation of the controller, or integer

$p(\ )$ probability density

q parameters of the difference equation of the controller

r weighting factor of the manipulated variable; or integer

s variable of the Laplace transform $s = \delta + i\omega$; or signal

t continuous time

u input signal of the process, manipulated variable $u(k) = U(k) - U_{00}$

v nonmeasurable, virtual disturbance signal

w reference value, command variable, setpoint $w(k) = W(k) - W_{00}$

x state variable

y output signal of the process, controlled variable $y(k) = Y(k) - Y_{00}$

z variable of the z-transformation $z = e^{T_0 s}$

$\left.\begin{array}{c} a \\ b \end{array}\right\}$ parameters of the differential equations of the *process*

$A(s)$ denominator polynomial of $G(s)$

$B(s)$ numerator polynomial of $G(s)$

$A(z)$	denominator polynomial of the z-transfer function of the process model
$B(z)$	numerator polynomial of the z-transfer function of the process model
$C(z)$	denominator polynomial of the z-transfer function of the noise model
$D(z)$	numerator polynomial of the z-transfer function of the noise model
$G(s)$	transfer function for continuous-time signals
$G(z)$	z-transfer function
$H(\ )$	transfer function of a holding element
I	control performance criterion
K	gain
L	word length
M	integer
N	integer or discrete measuring time
$P(z)$	denominator polynomial of the z-transfer function of the controller
$Q(z)$	numerator polynomial of the z-transfer function of the controller
$R(\ )$	dynamical control factor
S	power density or sum criterion
T	time constant
T_{95}	settling time of a step response until 95% of final value
T_0	sample time
T_t	dead time
U	process input (absolute value)
V	loss function
W	reference variable (absolute value)
Y	process output variable (absolute value)
b	control vector
c	output vector
k	parameter vector of the state controller
n	noise vector $(r \times 1)$
u	input vector $(p \times 1)$
v	noise vector $(p \times 1)$
w	reference variable vector $(r \times 1)$
x	state variable vector $(m \times 1)$
y	output vector $(r \times 1)$
A	system matrix $(m \times m)$
B	input matrix $(m \times p)$
C	output matrix, observation matrix $(r \times m)$
D	input-output matrix $(r \times p)$, or diagonal matrix
F	noise matrix or $F = A - BK$
G	matrix of transfer functions
I	unity matrix
K	parameter matrix of the state controller
Q	weighting matrix of the state variables $(m \times m)$
R	weighting matrix of the inputs $(p \times p)$; or controller matrix
$\mathscr{A}(z)$	denominator polynomial of the z-transfer function, closed loop
$\mathscr{B}(z)$	numerator polynomial of the z-transfer function, closed loop
$\mathfrak{F}$	Fourier-transform

$\mathfrak{I}$	information
$\mathfrak{L}(\ )$	Laplace-transform
$\mathfrak{Z}(\ )$	z-transform
$\mathscr{L}(\ )$	correspondence $G(s){\rightarrow}G(z)$
α	coefficient
β	coefficient
γ	coefficient; or state variable of the reference variable model
δ	deviation, or error
ε	coefficient
ζ	state variable of the noise model
η	state variable of the noise model; or noise/signal ratio
κ	coupling factor; or stochastic control factor
λ	standard deviation of the noise $v(k)$
μ	order of $P(z)$
ν	order of $Q(z)$; or state variable of the reference variable model
π	3.14159 . . .
σ	standard deviation, σ^2 variance, or related Laplace variable
τ	time shift
ω	angular frequency $\omega = 2\pi/T_p$ (T_p period)
Δ	deviation; or change; or quantization unit
Θ	parameter
Π	product
Σ	sum
Ω	related angular frequency
$\dot{x}$	$= dx/dt$
x_0	exact quantity
$\hat{x}$	estimated or observed variable
$\tilde{x}, \Delta x$	$= \hat{x} - x_0$ estimation error
$\bar{x}$	average
X_{00}	value in steady state

Mathematical abbreviations

$\exp(x)$	$= e^x$
$E\{\ \}$	expectation of a stochastic variable
$var\,[\ ]$	variance
$cov\,[\ ]$	covariance
dim	dimension, number of elements
tr	trace of a matrix: sum of diagonal elements
adj	adjoint
det	determinant

Indices

P	process
Pu	process with input u
Pv	process with input v

R or C	feedback controller, feedback control algorithm, regulator
S or C	feedforward controller, feedforward control algorithm
o	exact value
oo	steady state, d.c.-value

Abbreviations for controllers or control algorithms (C)

$i-PC-j$	parameter optimized controller with i parameters and j parameters to be optimized
DB	Deadbeat-controller
$LC-PA$	linear controller with pole assignment
PREC	predictor controller
MV	minimum variance controller
SC	state controller (usually with an observer)

Abbreviations for parameter estimation methods

COR-LS	correlation analysis with LS parameter estimation
IV	instrumental variables
LS	least squares
ML	maximum-likelihood
STA	stochastic approximation

The letter R means recursive algorithm, i.e. RIV, RLS, RML.

Other abbreviations

ADC	analog-digital converter
CPU	central processing unit
DAC	digital-analog converter
PRBS	pseudo-random binary signal

Remarks

The vectors and matrices in the Figures are roman types and underlined. Hence it corresponds e.g. $x \rightarrow \underline{x}$; $K \rightarrow \underline{K}$.
The symbol for the dimension of time t in seconds is usually s and sometimes sec in order to avoid misinterpretation as the Laplace variable s.

1 Introduction

Digital control systems are often associated with other tasks of process automation. A brief survey will therefore be presented dealing with the various tasks and state of the art of process automation.

Process Automation

During the last three decades automation techniques have developed an increasing influence on both, operation and design of technical processes. This has been brought about by the increasing requirements of the processes themselves, the possibilities offered especially by the electronic equipment together with a deeper theoretical understanding of automatic control systems.

According to the transport of materials, energy or information, technical processes can be divided as follows:

Continuous Processes: information flow in continuous streams
Batch Processes: flow in interrupted streams (packets)
Piece-good Processes: transport in "pieces"

Fig. 1.1a) shows a few examples. Various automation techniques have been developed for these classes of processes. The type of automation system is, however, also dependent on the size of the processes, indicated by the number of variables (sensors, actuators) or by the dimension of their units. This is described in Fig. 1.1b):

Large Processes
Medium Processes
Small Processes

During recent years the proportional cost for the automation of all processes has increased steadily. This is clear according to the cost statistics for large processes in process engineering where cost for automation augmented from 8% in 1963 to 15% in 1980. Between 1965 and 1980 the number of sensors installed in power plants increased from about 400 to 4000, the number of actuators from about 500 to 2000. But also for small processes the relative cost for automation has considerably risen (e.g. heating systems, machine tools, electrical drives, vehicles).

Fig. 1.2 shows the various *tasks of process automation*. In *feedforward control* input signals influence according to certain laws other variables as output signals. These output signals of feedforward control systems then are input signals of the process. They are influenced by feedforward control in such a way that a certain

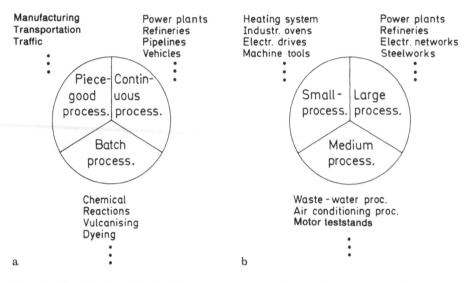

Fig. 1.1. Classification of technical processes. **a** flow of materials and energy; **b** size

causal process operation results. Feedforward control systems are characterized by an open-loop signal flow. One distinguishes between logic control for which input and output signals, according to Boolean Algebra, storages and time functions are related to each other and sequence control for which the switching

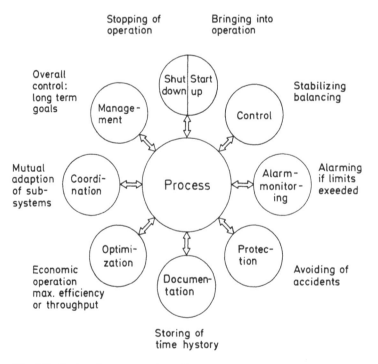

Fig. 1.2. Tasks of process automation

from one step to the following is performed dependent on the switching condition. Feedforward control systems are also used for automatic *start-up* and *shut-down* of processes. *Feedback-control systems* are meant to adjust certain variables (control variables) to given reference variables. They are characterized by a closed-loop structure resulting from the feedback. The reference variables are either constant or dependent on other variables (e.g. servo systems). If required, feedback control also ensures the stabilization of the process. Feedback control is e.g. used to maintain the energy-, mass- or momentum balance of processes independent of disturbances and in addition to keep various state variables at certain prescribed values. *Supervision* serves to indicate undesired or unpermitted process states (*monitoring*) and to take appropriate actions. In consequence one has to find out whether chosen state variables exceed or violate certain limits (limit value checking). This generally results in the triggering of an alarm signal. If this limit value violation signifies a dangerous state an appropriate action will be taken automatically. This is then called automatic *protection* of the process.

The storing of time history of process signals is performed by *documentation* (recorder, printer). Frequently this is combined with data processing and data reduction. The goal of *optimization* is the optimal economical control of the process. In this way efficiency and system throughput are being maximized by changing the input signals of feedforward and feedback control. This results in minimized operating costs. If several processes are to operate together in a compound system they have to be mutually adapted by *coordination*. The process *management* is responsible for the long-term adaptation of a process system (work, compound network) to planning, marketing, the raw material and to the personnel.

For small processes the minimum extend of automation consists in feedforward control, feedback control and supervision. For medium sized plants at least documentation has to be added. For modern large processes (e.g. power plants) almost all tasks are automatically performed excluding the process management.

The tasks of process automation are usually distributed to different levels. Fig. 1.3 gives an example. At the low level tasks are performed which act locally and need a fast action while the high levels are dedicated to tasks which act globally generally not requiring a fast response. They can also contain decisions.

Principles of feedforward and feedback control are used at all levels. In an analogous way to control loops one may also refer to feedback control as supervisory, optimization and coordination loops, or more generally just as *multilevel control*.

Conventional Control Systems

Conventional control systems are characterized by processing analog or binary signals with hardwired devices. Each task is performed by special equipment, such as a controller, alarm monitor, protector or recorder. Consequently, a totally decentralized system structure results, Fig. 1.4a). These systems call for a high planning effort, large cabling costs and are rather inflexible after installation. They only permit realization of the most important tasks at the lower automation levels. However, these conventional control systems are very reliable and easy to understand and operate.

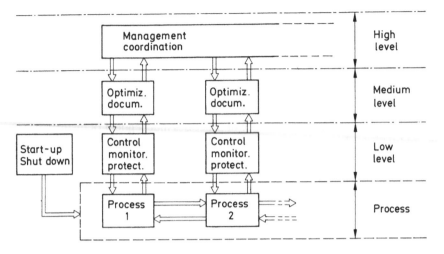

Fig. 1.3. Process automation in different levels

Process Computers (Minicomputers)

The appearance of digital process computers has essentially influenced process automation both structure and function. The following steps in developments could be observed. In 1959, for the first time process computers were used on-line, but mainly in an open-loop configuration for data-logging, data-reduction and process monitoring. The direct control of process variables was performed by analog equipment, principally because of the unsatisfactory reliability of digital process computers at that time. The reference values (set points) of analog controllers were given by the process computers (supervisory control), for example according to time schedules or for process optimization. Process computers have been used for direct digital control (DDC) in an on-line closed-loop configuration since 1962 for chemical and power plants [1.1.], [1.2], [1.3], [1.4], [1.5], [1.6].

As a result of the development of more powerful process computers and relevant software, the application of computers for process control has increased considerably. From 1965 on, computers have been standard components of process automation [1.5], [1.6]. For further details the reader is referred to the books [1.7] to [1.14]. Process computers have been used mainly for monitoring and coordination but also for data-logging and data-reduction [1.7] to [1.11]. Actual on-line optimization has been rarely applied, however.

The high cost for process computers caused a centralized system structure, Fig. 1.4b). From 1973 to 1978 the numbers of minicomputers increased from about 35,000 to 210,000, representing an annual increase of 35%. Almost a third to a half of this number has been used for process automation [1.16]. Because of the rather high cost of process computers the application was mainly restricted to large-scale and medium-scale processes. There was a tendency to overload the capacity. Software problems (reliability, maintenance) and the need of back-up computers then slowed down further expansion.

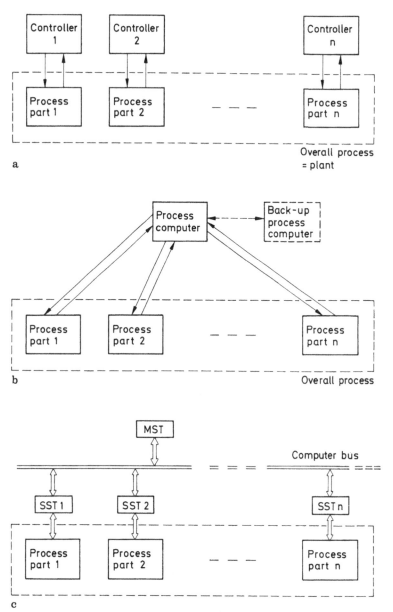

Fig. 1.4. Structures of process automation systems. **a** total decentralized system structure: conventional; **b** central system structure: process computers; **c** decentralized/centralized system structure: distributed microcomputer system (SST: substation; MST: main station)

Microcomputer Control Systems

The advent of the microprocessor in 1971 began to change the development into another direction. The functional construction of microcomputers is similar to minicomputers. The microprocessors and the memory units however, are integrated on single chips allowing cheaper mass production. Because of serial and parallel buses a range of external equipment can be connected by hardware and software. Hence, many different system structures are possible. For example, various sub-stations with a master station can be realized, Fig. 1.4c), thus forming a hierarchical system structure. A sub-station generally performs several tasks (e.g. 2–8 controllers and monitoring) thus leading to both, a *decentralized* and a *centralized structure* which on the lowest level is locally centralized yet globally decentralized, provided the microcomputers are distributed over the whole process. This is therefore referred to as a *distributed digital control system*. Some characteristics of these microcomputer systems are:

1. Sub-stations for basic functions as e.g. data processing, control, monitoring.
2. CRT (cathode ray tube) control panels.
3. Communication with data buses.
4. Master stations for higher level functions.

A great advantage is that the structure may be adapted to the different tasks of process automation and to the process structure. This makes the system more transparent, more reliable, easier to install, to program, to operate and to train. By using serial buses considerable cabling costs can be saved. For medium-scale and large-scale processes the development of *stand-alone microcomputers* is of interest. Having been on the market since about 1975, they indicate a continuous development. Some features are front panel push buttons with software-assignable functions, programmable setpoint and controller parameters as well as digitally driven displays and self-diagnosis. The pre-programed algorithms make them quite flexible. Self-tuning controller parameters tend to be the next step in development. Appropriate devices have been on the market since 1981.

Development of Digital Computers

Some characteristics are given in Table 1.1 showing the development of computer technology based on [1.17], [1.18], [1.19], [1.20]. Almost all of them show an exponential increase. The number of components per chip for dynamic memories increased by a factor of 20 to 30 per decade (doubling each year). In 1982 the 64-k-RAM was produced with about 50 million pieces. In 1984 the 256-k bit memories have been introduced, in 1986 the 1 M bit RAM on one chip and 4 M bit in 1987. Because of their more complex circuits microprocessors have a lower number of components per chip. This development, however, goes in parallel with the one of the memory systems. The cost to performance ratio of computer hardware expressed by million instructions per second (MIPS) demonstrates the many different technological developments. Within one decade a cost reduction of a factor 40 was achieved. Other characteristics given in Table 1.1 indicate similar important developments. Reliability is of special interest. In 1985 the MTBF (meantime between failures) of a 1 MIPS computer is about 10,000 hours (1.14 years).

The relative cost-to-performance ratio for complete systems, however, show a rather different trend. In the beginning hardware cost dominated. Between 1965

Table 1.1. Development of computers

	Compo-nents per chip	Computer hardware			Memories		Computer software cost
		Cost $/MIPS	Size FT^3/MIPS	Reliab. MTBF/h	Cost $/MByte	Size FT^3/MByte	%
1965	50	3000 000	3 000	40	2000 000	100	40
1975	1000	80 000	30	800	100 000	0.3	80
1985	30 000	2 000	0.2	10 000	1 000	0.008	90
Factor per decade							
65/75	20	37.5	100	20	20	33	0.5
75/85	30	40	150	12.5	100	37.5	0.9

and 1975 the cost for software increased from 40% to 80% indicating a complete change in the trade-off between software and hardware. This shows the great significance of software production.

On observing this rapid and far-reaching development which started with the microelectronic silicon circuits in the early sixties and resulted in powerful microprocessors and large memories in the beginning of the eighties, one may well refer to this as a technical revolution. Since about 28% of the IC (integrated circuit) market in 1980 was used for process automation [1.17] this will influence control and process engineering strongly. Automation systems therefore, will operate increasingly on a digital basis in the future. [1.20], [1.21].

Digital Control Systems

Signal processing with digital computers is not, unlike conventional analog control or feedforward control with binary elements, limited to some few basic functions. They are programable and can perform complex calculations. Therefore many new methods can be developed for digital process control, which for the low levels can be realized as programed *algorithms* and for the higher levels as programed *problem solving methods*. Since actions in all levels are performed by generalized feedforward and feedback control, multilevel control algorithms have to be designed, selected and adjusted to the processes. This book considers digital control at the lowest level of process automation, putting special emphasis on the application to continuous and batch processes. However, many of the methods for designing algorithms, for obtaining process models, for estimation of states and parameters, for noise filtering and actuator control can also be applied to the synthesis of digital monitoring, optimization and coordination systems. They can be also used for piece good processes.

The Contents

This book deals with the design of digital control systems with reference to process computers and microcomputers. Suitable design methods are described which are based on the basic theory of linear sampled-data control for deterministic and stochastic signals. Parameter-optimized (PID-), cancellation- and deadbeat control algorithms which derived from classical control theory as well as state- and minimum-variance control algorithms resulting from modern control theory through state variable representation and parametric stochastic process/signal models are considered. In order to investigate the behavior of the various feedforward and feedback control algorithms, almost all of the treated algorithms (and resulting control loops) were simulated on digital computers. They were designed by means of computer-aided techniques on process computers and microcomputers as well as tested on-line with simulated analogue processes, pilot processes and industrial processes and compared. The relation to the methods for process modelling, especially by identification methods are of great importance for all design methods and control algorithms.

A scheme to organize the contents is presented following the table of contents. The various chapters are concerned with the following aspects:

— design of control system structure
— design of control algorithms
— information on processes and signals
— realization with digital computers.

Some remarks on the contents and the choice of material are made in the sequel which also includes information on the newly added material of the second edition. The process and signal models used in this book are mainly *parametric* in the form of difference equations or vector difference equations, since the modern synthesis procedures are based on these models, the processes can be described compactly by a few parameters, and methods of time-domain synthesis are obtained with a small amount of calculations and provide structure optimal controllers. These models are the direct result of parameter estimation methods and can be used directly for state observers or state estimators. Non-parametric models such as transient functions or frequency responses in tabular form do not always offer these advantages. They restrict the possibilities for synthesis, particularly with regard to computer-aided design and to adaptive control algorithms. Yet, there are certain cases, in which also non-parametric models can be suitable, e.g. for analyzing the dynamic behaviour and for illustrating the results and for the design of parameter-optimized controllers.

Chapter 2 treats the general signal flow for *digital control systems* and explains briefly the steps taken for the design of digital control systems. A detailed introduction to the theoretical fundamentals of *linear sampled-data systems* follows in chapter 3, based on difference equations, z-transform and state-variable representation. (At this point the first edition presented only a summary of basic relations). A survey of control algorithms designed for deterministic and stochastic disturbances is provided in chapter 4.

Using conventional analogue controllers as a basis, chapter 5 discusses the derivation and design of *parameter-optimized control algorithms* with, for instance, P-, PI- or PID-behaviour as well as, separate from continuous signals, general discrete-time controllers of low order. Tuning rules for the controller parameters

Part A
Fundamentals

2 Control with Digital Computers
(Process Computers, Microcomputers)

Sampled-data Control Systems

In data processing with process computers signals are sampled and digitized, resulting in *discrete (discontinuous) signals*, which are quantized in *amplitude* and *time*, as shown in Fig. 2.1.

Unlike continuous signals these signals have discrete amplitude values at discrete time points. *Amplitude modulated* pulse series emerge, for which the pulse heights are rounded up or down, according to the quantization device.

The sampling is usually performed periodically with sampling time T_0 by a multiplexer which is constructed together with a measuring range selector and an analogue/digital converter. The digitized input data are sent to the central processor unit where the output data are calculated using programmed algorithms. If an analogue signal is required for the actuator, the output data emerge through a digital/analogue converter followed by a hold device. Fig. 2.2 shows a simplified block diagram.

The samplers of the input and output signal do not actually operate in synchronism but are displaced by an interval T_R. This interval results from the A/D-conversion and the data processing within the central processing unit. If this interval can be considered small in comparison with the time constants of the actuators, processes and sensors, it can often be neglected. Then synchronous

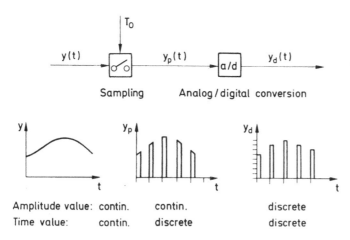

Fig. 2.1. An amplitude modulated, discrete-time and discrete-amplitude signal generated by sampling and analog/digital-conversion

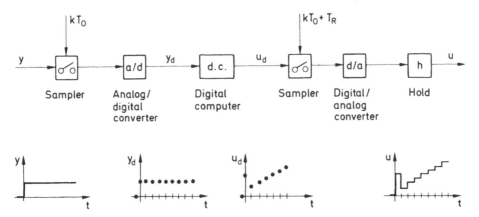

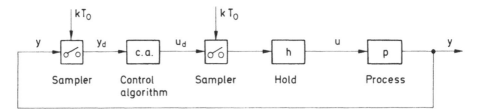

Fig. 2.2. The process computer as sampled-data controller. $k = 0, 1, 2, 3 \ldots$

Fig. 2.3. Control loop with a computer as a sampled-data controller

sampling at process computer inputs and outputs can be assumed. Also the quantization of the signal is small for computers with wordlengths of 16 bits and more and A/D-converters with at least 10 bits so that the signal amplitudes initially can be considered as continuous.

These simplifications lead to the block diagram in Fig. 2.3, which shows a control loop with process computer as a sampled-data controller. The samplers now operate synchronously and generate discrete-time signals. The manipulated variable u is calculated by a control algorithm using the control variable y and the reference value w as inputs. Such sampled-data control loops do not only exist in connection with process computers. Sampled-data also occur when:

— measured variables are only present at definite instants (e.g. rotating radar antenna, radar distance measurement, material sampling followed by later laboratory analysis, socio-economical, biological and meteorological data)
— multiplexing of expensive equipment (cables, channels)
— low-cost energy control switches

Design of Digital Control Systems

For electrical, pneumatic or hydraulic analogue controllers design is restricted mostly to single-purpose elements with proportional (P-), integral (I-) and differential (D-) behaviour for technical and economical reasons. The design possibilities of analogue controllers have thus been highly restricted. However, these restrictions are not valid for control algorithms in process computers. Because of high programming flexibility, much latitude can be used in the realization of sophisticated control algorithms. This enables the practical application of modern control theory, but also reinforces the question as to which control algorithms are best suited for a particular application. An answer to this question is possible only if enough is known about the mathematical models of the processes and their signals, and if it is known how the various control algorithms compare with each other with respect to control performance, manipulation effort, sensitivity, disturbance signals, design effort, computing effort and the process behaviour (linear, non-linear, position of poles and zeros, dead-times, structure of multivariable processes).

An extensive body of literature exists on the theory of sampled-data systems, and in many books on control engineering some chapters are dedicated to sampled-data control. Special books on the theory of sampled-data control systems were first based on difference equations [2.1], [2.2]. They were followed by books which also used the z-transformation [2.3] to [2.13], [2.15], [2.20]. Finally, the state representation was introduced [2.14], [2.17], [2.18], [2.19], [2.21]. Besides the first edition of this book [2.22], the revised English version [2.23] and translations into various languages [2.24], [2.25], other books appeared in the meantime treating digital control in detail: [2.25], [2.26], [2.27], [2.28], [2.29]. In the more application-oriented books [1.7]–[1.11] on process computer control, only those control algorithms which rely upon discretized analog PID-controllers are treated.

For the design of digital control systems as described in this book the following stages are considered (compare also the survey scheme, on page xiii).

1. *Information on the processes and the signals*

    The basis for the systemic design of control systems is the available information on the processes and their signals, which can be provided for example by:

    — direct measurable inputs, outputs, state variables,
    — process models and signal models,
    — state estimates of processes and signals.

    Process and signal models can be obtained by identification and parameter estimation methods, and in the case of process models, by theoretical modelling as well. Non-measurable state variables can be reconstructed by observers or state estimators.

2. *Control system structure*

    Depending on the process, and after suitable selection of appropriate manipulation variables and control variables, the control system structure has to be designed in the form of, for example:

    — single input/single output control systems
    — interconnected control systems
    — multi input/multi output control systems
    — decentralized control systems.

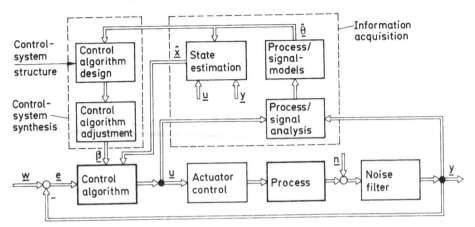

Fig. 2.4. Scheme for the design of digital control systems

3. *Feedforward and feedback control algorithms* (design and adjustment)
   Finally the feedforward and feedback control algorithms are to be designed and adjusted (or tuned) to the process. This can be done using:

   — simple tuning rules for the parameters
   — computer-aided design
   — self-optimizing adaptive control algorithms.

   Since several control algorithms with different properties are usually available, they have to be compared and selected according to various requirements.

4. *Noise filtering*
   High-frequency noise which contaminates the controlled variables and which cannot be controlled, has to be filtered by analogue and digital filters

5. *Feedforward or feedback control of actuators*
   Depending on the construction of the actuator, various feedforward or feedback controls of the actuator are possible. The control algorithms for the process have to be adjusted to the actuator control.

Finally, for all control and filter algorithms the effects of amplitude quantization have to be considered.

In Fig. 2.4 a scheme for the design of digital control systems is given. If *tuning rules* are applied to the adjustment of simple parameter-optimized control algorithms, simple process models are sufficient. For a single *computer-aided design*, exact process/signal models are required which can most appropriately be obtained by identification and parameter estimation. If the acquisition of information and the control algorithms design are performed continuously (on-line, real-time) then *adaptive control systems* can be realized.

3 Fundamentals of Linear Sampled-data Systems (Discrete-time Systems)

In this chapter an introduction to the mathematical treatment of linear discrete-time systems (sampled-data systems) is given. Basic relationships which are fundamental for the design of digital control systems examined in this book are presented. To assist in providing a better understanding examples are frequently given. Exercises and their solutions are given in the appendix.

3.1 Discrete-time Signals

3.1.1. Discrete Functions, Difference Equations

Discrete Signals

Discrete (*discontinuous*) *signals* are quantized in *amplitude* or *time*. In contrast to continuous signals which describe any amplitude value for any time instant, discrete signals contain only values of discrete amplitudes for discrete-time points. In the following, only *discrete-time* signals are considered. They consist of trains of pulses at certain time points. Quite often discrete-time signals can be found which are generated by *sampling* continuous signals at constant time intervals. The single pulses of the series can be modulated in several ways dependent on the respective value of the continuous signal. Pulse amplitude, pulse width and pulse frequency modulation can be distinguished. For digital control systems pulse amplitude modulation is usually of interest, especially for the case when the pulse height is proportional to the continuous signal value, the pulse width is constant and the pulses occur in equi-distant sampling instants, Fig. 3.1. This type of discrete-time signals leads to linear relationships in the treatment of linear dynamic systems. Since, with the use of digital computers, mainly amplitude-modulated signals are of interest the most important relationships of these signals will be derived in the subsequent discussion. Figure 3.1 shows the generation of discrete-time amplitude modulated pulse trains through periodic detection of the continuous signal $x(t)$ with a *switch* which closes with sampling time T_0 for the time period h. If the switch duration h is very small in comparison to the sampling time T_0 and if the switch is followed by a linear transfer element with time constants $T_i \gg h$, the pulse trains $x_p(t)$ can be represented by the discrete-time signal $x(kT_0)$, Fig. 3.2. Then $x(kT_0)$ are the amplitudes for the sampling instants and the switch becomes a *sampler*.

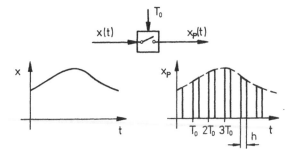

Fig. 3.1. Generation of an amplitude modulated discrete-time signal through a switch with duration h and sampling time T_0

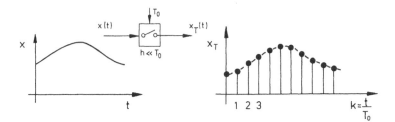

Fig. 3.2. Discrete-time signal $x_T(k)$ for $h \ll T_0$, generated by a sampler

Discrete-time Functions

An amplitude modulated discrete-time function $x_T(t)$ which is generated through equi-distant sampling of a continuous function $x(t)$ with sampling time T_0 is defined as:

$$\left. \begin{array}{ll} x_T(t) = x(kT_0) & \text{for} \quad t = kT_0 \\ x_T(t) = 0 & \text{for} \quad kT_0 < t < (k+1)T_0 \end{array} \quad k = 0,1,2,... \right\} \tag{3.1.1}$$

Various methods of generating discrete-time functions are given in the following examples. Signals as well as input/output relations are considered.

Example 3.1 Discrete-time Functions
a) If the·continuous function (*elementary function*)

$$x(t) = e^{-at}$$

is sampled, the discrete-time function becomes, with $t = kT_0$

$$x(kT_0) = e^{-akT_0} \quad k = 0,1,2...$$

Hence an explicit function results as

$$x_T(t) = x(kT_0)$$

b) The integration (*non-elementary function*)

$$x(t) = \frac{1}{T_I} \int_0^t w(t') dt'$$

can be performed numerically by a staircase approximation of the function $w(t')$, Figure 3.3a). This leads to:

$$x(kT_0) = \frac{1}{T_I} \sum_{v=0}^{k-1} T_0 w(vT_0).$$

Through this, the integral is approximated up to the time instant kT_0 by the area of rectangulars, the heights of which result from the former value $w((k-1)T_0)$ (delayed staircase function). Hence $x(kT_0)$ depends on a second function $w(vT_0)$

$$x_T(t) = x(kT_0) = f[w(vT_0), kT_0].$$

c) In the last example the numerical integration performed for one step ahead yields:

$$x((k+1)T_0) = \frac{1}{T_I} \sum_{v=0}^{k} T_0 w(vT_0).$$

The subtraction of the two equations results in the *recursive relation*

$$x((k+1)T_0) - x(kT_0) = \frac{T_0}{T_I} w(kT_0).$$

If kT_0 is replaced by k, it follows that

$$x(k+1) + a_1 x(k) = b_1 w(k)$$

with $a_1 = -1$ and $b_1 = T_0/T_I$.
With a backward shift of one sampling period

$$x(k) + a_1 x(k-1) = b_1 w(k-1).$$

results. This is a *first order difference equation*.
d) If the integral of b) is approximated until the time point kT_0 by the area of rectangles whose heights result from the current value $w(kT_0)$ (advanced staircase function), Figure 3.3b), then

$$x(kT_0) = \frac{1}{T_I} \sum_{v=1}^{k} T_0 w(vT_0)$$

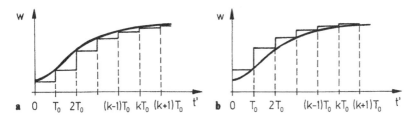

Fig. 3.3. Staircase approximation of a continuous function $w(t)$. **a** delayed staircase function **b** advanced staircase function

follows and the *recursive equation* according to c) yields

$$x(k+1) + a_1 x(k) = b_1 w(k+1)$$

or

$$x(k) + a_1 x(k-1) = b_1 w(k).$$

For small sampling times this approximation gives almost the same result as c).

If discrete-time functions which depend on other discrete-time functions can be written in recursive form, *difference equations* are obtained.
    An mth-order difference equation is:

$$x(k) + a_1 x(k-1) + \ldots + a_m x(k-m)$$
$$= b_0 w(k) + b_1 w(k-1) + \ldots + b_m w(k-m). \tag{3.1.2}$$

The current value of the output signal at time k can be determined recursively from:

$$x(k) = - a_1 x(k-1) - \ldots - a_m x(k-m)$$
$$+ b_0 w(k) + b_1 w(k-1) + \ldots + b_m w(k-m) \tag{3.1.3}$$

if the current input and m past inputs $w(k-1), \ldots, w(k-m)$ and m past outputs $x(k-1), \ldots, x(k-m)$ are known.

Difference Equations from Differential Equations

Difference equations can also be obtained by *discretizing differential equations*. Here a first order differential is approximated by a first order difference, a second order differential by a second order difference, etc. The following relationships result if backward differences are used:

Continuous function	Discrete-time function
· first order differential	· first order difference

$$\frac{dx(t)}{dt} = \lim_{\Delta t \to 0} \frac{x(t) - x(t - \Delta t)}{\Delta t} \qquad \Delta x(k) = x(k) - x(k-1)$$

· second order differential   · second order difference

$$\frac{d^2 x(t)}{dt^2} = \lim_{\Delta t \to 0} \frac{\dfrac{dx(t)}{dt} - \dfrac{dx(t - \Delta t)}{dt}}{\Delta t} \qquad \begin{aligned} \Delta^2 x(k) &= \Delta x(k) - \Delta x(k-1) \\ &= x(k) - 2x(k-1) \\ &\quad + x(k-2) \end{aligned}$$

$$\frac{d^3 x(t)}{dt^3} = \lim_{\Delta t \to 0} \frac{\dfrac{d^2 x(t)}{dt^2} - \dfrac{d^2 x(t - \Delta t)}{dt^2}}{\Delta t} \qquad \begin{aligned} \Delta^3 x(k) &= \Delta^2 x(k) - \Delta^2 x(k-1) \\ &= x(k) - 3x(k-1) \\ &\quad + 3x(k-2) + x(k-3) \end{aligned}$$

$$\vdots \qquad\qquad\qquad\qquad\qquad\qquad\qquad \vdots$$

In order to discretize a differential equation the following terms are used instead of the differentials

$$\left.\begin{array}{l}\dfrac{dx(t)}{dt} \approx \dfrac{x(k) - x(k-1)}{T_0} \\[2mm] \dfrac{d^2x(t)}{dt^2} \approx \dfrac{x(k) - 2x(k-1) + x(k-2)}{T_0^2} \\[2mm] \dfrac{d^3x(t)}{dt^3} \approx \dfrac{x(k) - 3x(k-1) + 3x(k-2) + x(k-3)}{T_0^3} \\[2mm] \vdots \qquad\qquad\qquad\qquad \vdots \end{array}\right\}$$
(3.1.4)

Example 3.2: Discretization of a differential

Unlike example 3.1b), c) and d) the function $w(t)$ is not to be integrated but differentiated

$$x(t) = T_D \dfrac{dw(t)}{dt}.$$

Backwards differencing yields

$$x(kT_0) = \dfrac{T_D}{T_0} [w(kT_0) - w((k-1)T_0)]$$

and thus the difference equation is

$$x(k) = b_0 w(k) - b_1 w(k-1)$$

with $b_0 = b_1 = T_D/T_0$.

Example 3.3: Discretization of a first order differential equation

A differential equation is given by

$$a_1 \dfrac{dx(t)}{dt} + x(t) = b_0 w(t)$$

On applying backwards differencing by inserting (3.1.4) one obtains:

$$a_0 x(k) + a_1 x(k-1) = b_0 w(k)$$

with:

$$a_0 = \dfrac{a_1}{T_0} + 1; \; a_1 = -\dfrac{a_1}{T_0}; \; b_0 = b_0.$$

Example 3.4: Discretization of a second order differential equation

The differential equation is given as

$$a_2 \ddot{x}(t) + a_1 \dot{x}(t) + x(t) = b_0 w(t).$$

Using backwards differencing by inserting (3.1.4) results in

$$a_0 x(k) + a_1 x(k-1) + a_2 x(k-2) = b_0 w(k)$$

with

$$a_0 = \dfrac{a_2}{T_0^2} + \dfrac{a_1}{T_0} + 1; \; a_1 = -\dfrac{a_1}{T_0} - \dfrac{2a_2}{T_0^2}; \; a_2 = \dfrac{a_2}{T_0^2}$$

$$b_0 = b_0.$$

These approximations are only satisfactory if the sampling time T_0 is small. In order to describe the function $x(t)$ by differentiating backwards only the function value $x(k)$ and its differences to the next value $x(k-1)$ and other past values $x(k-2)$, $x(k-3)$, ... is used. This is equivalent to a truncation of a Taylor-series expansion for each derivative of $x(t)$ after the first term. Hence, for the treatment of dynamic systems with discrete input and output signals other difference equations are derived in sequence which are also valid for large sample times T_0.

Eq. (3.1–2) is the standard form of a difference equation. A form corresponding to a differential equation results from introducing differences up to n^{th} order

$$\alpha_n \Delta^n x(k) + \alpha_{n-1} \Delta^{n-1} x(k) + ... + \alpha_1 \Delta x(k) + x(k)$$
$$= \beta_m \Delta^m w(k) + ... + \beta_1 \Delta w(k) + \beta_0 w(k) \qquad (3.1.5)$$

3.1.2 Impulse Trains

An expedient mathematical treatment of discrete-time functions is obtained if the pulse train $x_p(t)$ is approximated by an δ-impulse train, where an δ-impulse is defined by

$$\delta(t) = \begin{cases} 0 & t \neq 0 \\ \infty & t = 0 \end{cases} \qquad (3.1.6)$$

and has the area given by the integral

$$\int_{-\infty}^{\infty} \delta(t) dt = 1 \text{ s.} \qquad (3.1.7)$$

If the switch duration is very small compared with the sample time, $h \ll T_0$, the pulses of the pulse train $x_p(t)$ with the area $x(t) h$ can be approximated by impulses $\delta(t)$ with the same area

$$x_p(t) \approx x_\delta(t) = x(t) \frac{h}{1 \text{ s}} \sum_{k=0}^{\infty} \delta(t - kT_0). \qquad (3.1.8)$$

The resulting impulse train $x_\delta(t)$ is not a realizable signal but an approximation of the pulse train $x_p(t)$. This assumption of an ideal sampler, however, leads to a considerably simplified mathematical description of the transfer behaviour of systems with discrete-time signals, if the switches are followed by linear transfer elements. Figure 3.4 illustrates this approximation. The lengths of the arrows correspond to the area of the impulses.

Since the impulse train only exists for kT_0 Eq. (3.1–7) becomes

$$x_\delta(t) = \frac{h}{1 \text{ s}} \sum_{k=0}^{\infty} x(kT_0) \delta(t - kT_0). \qquad (3.1.9)$$

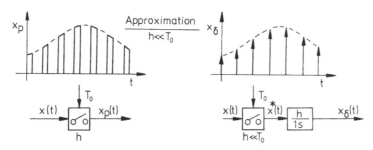

Fig. 3.4. Approximation of the pulse train $x_p(t)$ by an impulse train $x_\delta(t)$

Thus the switch with the impulse train $x(t)$ is replaced by an ideal sampler which results in an δ-impulse train amplified by $h/1$ sec. The switch duration h cancels for transfer systems with identical synchronously operating switches at the input and the output. Furthermore, different values for h do not affect the result if a data-hold follows the switch. Therefore, for simplicity, the switch duration will be ignored (equivalent to choosing $h = 1$ s) which leads to the "starred" impulse train:

$$x^*(t) = \sum_{k=0}^{\infty} x(kT_0)\delta(t-kT_0) \qquad (3.1.10)$$

Through this approximation and normalization, the output of the sampler, Fig. 3.2 is just multiplied by impulses $\delta(t - kT_0)$. Note that this approximation assumes that:

a) $h \ll T_0$,
b) the sampler is followed by a linear realizable system $G(s) = Z(s)/N(s)$.

3.1.3 Fourier-Transform of the Impulse Train

a) Limited Switch Duration

The pulse train $x_p(t)$ and its approximation $x_\delta(t)$ are to be considered in the frequency range. The pulse train $x_p(t)$ is assumed to consist of a train of rectangular pulses with the width h and the height $x(kT_0)$. This can also be considered as a switch generating a carrier signal in the form of a rectangular pulse train $p(t)$ of height 1 as follows

$$p(t) = \sum_{k=0}^{\infty} [1(t-kT_0) - 1(t-kT_0-hT_0)] = \sum_{k=0}^{\infty} p(t-kT_0) \qquad (3.1.11)$$

Thus $1(t)$ represents a step function of height 1 and this carrier signal is modulated with the input signal in such a way that

$$x_p(t) = x(t)p(t). \qquad (3.1.12)$$

results.

Since $p(t)$ is periodic with the period T_0 the *carrier signal* can be developed into a Fourier-series

$$p(t) = \sum_{v=-\infty}^{\infty} c_v e^{iv\omega_0 t} \tag{3.1.13}$$

where ω_0 is the sampling frequency given as

$$\omega_0 = \frac{2\pi}{T_0}$$

and the Fourier-coefficients are

$$c_v(iv\omega_0) = \frac{1}{T_0} \int_0^{T_0} p(t) e^{-iv\omega_0 t} dt \tag{3.1.14}$$

With $p(t) = 1$ one obtains

$$c_v(iv\omega_0) = \frac{1}{T_0} \int_0^{h} e^{-iv\omega_0 t} dt = \frac{1 - e^{-iv\omega_0 h}}{iv\omega_0 T_0}. \tag{3.1.15}$$

This is the Fourier-transform of a rectangular impulse which can also be derived by trigonometrical transformation

$$c_v(iv\omega_0) = \frac{h}{T_0} \frac{\sin \frac{v\omega_0 h}{2}}{\frac{v\omega_0 h}{2}} \cdot e^{-\frac{iv\omega_0 h}{2}}, \tag{3.1.16}$$

see Fig. 3.5a. Hence:

$$p(t) = \frac{h}{T_0} \sum_{v=-\infty}^{\infty} \frac{\sin \frac{v\omega_0 h}{2}}{\frac{v\omega_0 h}{2}} e^{-\frac{iv\omega_0 h}{2}} \cdot e^{iv\omega_0 t}$$

$$= \sum_{v=-\infty}^{\infty} c_v(iv\omega_0) e^{iv\omega_0 t}. \tag{3.1.17}$$

A Fourier-transform application to the *pulse train* $x_p(t)$ yields

$$\mathcal{F}\{x_p(t)\} = x_p(i\omega) = \int_{-\infty}^{\infty} x_p(t) e^{-i\omega t} dt = \int_{-\infty}^{\infty} x(t) p(t) e^{-i\omega t} dt$$

$$= \sum_{v=-\infty}^{\infty} c_v(iv\omega_0) \int_{-\infty}^{\infty} x(t) e^{-i(\omega - v\omega_0)t} dt$$

$$= \sum_{v=-\infty}^{\infty} c_v(iv\omega_0) x(i(\omega - v\omega_0))$$

$$= c_0 x(i\omega) + c_1(i\omega_0) x(i(\omega - \omega_0))$$

$$+ c_2(i2\omega_0) x(i(\omega - 2\omega_0)) + ... \tag{3.1.18}$$

Hence higher frequency signal components are introduced by the sampling. The Fourier-transform results from the *fundamental spectrum*

$$x_p(i\omega)|_{v=0} = \frac{h}{T_0} x(i\omega) \tag{3.1.19}$$

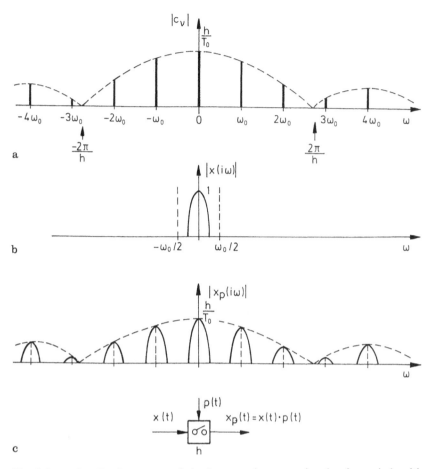

Fig. 3.5a–c. Amplitude spectra of the input and output signals of a switch with switch duration h. **a** amplitude spectrum of rectangular pulse train $p(t)$; **b** amplitude spectrum of the continuous input signal $x(t)$; **c** amplitude spectrum of the switch output $x_p(t)$

and the *complementary spectra* $x(i(\omega - v\omega_0))$, $v = \pm 1, \pm 2, \ldots$ multiplied with the Fourier-coefficients c_v, Fig. 3.5.

Thus the sampled signal also contains the continuous spectrum of the input signal $x(t)$ which is, however, multiplied by the factors c_v.

b) Small Switch Duration

For small pulse-width h it follows that

$$x_p(t) = \sum_{k=0}^{\infty} x(kT_0)[1(t-kT_0) - 1(t-kT_0-hT_0)] \qquad (3.1.20)$$

and with Laplace-transform one obtains

$$x_p(s) = \sum_{k=0}^{\infty} x(kT_0) \left[\frac{1-e^{-hs}}{s} \right] e^{-kT_0 s}. \tag{3.1.21}$$

With $h \to 0$ becomes

$$1 - e^{-hs} = 1 - \left[1 - hs + \frac{(hs)^2}{2!} - \ldots \right] \approx hs \tag{3.1.22}$$

and

$$x_\delta(s) \approx h \sum_{k=0}^{\infty} x(kT_0) e^{-kT_0 s}. \tag{3.1.23}$$

Using inverse Laplace-transform yields

$$x_\delta(t) = \frac{h}{1\,\text{s}} \sum_{k=0}^{\infty} x(kT_0) \delta(t-kT_0), \tag{3.1.24}$$

which is the same term as Eq. (3.1.9). This means that for small pulse durations h the resulting output signal can be approximated by δ-impulses which are multiplied by $x(kT_0)$ and $h/1$ sec. Instead of rectangular pulses the carrier signal now consists of δ-impulses.

In order to obtain the Fourier-transform for $h \to 0$ the δ-function is approximated by a rectangular pulse of height $1/h$

$$\delta(t) = \lim_{h \to 0} \frac{1}{h} p(t). \tag{3.1.25}$$

Eq. (3.1.18) and (3.1.15) lead to

$$\mathfrak{F}\{x_\delta(t)\} = x_\delta(i\omega) = \lim_{h \to 0} \frac{1}{h} \mathfrak{F}\{x_p(t)\}$$

$$= \lim_{h \to 0} \frac{1}{h} \sum_{v=-\infty}^{\infty} \frac{1-e^{-iv\omega_0 h}}{iv\omega_0 T_0} x(i(\omega - v\omega_0))$$

$$= \frac{1}{T_0} \sum_{v=-\infty}^{\infty} x(i(\omega - v\omega_0)). \tag{3.1.26}$$

Hence the Fourier-transform of the impulse train $x_\delta(t)$ approximated by δ-functions is composed of the basic spectrum of the continuous signal $x(t)$

$$x_\delta(i\omega) = \frac{1}{T_0} x(i\omega) \tag{3.1.27}$$

and the complementary spectra $x(i(\omega - v\omega_0))$, $v = \pm 1, \pm 2, \ldots$ which are multiplied by $1/T_0$.

If $x(t)$ has no jump discontinuity then for $t=0$

$$x_\delta(i\omega) = x(0^+)/2 + (1/T_0) \sum_{\nu=-\infty}^{\infty} x(i(\omega-\nu\omega_0))$$

as shown in [2.25.]

3.2 Laplace-transformation of Discrete-time Functions and Shannon's Sampling Theorem

3.2.1 Laplace-transformation

The Laplace-transformation

$$x(s) = \mathfrak{L}\{x(t)\} = \int_0^\infty x(t)e^{-st}dt \tag{3.2.1}$$

where $s = \delta + i\omega$, applied to an δ-impulse, yields

$$\mathfrak{L}\{\delta(t)\} = \int_0^\infty \delta(t)e^{-st}dt = 1 \text{ s.} \tag{3.2.2}$$

and for shifted impulses

$$\mathfrak{L}\{\delta(t-kT_0)\} = e^{-kT_{0}s} \cdot 1 \text{ s.} \tag{3.2.2}$$

With (3.1.10) the δ-impulse train becomes

$$\mathfrak{L}\{x^*(t)\} = x^*(s) = \sum_{k=0}^{\infty} x(kT_0)e^{-kT_{0}s} \cdot 1 \text{ s.} \tag{3.2.4}$$

Note that this Laplace-transform contains the dimension s. The Laplace-transform of a discrete-time function $x(kT_0)$ is now periodic with frequency given as

$$\omega_0 = 2\pi/T_0 \tag{3.2.5}$$

since

$$x^*(s+i\nu\omega_0) = x^*(s) \quad \nu=0, \pm 1, \pm 2, \dots \tag{3.2.6}$$

This results from

$$x^*(s+i\nu\omega_0) = \sum_{k=0}^{\infty} x(kT_0)e^{-kT_{0}s} \cdot e^{-kT_{0}\nu\omega_0 i}$$

$$= \sum_{k=0}^{\infty} x(kT_0)e^{-kT_{0}s} \cdot \underbrace{e^{-k\nu 2\pi i}}_{1}$$

$$= x^*(s) \tag{3.2.7}$$

Then with $s = \delta + i\omega$

$$x^{*}(\delta + i(\omega + v\omega_0)) = x^{*}(\delta + i\omega); \; v = \pm 1, \pm 2, \ldots \tag{3.2.8}$$

which means that $x^{*}(s)$ is repeated for all $v\omega_0$, as shown for the Fourier-transform $x_\delta(i\omega)$, (3.1.26). If $x^{*}(\delta + i\omega)$ is known for all δ and $-\omega_0/2 \leq \omega < \omega_0/2$, i.e. one strip in the s-plane parallel to the real axis then it is known completely for all ω.

3.2.2 Shannon's Sampling Theorem

If the continuous signal $x(t)$ is sampled with a small sample time $T_0 = \Delta t$, and approximated by a staircase function, the Laplace-transform of the continuous signal, (3.2.1), can be approximated by

$$x(s) \approx \sum_{k=0}^{\infty} x(k\Delta t) e^{-k\Delta t s} \Delta t. \tag{3.2.9}$$

A comparison with (3.2.4) for small $T_0 = \Delta t$ yields

$$T_0 x^{*}(s) \approx x(s) \text{ bzw. } T_0 x^{*}(\delta + i\omega) \approx x(\delta + i\omega). \tag{3.2.10}$$

Compare also (3.1.26).

It is now assumed that a continuous signal $x(t)$ contains a Fourier-transform for which

$$x(i\omega) \neq 0 \quad \text{für } -\omega_{max} \leq \omega \leq \omega_{max}$$

$$x(i\omega) = 0 \quad \text{für } \omega < -\omega_{max} \text{ und } \omega > \omega_{max},$$

is valid, see Fig. 3.6a). This band-limited signal is now to be sampled with sample time T_0 and approximated by an δ-impulse train $x^{*}(t)$.

If T_0 is sufficiently small, the Fourier transform then exists in a "basic spectrum"

$$T_0 x^{*}(i\omega) \approx x(i\omega) \quad -\omega_{max} \leq \omega \leq \omega_{max}$$

and periodically with ω_0 repeating "complementary spectra" (sidebands)

$$T_0 x^{*}(i(\omega + v\omega_0)) \approx x(i\omega) \quad v = \pm 1, \pm 2, \ldots$$

compare Fig. 3.6b). In comparison with the continuous signal, additional higher frequency components arise from the sampled signal. This was already discussed in section 3.1.3.

If the originally continuous signal has to be reconstructed from the sampled signal this can be done by filtering with an ideal bandpass filter with

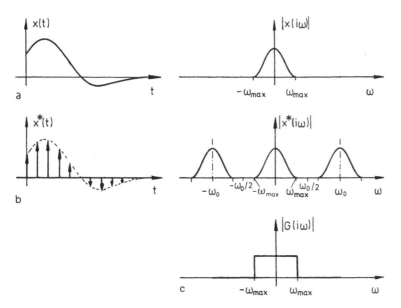

Fig. 3.6a–c. Fourier transform of a sampled signal. **a** Magnitude of the Fourier-transform of a continuous signal; **b** Magnitude of the Fourier-transform of a sampled system; **c** Magnitude of the frequency response of an ideal bandpass filter.

$$|G(i\omega)| = 1 \quad -\omega_{max} \leqq \omega \leqq \omega_{max}$$

$$|G(i\omega)| = 0 \quad \omega < -\omega_{max} \text{ and } \omega > \omega_{max}$$

see Fig. 3.6). This can only be achieved exactly if

$$\omega_0/2 = \pi/T_0 > \omega_{max}$$

If the sample time is too large i.e. the sampling frequency is too small then

$$\omega_0/2 < \omega_{max},$$

and overlapping fundamental and side spectra will arise, and an error-free filtering of the fundamental spectrum is not possible. Therefore with this incorrectly chosen sampling frequency the band-limited continuous signal cannot be recovered without errors.

In order to recover the continuous band-limited signal which has a maximum frequency ω_{max} from the sampled signal, the sampling frequency ω_0 has to be chosen as

$$\omega_0 > 2\,\omega_{max} \tag{3.2.11}$$

or, with (3.2.5)

$$T_0 < \pi/\omega_{max}\,. \tag{3.2.12}$$

This is *Shannon's sampling theorem*. The oscillation of the highest frequency ω_{max} with period $T_p = 2\pi/\omega_{max}$ has to be sampled at least twice per period.

It should be mentioned that band-limited continuous signals in reality do not arise in communication or control systems. However, in sampled data control, the Shannon frequency (also called Nyquist frequency)

$$\omega_{Sh} = \frac{\omega_0}{2} = \frac{\pi}{T_0} \qquad (3.2.13)$$

plays an important role, at least as a reference frequency.

3.2.3 Holding Element

If a sampler is followed by a *zero-order hold* which holds the sampled signals $x(kT_0)$ for one sampling interval, then a staircase signal results, Fig. 3.7.

The transfer function of a zero-order hold can be derived as follows. For the impulse train as input it is

$$x^*(t) = \sum_{k=0}^{\infty} x(kT_0)\delta(t-kT_0)$$

with Laplace-transformed function

$$x^*(s) = \sum_{k=0}^{\infty} x(kT_0)e^{-kT_0 s}.$$

The staircase function can be described by superposition of time shifted step functions $1(t)$

$$m(t) = \sum_{k=0}^{\infty} x(kT_0)[1(t-kT_0) - 1(t-(k+1)T_0)]. \qquad (3.2.14)$$

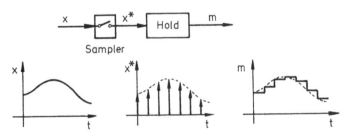

Fig. 3.7. Sampler with zero-order hold

with on taking the Laplace transform

$$m(s) = \underbrace{\sum_{k=0}^{\infty} x(kT_0)e^{-kT_0 s}.\frac{1}{s}[1 - e^{-T_0 s}]}_{x^*(s)}.$$

The transfer function of the zero-order hold is then

$$H(s) = \frac{m(s)}{x^*(s)} = \frac{1}{s}[1 - e^{-T_0 s}]. \tag{3.2.15}$$

The zero-order hold can be modelled as an integrator which is only effective for time period T_0. Its impulse response equals a rectangular pulse of height 1 and duration T_0. The gain results from (3.2.15) for $s \to 0$ to $H(0) = T_0/1$ s.

In an attempt to recover the continuous signal $x(t)$ from the staircase signal $m(t)$ by averaging, involves time shifting $x(t)$ by time $T_0/2$. Hence the sample and holding element introduce a phase shift which approximately equals a dead time element with the dead time $T_t = T_0/2$.

A series expansion of (3.2.15) yields for $s = i\omega$

$$H(i\omega) = \frac{1}{i\omega}\left[1 - \frac{1}{1 + \dfrac{T_0 i\omega}{1!} + \dfrac{(T_0 i\omega)^2}{2!} + \dots}\right]. \tag{3.2.16}$$

This leads to

$$\lim_{\omega \to 0} H(i\omega) = \frac{T_0}{1 + T_0 i\omega}. \tag{3.2.17}$$

For low frequencies the zero-order hold behaves like a first order low-pass filter.

The sampler and zero-order hold operations generate a staircase approximation of the continuous input signal $x(t)$. If this signal, after sampling, is to be even better approximated the slope $\dot{x}(t)$ can be held at the sample time point so that

$$m(t) = \sum_{k=0}^{\infty} x(kT_0)[1(t - kT_0) - 1(t - (k+1)T_0)]$$

$$+ \frac{x(kT_0) - x((k-1)T_0)}{T_0}[(t - kT_0) - (t - (k+1)T_0]$$

$$\tag{3.2.18}$$

The transfer function of this *first-order hold* follows by Laplace-transformation as

$$H'(s) = \frac{m(s)}{x^*(s)} = \frac{1 + T_0 s}{T_0}\left[\frac{1 - e^{-T_0 s}}{s}\right]^2. \tag{3.2.19}$$

The gain is $H'(0) = T_0$ and for low frequencies also (3.2.17) is valid. A noticeable advantage of a first-order hold compared with a zero-order hold can be therefore only realized for higher frequencies.

Because of easier realization, a zero-order hold is mostly used for digital control systems.

3.2.4 Frequency Response of Sampled Systems

If the sampler is followed by a linear transfer element with frequency response

$$G(i\omega) = \frac{y(i\omega)}{u(i\omega)},$$
(3.2.20)

then according to (3.1.26)

$$u_\delta(i\omega) = \frac{1}{T_0} \sum_{v=-\infty}^{\infty} u(i(\omega-v\omega_0))$$
(3.2.21)

becomes valid for the input signal and

$$y(i\omega) = G(i\omega)u_\delta(i\omega).$$
(3.2.22)

for the continuous output signal.
    If $y(t)$ is also sampled then

$$y_\delta(i\omega) = \frac{1}{T_0} \sum_{v=-\infty}^{\infty} y(i(\omega-v\omega_0)).$$
(3.2.23)

results.
    This shows that side-bands emerge also in the output signal. From (3.2.22) follows with (3.2.6)

$$y_\delta(i\omega) = \frac{1}{T_0} \sum_{v=-\infty}^{\infty} G(i(\omega-v\omega_0)) \cdot u_\delta(i(\omega-v\omega_0))$$
(3.2.24)

$$= u_\delta(i\omega) \frac{1}{T_0} \sum_{v=-\infty}^{\infty} G(i(\omega-v\omega_0))$$

If (3.1.9) is inserted the factor $h/1$ s cancels leading to

$$y^*(i\omega) = u^*(i\omega)G^*(i\omega)$$
(3.2.25)

The frequency response of a transfer element with a sampler at the input and the output then is [2.11, 2.20]

$$G^*(i\omega) = \frac{y^*(i\omega)}{u^*(i\omega)} = \frac{1}{T_0} \sum_{v=-\infty}^{\infty} G(i(\omega-v\omega_0))$$
(3.2.26)

This means that the frequency response defined in this way also contains side-bands. For the range of the basic spectrum of the input signal it follows that

$$G^*(i\omega) = \frac{1}{T_0}G(i\omega) \quad 0 \leqq \omega < \omega_0/2. \tag{3.2.27}$$

becomes valid for large ω_0.

For the Nyquist-frequency $\omega = \omega_0/2$ magnitude and phase of the frequency response depend on the phase length with which the output oscillation is sampled.

Similarly for continuous systems, the frequency response of sampled systems in the range of $\omega < \omega_0/2$ can be determined experimentally. If the sampling time becomes too large, however, the overlapping of the side-bands causes an inability to separate the components. Because of the inconvenient manipulation for discrete-time systems and the more expedient treatment in the time- and z-domain the frequency response is not often used.

3.3 z-Transform

3.3.1 Introduction of z-Transform

By introducing the abbreviation

$$z = e^{T_0 s} = e^{T_0(\delta + i\omega)} = e^{T_0 \delta}[\cos T_0 \omega + i \sin T_0 \omega] \tag{3.3.1}$$

into (3.2.4), the z-transform $x(t)$ follows as

$$x(z) = \mathfrak{Z}\{x(kT_0)\} = \sum_{k=0}^{\infty} x(kT_0)z^{-k} \cdot 1 \text{ s}$$

$$= [x(0) + x(T_0)z^{-1} + x(2T_0)z^{-2} + ...] \cdot 1 \text{ s} \tag{3.3.2}$$

Note, that like the Laplace-transform the z-transform also contains the dimension s. The definition (3.3.2) however, only makes sense, if $x(z)$ converges. If $x(kT_0)$ is limited, $x(z)$ converges for $|z| = |\exp(T_0\delta)| > 1$. Just as in the Laplace-transform case δ can be appropriately chosen so that for most discrete functions convergence can be enforced. The assumptions made for the Laplace-transform, especially $x(kT_0) = 0$ for $k < 0$, are also valid for the z-transform.

Generally, $x(z)$ is an infinite series. Many functions, however, can be represented in closed form if the properties of power series are used.

Example 3.5:
a) Step function: $x(kT_0) = 1(kT_0)$

$$x(z) = 1 + z^{-1} + z^{-2} + ... = \frac{1}{1 - z^{-1}} = \frac{z}{z - 1} \tag{3.3.3}$$

provided $|z| > 1$.
b) Exponential function: $x(kT_0) = e^{-akT_0}$ (a is real)

$$x(z) = 1 + (e^{+aT_0}z)^{-1} + (e^{+aT_0}z)^{-2} + ... \tag{3.3.4}$$

$$= \frac{1}{1 - (e^{+aT_0}z)^{-1}} = \frac{z}{z - e^{-aT_0}} = \frac{z}{z + a_1}$$

provided that $|e^{+aT_0} z)| > 1$.

c) Sine function: $x(kT_0) = \sin \omega_1 kT_0$

With $\sin \omega_1 kT_0 = \dfrac{1}{2i} [e^{i\omega_1 kT_0} - e^{-i\omega_1 kT_0}]$

this follows, using the result of b) with $a = i\omega_1$

$$x(z) = \frac{1}{2i} \left[\frac{z}{z - e^{i\omega_1 T_0}} - \frac{z}{z - e^{-i\omega_1 T_0}} \right]$$

(3.3.5)

$$= \frac{z \sin \omega_1 T_0}{z^2 - 2z\cos \omega_1 T_0 + 1} = \frac{b_1 z}{z^2 + a_1 z + 1}.$$

These examples show how the z-transforms of some simple functions can be obtained. A table of corresponding entries (continuous time functions, Laplace-transform and z-transform) is given in the Appendix A1. This table shows:

a) There is a direct correspondence between the denominators of $x(s)$ and $x(z)$

$$\frac{\cdots}{\displaystyle\prod_{j=1}^{n}(s - s_j)} \leftrightarrow \frac{\cdots}{\displaystyle\prod_{j=1}^{n}(z - z_j)} \quad \text{with } z_j = e^{T_0 S_j}.$$

b) There is no direct correspondence between the numerators of $x(s)$ and $x(z)$. For example, $x(z)$ can possess numerator-polynomials even if $x(s)$ does not.

3.3.2 z-Transform Theorems

Some important theorems for the application of the z-transform are given in the following:

Linearity

$$3\{a\, x_1(k) + b\, x_2(k)\} = a\, 3\{x_1(k)\} + b\, 3\{x_2(k)\}$$

(3.3.6)

Proof: If (3.3.6) is introduced in (3.3.2) the given superposition law results.

Example: If $x_1(k) = c_1 k$ and $x_2(k) = c_2$
then

$$3\{c_1 k + c_2\} = c_1 \frac{T_0 z}{(z-1)^2} + c_2 \frac{z}{(z-1)} = \frac{(c_1 T_0 - c_2)z + c_2 z^2}{(z-1)^2}$$

Effect of Shifting to the Right:

The discrete-time function $x(k)$ is shifted to the right by d sampling times T_0 (shift into the past)

$$\mathfrak{Z}\{x(k-d)\} = z^{-d}x(z) \qquad d \geq 0 \tag{3.3.7}$$

Proof:

$$\mathfrak{Z}\{x(k-d)\} = \sum_{k=0}^{\infty} x(k-d)z^{-k} = z^{-d}\sum_{k=0}^{\infty} x(k-d)z^{-(k-d)}$$

Substitution $q = k - d$ and notice of $x(q) = 0$ for $q < 0$ yields

$$\mathfrak{Z}\{x(k-d)\} = z^{-d}\sum_{q=-d}^{\infty} x(q)z^{-q} = z^{-d}\sum_{q=0}^{\infty} x(q)z^{-q}$$

$$= z^{-d}x(z).$$

Example: The z-transform for a step function of height 1 which is delayed by 3 sample times is

$$\mathfrak{Z}\{1(k-3)\} = z^{-3}\frac{z}{z-1} = \frac{1}{z^2(z-1)}.$$

Effect of Shifting to the Left:

The discrete-time function $x(k)$ is shifted to the left (shifted into the future).

$$\mathfrak{Z}\{x(k+d)\} = z^d\left[x(z) - \sum_{q=0}^{d-1} x(q)z^{-q}\right] \qquad d \geq 0 \tag{3.3.8}$$

Proof:

$$\mathfrak{Z}\{x(k+d)\} = \sum_{k=0}^{\infty} x(k+d)z^{-k} = z^d\sum_{k=0}^{\infty} x(k+d)z^{-(k+d)}$$

Substitution $q = k + d$

$$\mathfrak{Z}\{x(k+d)\} = z^d\sum_{q=d}^{\infty} x(q)z^{-q} = z^d\left[\sum_{q=0}^{\infty} x(q)z^{-q} - \sum_{q=0}^{d-1} x(q)z^{-q}\right]$$

$$= z^d\left[x(z) - \sum_{q=0}^{d-1} x(q)z^{-q}\right].$$

By shifting to the left the function values of the original non-shifted function $x(k)$ for $k = 0, 1, \ldots, d-1$ vanish because the z-transform is defined only for $q > 0$.

Example: The z-transform for a step function of height 1 which is shifted into the future by 3 sample times is

$$\mathfrak{Z}\{1(k+3)\} = z^3\left[\frac{z}{z-1} - (1 + z^{-1} + z^{-2})\right] = \frac{z}{z-1}$$

Effect of Damping:

$$3\{x(k)e^{-akT_0}\} = x(ze^{aT_0}) \tag{3.3.9}$$

Proof: With the substitution $z_s = ze^{aT_0}$

$$3\{x(k)e^{-akT_0}\} = \sum_{k=0}^{\infty} x(k)e^{-akT_0}z^{-k} = \sum_{k=0}^{\infty} x(k)(ze^{aT_0})^{-k}$$

$$= \sum_{k=0}^{\infty} x(k)z_s^{-k} = x(z_s) = x(ze^{aT_0})$$

Example: Given the z-transform of a step function $x(z) = z/(z-1)$. Find the z-transform of e^{-2kT_0}.

$$3\{1(k)e^{-2kT_0}\} = \frac{ze^{2T_0}}{ze^{2T_0}-1} = \frac{z}{z-e^{-2T_0}}.$$

is used.

Effect of Initial Value:

$$x(0) = \lim_{z\to\infty} x(z)$$

Proof:

$$x(z) = \sum_{k=0}^{\infty} x(k)z^{-k} = x(0) + x(1)z^{-1} + x(2)z^{-2} + \dots$$

$$\lim_{z\to\infty} x(z) = x(0)$$

Example: The initial value of the discrete-time function is to be calculated, which has the z-transform in example *Linearity*

$$x(0) = \lim_{z\to\infty} \frac{(c_1T_0-c_2)z + c_2z^2}{(z-1)^2} \cdot \frac{z^{-2}}{z^{-2}}$$

$$= \lim_{z\to\infty} \frac{(c_1T_0-c_2)z^{-1} + c_2}{1 - 2z^{-1} - z^{-2}} = c_2$$

Final Value:

$$\lim_{k\to\infty} x(k) = \lim_{z\to1} \frac{z-1}{z} x(z) = \lim_{z\to1} (z-1)x(z).$$

This is only valid if the final value $x(\infty)$ exists. This is the case if, for the poles of $(z-1)x(z)$, $|z| < 1$ holds.

Proof: The discrete-time function $x(k)$ can be written as a sum of its first-order difference

$$x(k) = \sum_{q=0}^{k} [x(q) - x(q-1)] = \sum_{q=0}^{k} \Delta x(q).$$

Then the final value

$$\lim_{k \to \infty} x(k) = \sum_{q=0}^{\infty} \Delta x(q) = \Delta x(0) + \Delta x(1) + \Delta x(2) + \dots$$

is valid.

This series also results from the z-transform of the differences $\Delta x(q)$ if one applies

$$\lim_{z \to 1} 3\{\Delta x(q)\} = \lim_{z \to 1} \sum_{q=0}^{\infty} \Delta x(q) z^{-q} = \Delta x(0) + \Delta x(1) + \Delta x(2) + \dots$$

This leads to

$$\lim_{k \to \infty} x(k) = \lim_{z \to 1} 3\{x(q) - x(q-1)\} = \lim_{z \to 1} (1 - z^{-1}) x(z)$$

$$= \lim_{z \to 1} \frac{z-1}{z} x(z).$$

Example: The final value of the function $x(k)$ with the z-transform $x(z) = 2z/(z-1)(z-0.5)$ is to be determined.

$$\lim_{k \to \infty} x(k) = \lim_{z \to 1} (z-1) \frac{2z}{(z-1)(z-0,5)} = 4.$$

3.3.3 Inverse z-Transform

Equation for the Inverse z-Transform

In order to derive the inverse z-transform, one uses the fact that $x(z)$ and hence $x^*(s)$ is periodic with $\omega_0 = 2\pi/T_0$ and is furthermore symmetrical with the real axis of the s-plane. That is why $x^*(s)$ has to have the inverse transform operation applied to it only in the domain $\delta - i\pi/T_0 \le s \le \delta + i\pi/T_0$. In this case δ has to be larger than the convergence abscissa.

On multiplying

$$x^*(s) = \sum_{k=0}^{\infty} x(k) e^{-kT_0 s} \tag{3.3.10}$$

by $e^{lT_0 s}$ and integrating this equation ($a = \pi/T_0$) leads to

$$\int_{\delta-ia}^{\delta+ia} x^*(s)e^{\ell Tos}ds = \int_{\delta-ia}^{\delta+ia}\left[\sum_{k=0}^{\infty} x(k)e^{-kTos}\right]e^{\ell Tos}ds$$

$$= \sum_{k=0}^{\infty} x(k)\underbrace{\int_{\delta-ia}^{\delta+ia} e^{-(k-\ell)Tos}ds}$$

$$=0 \text{ für } k\neq\ell$$
$$=2ia \text{ für } k=\ell$$

$$=2ia\ x(\ell).$$

Hence

$$x(\ell) = \Im^{-1}\{x^*(s)\} = \frac{T_0}{2\pi i}\int_{\delta-i\pi/T_0}^{\delta+i\pi/T_0} x^*(s)e^{\ell Tos}ds. \qquad (3.3.11)$$

If $z = e^{Tos}$ and $dz = T_0 e^{Tos}\ ds$, then

$$x(\ell) = \Im^{-1}\{x(z)\} = \frac{1}{2\pi i}\oint x(z)z^{\ell-1}dz. \qquad (3.3.12)$$

results.

Thus the integration has to be performed on a unit circle with radius e^{δ}. For a rational fraction of $x(z)$ the integral can be determined by applying Cauchy's residual law [2.11, 2.13].

Inverse Transformation by Expansion into Simple Terms

A more simple procedure can be achieved by expanding the z-transform into a sum of the low-order terms listed in the z-transform table and by observing the linearity law (3.3.6). According to the rules of partial fraction expansion, a given rational fraction term $x(z)$ can be written as follows

$$x(z) = \frac{Az}{(z-1)} + \frac{Bz}{(z-1)^2} + \frac{Cz}{(z-a)} + \frac{Dz}{(z^2-cz+d)} + \dots \qquad (3.3.13)$$

Using corresponding entries from the table in Appendix A1 an explicit function of the continuous signal in the time domain is obtained as

$$x(t) = A + Bt/T_0 + Ce^{-at} + De^{-at}\sin\omega_1 t + \dots \qquad (3.3.14)$$

and with $t = kT_0$ the discrete-time function is

$$x(k) = A + Bk + Ce^{-akT_0} + De^{-akT_0}\sin\omega_1 kT_0 + \dots \qquad (3.3.15)$$

Inverse Transform by Division

If one is not interested in the explicit function $x(k)$ but only in the numerical values, $x(z)$ can be defined by dividing through as follows

$$(b_0z^n + b_1z^{n-1} + ... + b_n):(z^n + a_1z^{n-1} + ... + a_n)$$

$$= c_0 + c_1z^{-1} + c_2z^{-2} + ... \qquad (3.3.16)$$

On comparing the coefficients with the defining equation

$$x(z) = \sum_{k=0}^{\infty} x(k)z^{-k} = x(0) + x(1)z^{-1} + x(2)z^{-2} + ... \qquad (3.3.17)$$

it follows directly that

$$x(0) = c_0; x(1) = c_1; x(2) = c_2; ... \qquad (3.3.18)$$

Example 3.6:
Given the z-transform

$$x(z) = \frac{0,1(z+1)z}{(z-1)^2(z-0,6)}$$

with sampling time $T_0 = 1$ s. Find the discrete-time function $x(k)$.

a) Expansion into simple terms

$$x(z) = \frac{Az}{(z-1)} + \frac{Bz}{(z-1)^2} + \frac{Cz}{(z-0,6)}$$

$$= \frac{(A+C)z^3 + (B-2C-1,6A)z^2 + (0,6A-0,6B+C)z}{(z-1)^2(z-0,6)}.$$

By comparison of the coefficients, it follows that:
$A = -1; B = 0.5; C = 1$.
This leads to

$$x(z) = \frac{-z}{(z-1)} + \frac{0,5z}{(z-1)^2} + \frac{z}{(z-0,6)}.$$

Resulting from the correspondences of the z-transform table and $e^{-aT_0} = 0.6$ respectively $a = -(\ln 0.6)/T_0 = 0.5198/T_0$, it follows that

$$x(t) = -1 + 0,5\ t/T_0 + e^{-at}$$

and with $t = kT_0$

$$x(k) = -1 + 0,5\ k + e^{-0,5198\ k}.$$

The numerical values are

$$
\begin{array}{ll}
x(0)=0 & x(3)=0,716 \\
x(1)=0,1 & x(4)=1,1296 \\
x(2)=0,36 &
\end{array}
$$

b) Division

$$(0,1z^2 + 0,1z) : (z^3 - 2,6z^2 + 2,2z - 0,6)$$

$$= 0,1z^{-1} + 0,36z^{-2} + 0,716z^{-3} + 1,1296z^{-4} + \dots$$

Consequently, the numerical values are

$$
\begin{array}{ll}
x(0) = 0 & x(3) = 0,716 \\
x(1) = 0,1 & x(4) = 1,1296 \\
x(2) = 0,36 &
\end{array}
$$

.

.

.

3.4 Convolution Sum and z-Transfer Function

3.4.1 Convolution Sum

As already shown in Fig. 2.3, samplers are connected before and after the process which is to be controlled. This is once again described in Fig. 3.8. A sampler operates at the input of a linear system with transfer function $G(s)$ or impulse response $g(t)$. The impulse train approximation of the input to the system is

$$u^*(t) = \sum_{k=0}^{\infty} u(kT_0)\delta(t-kT_0). \tag{3.4.1}$$

With the impulse response $g(t)$ as response to a δ-impulse, the *convolution sum*

$$y(t) = \sum_{k=0}^{\infty} u(kT_0)g(t-kT_0). \tag{3.4.2}$$

results for the output variable $y(t)$.
If also the input and output are sampled synchronously, then yields for $t = nT_0$

$$y(nT_0) = \sum_{k=0}^{\infty} u(kT_0)g((n-k)T_0)$$

$$= \sum_{v=0}^{\infty} u((n-v)T_0)g(vT_0). \tag{3.4.3}$$

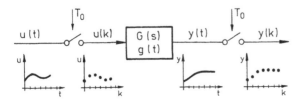

Fig. 3.8. A linear process with sampled input and output

Like for the convolution integral for continuous-time systems, the value of the output signal at time point nT_0 is obtained by the product sum of $u(kT_0)$ and $g((n-k)T_0)$ which again are timeshifted by nT_0 in opposite direction. Thereby the timeshift can be realized either in the impulse function, that is $g((n-k)T_0)$ or in the input signal, i.e. $u((n-v)T_0)$ as can be immediately seen by writing the addends in (3.4.3).

3.4.2 Pulse Transfer Function and z-Transfer Function

Introducing (3.4.3) the Laplace-transform of the sampled output $y(nT_0)$ yields

$$y^*(s) = \sum_{n=0}^{\infty} y(nT_0)e^{-nT_0 s}$$

$$= \sum_{n=0}^{\infty} \sum_{k=0}^{\infty} u(kT_0)g((n-k)T_0)e^{-nT_0 s}. \tag{3.4.4}$$

and by substituting $q = n-k$ (thus also $n = q+k$)

$$y^*(s) = \sum_{q=-k}^{\infty} \sum_{k=0}^{\infty} u(kT_0)g(qT_0)e^{-qT_0 s}e^{-kT_0 s}$$

$$= \underbrace{\sum_{q=0}^{\infty} g(qT_0)e^{-qT_0 s}}_{G^*(s)} \underbrace{\sum_{k=0}^{\infty} u(kT_0)e^{-kT_0 s}}_{u^*(s)} \tag{3.4.5}$$

considering that $g(qT_0) = 0$ for $q < 0$. Hence, the Laplace-transform $u^*(s)$ of the input signal is expressed as a separate factor and according to the case of continuous signals the *pulse transfer function* is defined as follows

$$G^*(s) = \frac{y^*(s)}{u^*(s)} = \sum_{q=0}^{\infty} g(qT_0)e^{-qT_0 s}. \tag{3.4.6}$$

Introducing the abbreviation $z = e^{T_0 s}$ leads to the *z-transfer function*

$$G(z) = \frac{y(z)}{u(z)} = \sum_{q=0}^{\infty} g(qT_0)z^{-q} = \mathfrak{Z}\{g(q)\}. \tag{3.4.7}$$

Hence, the z-transfer function is the relation between the z-transform of the sampled output and the z-transform of the sampled input and equals furthermore the z-transform of the impulse function. The following example shows the evaluation of a z-transfer function.

Example 3.7: A first order lag without holding element
A first order lag with the s-transfer function

$$G(s) = \frac{K}{1 + Ts} = \frac{K'}{a + s}$$

with $a = 1/T$ and $K' = K/T$, has, according Laplace-function the impulse function

$$g(t) = K'e^{-at} \cdot 1 \text{ s } [-]$$

(In order that $g(t)$ maintains the correct dimension, respectively none, one has to correct with the time unit. The time unit was introduced by the inverse Laplace-transformation.) For the sampling instants follows

$$g(kT_0) = K'e^{-akT_0} \cdot 1 \text{ s.}$$

This leads with (3.4.7) and example 3.5b) to

$$G(z) = K' \sum_{q=0}^{\infty} (e^{+aT_0}z)^{-q} \cdot 1 \text{ s} = \frac{K'z}{z - e^{-aT_0}} \cdot 1 \text{ s}$$

$$= \frac{K' \cdot 1 \text{ s}}{1 - e^{-aT_0}z^{-1}} = \frac{b_0}{1 + a_1z^{-1}}.$$

Taking $K = 1$, $T = 7.5$ s and $T_0 = 4$ s yields the parameters

$$b_0 = K' \cdot 1 \text{ s} = (K/T) \cdot 1 \text{ s} = 0{,}1333$$

$$a_1 = -e^{-aT_0} = -0{,}5866.$$

The operations in this example consist of

$$G(s) \rightarrow g(t) \rightarrow g(kT_0) \rightarrow G(z)$$

or

$$G(z) = 3\{[\mathcal{L}^{-1}\{G(s)\}]_{t = kT_0}\}. \tag{3.4.8}$$

However, using simple terms from the z-transform table it follows that

$$G(s) \rightarrow G(z),$$

Thus the solution can be derived without calculating the impulse function. If one looks for the corresponding $x(z)$ in the z-transform table directly, it can be written as

$$G(z) = \mathcal{L}\{G(s)\}. \tag{3.4.9}$$

For example 3.7 this yields

$$G(z) = \mathcal{L}\left\{\frac{K'}{a+s}\right\} \cdot 1 \text{ s} = \frac{K'z}{z - e^{-aT_0}} 1 \text{s.}$$

If the sampler is followed by a hold, as shown in Fig. 3.9, then becomes

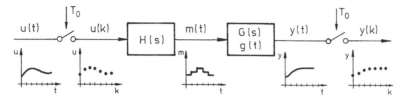

Fig. 3.9. Linear process with connected hold and sampled input and output signals

$$HG(z) = \mathscr{Z}\{H(s)G(s)\}. \tag{3.4.10}$$

For a zero-order hold this then yields

$$HG(z) = \mathscr{Z}\left\{\frac{1-e^{-T_0 s}}{s}G(s)\right\} = \mathscr{Z}\left\{\frac{G(s)}{s}\right\} - \mathscr{Z}\left\{\frac{G(s)}{s}e^{-T_0 s}\right\}$$

$$= (1 - z^{-1})\mathscr{Z}\left\{\frac{G(s)}{s}\right\} = \frac{z-1}{z}\mathscr{Z}\left\{\frac{G(s)}{s}\right\} \tag{3.4.11}$$

and for a first-order hold it follows that

$$HG(z) = (1 - z^{-1})^2 \mathscr{Z}\left\{\frac{1+T_0 s}{T_0}\frac{G(s)}{s^2}\right\}. \tag{3.4.12}$$

The transfer function with a first-order hold, however, generally brings no advantages, compare [2.17]. Consequently, this hold will not be used.

Example 3.8: A first-order lag with zero-order hold
According (3.4.11) and the z-transform table (Appendix) becomes

$$HG(z) = \frac{z-1}{z}\mathscr{Z}\left\{\frac{K'}{s(a+s)}\right\} = \frac{z-1}{z}\frac{(1-e^{-aT_0})z}{(z-1)(z-e^{-aT_0})}\frac{K'}{a}$$

$$= \frac{(1-e^{-aT_0})}{(z-e^{-aT_0})}\frac{K'}{a} = \frac{b_1 z^{-1}}{1+a_1 z^{-1}}.$$

Using the same parameters as in example 3.7, one obtains

$$b_1 = (1-e^{-aT_0})\frac{K'}{a} = 0{,}4134$$

$$a_1 = -e^{-aT_0} = -0{,}5866.$$

As shown by the comparison with example 3.7, the hold operation introduces into the numerator a lag term z^{-1} together with a second parameter. The denominator does not change.

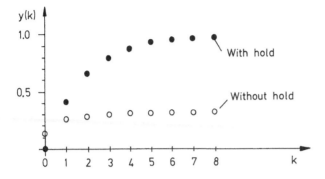

Fig. 3.10. Transient functions of a first-order lag with and without zero-order hold.

Now, the transient functions of both lag elements are to be considered.
Without hold according example 3.7.

$$y(k) = -a_1 y(k-1) + b_0 u(k).$$

With $u(k) = 1(k)$ and the numerical values for a_1 and b_0 it follows that

$$y(0) = 0{,}1333 \quad y(2) = 0{,}2897$$
$$y(1) = 0{,}2666 \quad y(3) = 0{,}3032.$$

and with zero-hold operation included

$$y(k) = -a_1 y(k-1) + b_1 u(k-1)$$

The corresponding numerical values yield

$$y(0) = 0 \qquad y(2) = 0{,}6559$$
$$y(1) = 0{,}4134 \quad y(3) = 0{,}7891$$

Figure 3.10 shows both transient functions, illustrating that a hold has to be connected before, if the discrete-time model is desired to have a behaviour equivalent to the continuous time element.

Further examples which can be derived from the z-transform table are given in Table 3.1 and also in Appendix A2. The determination of z-transfer functions via the s-transfer function are treated separately in section 3.7.3.

If a description of a linear dynamic system in the form of a difference equation according (3.1.2) already exists then the following can be written directly

$$y(k) + a_1 y(k-1) + \dots + a_m y(k-m)$$

$$= b_0 u(k) + b_1 u(k-1) + \dots + b_m u(k-m) \tag{3.4.13}$$

On applying the shifting theorem to the right hand-side results in

$$y(z)[1 + a_1 z^{-1} + \dots + a_m z^{-m}] = u(z)[b_0 + b_1 z^{-1} + \dots + b_m z^{-m}]$$

thus leading to the z-domain transfer function

$$G(z) = \frac{y(z)}{u(z)} = \frac{b_0 + b_1 z^{-1} + \dots + b_m z^{-m}}{1 + a_1 z^{-1} + \dots + a_m z^{-m}} = \frac{B(z^{-1})}{A(z^{-1})}. \tag{3.4.14}$$

Table 3.1. z-transfer function without and with zero-order hold

$G(s)$	$G(z)=\mathscr{L}\{G(s)\}$	$HG(z)=\dfrac{z-1}{z}\mathscr{L}\left\{\dfrac{G(s)}{s}\right\}$
$\dfrac{1}{s}$	$\dfrac{z}{z-1}$	$\dfrac{T_0}{z-1}$
$\dfrac{1}{s^2}$	$\dfrac{T_0 z}{(z-1)^2}$	$\dfrac{T_0^2(z+1)}{2(z-1)^2}$
$\dfrac{1}{s^3}$	$\dfrac{T_0^2 z(z+1)}{(z-1)^3}$	$\dfrac{T_0^3(z^2+4z+1)}{6(z-1)^3}$
$\dfrac{1}{s^4}$	$\dfrac{T_0^3 z(z^2+4z+1)}{6(z-1)^4}$	$\dfrac{T_0^4(z^3+11z^2+11z+1)}{24(z-1)^4}$
$\dfrac{1}{s+a}$	$\dfrac{z}{z-e^{-aT_0}}$	$\dfrac{(1-e^{-aT_0})}{a(z-e^{-aT_0})}$
$\dfrac{1}{(s+a)^2}$	$\dfrac{T_0 z e^{-aT_0}}{(z-e^{-aT_0})^2}$	$\dfrac{(1-e^{-aT_0}(1+aT_0))z+e^{-aT_0}(e^{-aT_0}-1+aT_0)}{a^2(z-e^{-aT_0})^2}$
$\dfrac{1}{(s+a)(s+b)}$	$\dfrac{1}{(b-a)}\dfrac{(e^{-aT_0}-e^{-bT_0})z}{(z-e^{-aT_0})(z-e^{-bT_0})}$	$\dfrac{1}{ab(a-b)}\dfrac{(Az+B)}{(z-e^{-aT_0})(z-e^{-bT_0})}$

$$A=a-b-ae^{-bT_0}+be^{-aT_0}$$
$$B=(a-b)e^{-(a+b)T_0}-ae^{-aT_0}+be^{-bT_0}$$

3.4.3 Properties of the z-Transfer Function and Difference Equations

Proportional Behaviour

For processes with proportional behaviour, the *gain* can be obtained by using the finite value theorem as

$$K=\frac{y(k\to\infty)}{u(k\to\infty)}=\frac{\lim_{z\to1}(z-1)y(z)}{\lim_{z\to1}(z-1)u(z)}=\lim_{z\to1}\frac{y(z)}{u(z)}$$

$$=\lim_{z\to1}G(z)=\frac{b_0+b_1+...+b_m}{1+a_1+...+a_m}. \tag{3.4.15}$$

Unlike $G(s)$ the sums of all parameters in the numerator and denominator have to be evaluated.

Integral Behaviour

As for instance demonstrated with $G(s)=1/T_I s$ and from the z-transform table for $G(z)=\mathscr{L}\{1/T_I s\}=z/T_I(z-1)=1/T_I(1-z^{-1})$ processes with integral behaviour have a pole at $z = 1$

$$G(z) = \frac{y(z)}{u(z)} = \frac{1}{(1-z^{-1})} \frac{b_0 + b_1 z^{-1} + ... + b_m z^{-m}}{1 + a_1' z^{-1} + ... + a_m' z^{-(m-1)}}. \qquad (3.4.16)$$

The "steady-state" gradient after a step input of height u_0 then is

$$\lim_{k \to \infty} \Delta y(k) = \lim_{k \to \infty} (y(k) - y(k-1))$$

$$= \lim_{z \to 1} y(z) (1 - z^{-1})$$

$$= \frac{b_0 + b_1 + ... + b_m}{1 + a_1' + ... + a_m'} u_0. \qquad (3.4.17)$$

If $b_0 \neq 0$ the system has a jump discontinuity at $k = 0$. However, for most real processes b_0 is zero.

Differentiating Behaviour

With finite sampling time a discrete-time transfer element with perfectly differentiating behaviour cannot be expected. Consider the use of the continuous-time smoothed differentiating element

$$G(s) = \frac{y(s)}{u(s)} = \frac{T_D s}{1 + T_1 s} \qquad (3.4.18)$$

Then together with the zero-order hold element one has

$$HG(z) = \frac{z-1}{z} \mathscr{L} \left\{ \frac{T_D}{1 + T_1 s} \right\} = \frac{b_0(z-1)}{(z+a_1)} = \frac{b_0(1-z^{-1})}{1 + a_1 z^{-1}} \qquad (3.4.19)$$

with $a_1 = e^{-T_0/T_1}$ and $b_0 = T_D/T_1$.

Hence, there is a pole at $z = -a_1$ and a zero at $z_{D1} = 1$. This implies the difference equation

$$y(k) + a_1 y(k-1) = b_0[u(k) - u(k-1)] = b_0 \Delta u(k) \qquad (3.4.20)$$

thus forming a first-order difference of the input signal. For $T_1 \to 0$

$$a_1 \to 0 \quad \text{und} \quad b_0 \to \infty$$

which can no longer be realized physically.

Deadtime

A deadtime element can be described by

$$y(t) = u(t - T_t)$$

and the s-transfer function

$$G(s) = \frac{y(s)}{u(s)} = e^{-T_t s}.$$

If the deadtime can be expressed by integers $d = T_t/T_0 = 1, 2, 3, \ldots$, then according to the shifting theorem

$$y(z) = z^{-d} u(z)$$

and thus the following results

$$G(z) = D(z) = \frac{y(z)}{u(z)} = z^{-d}. \tag{3.4.21}$$

For a time lag system $G(z)$ preceded or followed by a deadtime therefore yields

$$DG(z) = \frac{y(z)}{u(z)} = G(z) z^{-d}. \tag{3.4.22}$$

Hence, for discrete-time systems deadtime elements have the same type of transfer function as other dynamic elements. Contrary to continuous-time systems deadtime elements can be rather easily included.

Realizability

A discrete-time transfer system is realizable if the causality principle is satisfied, that is the output variable $y(k)$ does not depend on future values of the input signal $u(k+j)$, $j = 1, 2, \ldots$. This leads to:

a) a difference equation

$$a_0 y(k) + a_1 y(k-1) + \ldots + a_n y(k-n)$$
$$= b_0 u(k) + b_1 u(k-1) + \ldots + b_m u(k-m) \tag{3.4.23}$$

and the corresponding z-transfer function

$$G(z) = \frac{y(z)}{u(z)} = \frac{b_0 + b_1 z^{-1} + \ldots + b_m z^{-m}}{a_0 + a_1 z^{-1} + \ldots + a_n z^{-n}} \tag{3.4.24}$$

have the following realizability conditions:

— if $b_0 \neq 0$ then $a_0 \neq 0$,
— if $b_0 = 0$, $a_0 = 0$ and $b_1 \neq 0$ then $a_1 \neq 0$, etc. $\tag{3.4.25}$

This results in

$$m \lesseqgtr n.$$

b) a difference equation

$$a'_0 y(k) + a'_1 y(k+1) + \dots + a'_n y(k+n)$$
$$= b'_0 u(k) + b'_1 u(k+1) + \dots + b'_m u(k+m) \qquad (3.4.26)$$

and the corresponding z-transfer function

$$G(z) = \frac{y(z)}{u(z)} = \frac{b'_0 + b'_1 z + \dots + b'_m z^m}{a'_0 + a'_1 z + \dots + a'_n z^n} \qquad (3.4.27)$$

are realizable, if

$$m \leq n \quad \text{für} \quad a'_n \neq 0. \qquad (3.4.28)$$

Correspondence with the Impulse Response

The impulse response results from the difference equation, (3.4.23), with

$$y(k) = b_0 u(k) + b_1 u(k-1) + \dots + b_m u(k-m)$$
$$- a_1 y(k-1) - \dots - a_n y(k-n) \qquad (3.4.29)$$

introducing a δ-impulse described by

$$\left. \begin{aligned} u(0) &= 1 \\ u(k) &= 0 \quad \text{für} \quad k > 0. \end{aligned} \right\} \qquad (3.4.30)$$

This yields

$$\left. \begin{aligned} g(0) &= b_0 \\ g(1) &= b_1 - a_1 g(0) &&= b_1 - a_1 b_0 \\ g(2) &= b_2 - a_1 g(1) - a_2 g(0) = b_2 - a_1 b_1 + a_1^2 b_0 - a_2 b_0 \\ &\quad \cdot \\ &\quad \cdot \\ &\quad \cdot \\ g(k) &= b_k - a_1 g(k-1) - \dots - a_k g(0) && \text{für} \quad k \leq m \\ g(k) &= \quad - a_1 g(k-1) - \dots - a_k g(k-n) && \text{für} \quad k > m \end{aligned} \right\} \qquad (3.4.31)$$

Hence, the parameter b_0 equals $g(0)$. If $b_0 = 0$ then $g(1) = b_1$. The last equation of (3.4.31) shows that with increasing k the impulse response is more and more characterized by the parameters a_i.

Relationship with the Transient Function

From (3.4.29) follows, with

$$u(k) = 1 \; f\ddot{u}r \; k \geq 0 \qquad (3.4.32)$$

for the transient function (step response)

$$h(0) = b_0$$
$$h(1) = b_0 + b_1 - a_1 h(0) = b_0 + b_1 - a_1 b_0$$
$$h(2) = b_0 + b_1 + b_2 - a_1 h(1) - a_2 h(0)$$
$$= b_0 + b_1 + b_2 - a_1 b_0 + a_1 b_1 + a_1^2 b_0 - a_2 b_0$$

.
.
.

$$h(k) = b_0 + b_1 + ... + b_k - a_1 h(k-1) - a_2 h(k-2) - ... - a_k h(0)$$
$$\text{für } k \leq m$$
$$h(k) = b_0 + b_1 + ... + b_m - a_1 h(k-1) - a_2 h(k-2) - ... - a_n h(k-n)$$
$$\text{für } k > m$$
$$(3.4.33)$$

Relationship Between the Impulse Response and Step Response

The convolution sum, (3.4.3), leads with the unit step function $u(kT_0) = 1, k \geq 0$ as input signal together with the impulse response to the convolved transient response

$$h(n) = \sum_{k=0}^{\infty} g(n-k).$$
$$(3.4.34)$$

This shows that the step response follows from the superposition of time-shifted impulse responses.

Writing (3.4.34) for $n - 1$, yields

$$h(n-1) = \sum_{k=0}^{\infty} g(n-1-k)$$

and

$$h(n) = \sum_{k=0}^{\infty} g(n-1-k) + g(n).$$

For the evaluation of the impulse response with given transient function this results in

$$g(n) = h(n) - h(n-1) = \Delta h(n)$$
$$(3.4.35)$$

Cascaded Linear Sampled Data Systems

For a linear transfer system $G(s)$ with a sampled input and a sampled output the correspondence of the sampled signals (discrete-time signals following the sampler) $u(k)$ and $y(k)$, compare Fig. 3.8 and (3.4.7) and (3.4.9), is described by

$$y(z) = u(z) \cdot G(z) = u(z) \cdot \mathcal{Z}\{G(s)\}.$$
$$(3.4.36)$$

The multiplication sign emphasizes that there is a sampler between the continuous signal $u(t)$ and the transfer function $G(s)$. Equation (3.4.36) indicates that all

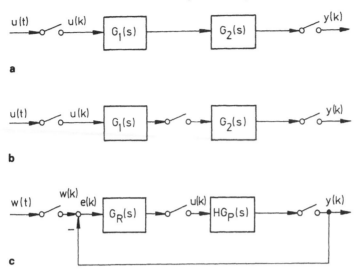

Fig. 3.11. Examples of cascaded systems with samplers. **a** Cascaded system not separated by a sampler; **b** Cascaded system separated by a sampler; **c** Sampled data control loop

elements not separated by a sampler must be multiplied out first. Only then can the operation $G(z) = \mathscr{Z}\{G(s)\}$ be formed.

According to Fig. 3.11a, the following equation becomes valid for cascaded systems not separated by a sampler

$$y(z) = u(z) \cdot \mathscr{Z}\{G_1(s)G_2(s)\} = u(z) \cdot G_1G_2(z) \qquad (3.4.37)$$

and for Fig. 3.11b with separating sampler

$$y(z) = u(z) \cdot \mathscr{Z}\{G_1(s)\} \cdot \mathscr{Z}\{G_2(s)\} = u(z) \cdot G_1(z) \cdot G_2(z). \qquad (3.4.38)$$

The separating sampler is thus described by the multiplication sign between $G_1(z)$ and $G_2(z)$. Hence

$$G_1G_2(z) \neq G_1(z) \cdot G_2(z). \qquad (3.4.39)$$

(Notice: For pure dead time elements in (3.4.39) the equality holds, as the sampler does not change the behavior.)

The reference transfer behaviour of a sampled data control loop, Fig. 3.11c), can be evaluated as follows

$$y(z) = HG_P(z) \cdot G_R(z) \, [w(z) - y(z)]$$

$$G_w(z) = \frac{y(z)}{w(z)} = \frac{HG_P(z) \cdot G_R(z)}{1 + HG_P(z) \cdot G_R(z)}.$$

Note that each sampler between the z-transformed signals and the z-transfer functions lead to a multiplication sign.

Example 3.9: Cascaded systems

The z-transfer function of two transfer elements according to Figure 3.11a and 3.11b are to be determined for

$$G_1(s) = \frac{K_1}{1 + T_1 s} \quad \text{und } G_2(s) = \frac{K_2}{1 + T_2 s}$$

a) $G(z) = \mathscr{Z}\{G_1(s)G_2(s)\} = \mathscr{Z}\left\{\frac{c}{(s+a)\,(s+b)}\right\}$

$$= \frac{c}{b-a} \frac{z(e^{-aT_0} - e^{-bT_0})}{(z - e^{-aT_0})(z - e^{-bT_0})}$$

$a = 1/T_1;\ b = 1/T_2;\ c = K_1 K_2\ a\ b$

b) $G(z) = \mathscr{Z}\{G_1(s)\} \cdot \mathscr{Z}\{G_2(s)\} = \mathscr{Z}\left\{\frac{aK_1}{s+a}\right\} \cdot \mathscr{Z}\left\{\frac{bK_2}{s+b}\right\}$

$$= \frac{a\ b\ K_1 K_2}{(z - e^{-aT_0})(z - e^{-bT_0})}$$

3.5 Poles and Zeros, Stability

A linear process is considered which for continuous signals is described by the rational transfer function

$$G(s) = \frac{y(s)}{u(s)} = \frac{B(s)}{A(s)} = \frac{b_0 + b_1 s + b_2 s^2 \ldots + b_m s^m}{1 + a_1 s + a_2 s^2 + \ldots + a_n s^n}$$

$$= \frac{(s - s_{01})(s - s_{02})\ldots(s - s_{0m})}{(s - s_1)(s - s_2)\ldots(s - s_n)} \frac{b_m}{a_n} \tag{3.5.1}$$

For the corresponding sampled process, according to section 3.4, without or with the zero-order hold, the rational z-transfer function results

$$G(z) = \frac{y(z)}{u(z)} = \frac{B(z^{-1})}{A(z^{-1})} = \frac{b_0 + b_1 z^{-1} + \ldots + b_m z^{-m}}{1 + a_1 z^{-1} + \ldots + a_n z^{-n}} \tag{3.5.2}$$

and after rewriting $(n > m)$

$$G(z) = \frac{B(z)}{A(z)} = \frac{(b_0 z^m + b_1 z^{m-1} + \ldots + b_m) z^{n-m}}{z^n + a_1 z^{n-1} + \ldots + a_n}$$

$$= \frac{(z - z_{01})(z - z_{02}) \ldots (z - z_{0n})}{(z - z_1)(z - z_2) \ldots (z - z_n)} \tag{3.5.3}$$

The roots z_i, $i = 1, 2, \ldots, n$ of the denominator polynominal $A(z) = 0$ are the *poles* and the roots z_{0i}, $i = 1, 2, \ldots, n$ of the numerator polynominal are the *zeros* of the z-transfer function.

In this way the poles describe the modal behaviour of the process and depend on its internal couplings. They are therefore relevant for stability. Zeros indicate how the input variable effects the internal variables and how those, in turn, influence the output variable.

3.5.1 Location of Poles in the z-Plane

In the following continuous transfer functions of first and second order are assumed. Forming the z-transfer function, it is investigated how real and complex-conjugate poles map into the z-plane.

Real Poles

It was shown in Example 3.7 that a first-order lag

$$G(s) = \frac{y(s)}{u(s)} = \frac{K'}{a + s} \tag{3.5.4}$$

with pole $s_1 = -a$ (continuous signals) and with no hold leads to the following transfer function

$$G(z) = \frac{y(z)}{u(z)} = \frac{K'z}{z - z_1} = \frac{K'}{1 - a_1 z^{-1}} \tag{3.5.5}$$

together with a pole

$$z_1 = a_1 = e^{-aT_0} \tag{3.5.6}$$

(Note: Sign of a_1 equals sign of z_1). The corresponding difference equation is

$$y(k) - a_1 y(k-1) = K'u(k). \tag{3.5.7}$$

For an initial value $y(0) \neq 0$ and for $u(k) = 0$ the homogeneous difference equation is

$$y(k) - a_1 y(k-1) = 0 \tag{3.5.8}$$

giving

$$\left.\begin{array}{l} y(1) = a_1 y(0) \\ y(2) = a_1 y(1) = a_1^2 y(0) \\ \quad \cdot \qquad\qquad\quad \cdot \\ \quad \cdot \qquad\qquad\quad \cdot \\ \quad \cdot \qquad\qquad\quad \cdot \\ y(k) = \qquad\quad\; a_1^k y(0). \end{array}\right\} \tag{3.5.9}$$

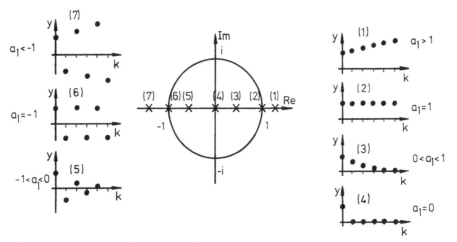

Fig. 3.12. Real poles and corresponding modal behaviour

This first-order system only converges to zero and is, only asymptotically stable for $|a_1| < 1$. The time behaviour of $y(k)$ for different positions of the pole a_1 in the z-plane is shown in Fig. 3.12. A negative value of a_1 results in a periodic behaviour with alternating signs, which is not possible for first order systems with continuous signals.

Since the poles in the s-plane and the z-plane are related by

$$z_1 = a_1 = e^{s_1 T_0} \tag{3.5.10}$$

the following correspondence results

$$\text{s-poles for } -\infty < s_1 < +\infty \rightarrow \text{z-poles for } 0 < z_1 < \infty.$$

This shows that positive real poles in the z-plane only result from real poles in the s-plane. Therefore, real negative poles $z_1 < 0$ have no corresponding poles on the real axis of the s-plane. (A negative real double pole for $z_1 < 0$, however, is equivalent to a special complex-conjugate pole-pair in the s-plane, as will be shown in Fig. 3.14).

Complex-conjugate Poles

The transfer function

$$G(s) = \frac{y(s)}{u(s)} = \frac{K}{T^2 s^2 + 2DTs + 1} = \frac{K\omega_{00}^2}{s^2 + 2D\omega_{00}s + \omega_{00}^2}$$

$$= \frac{K(a^2 + \omega_1^2)}{(s+a)^2 + \omega_1^2} = \frac{K(a^2 + \omega_1^2)}{(s-s_1)(s-s_2)} \tag{3.5.11}$$

with

$$a = D/T = D\omega_{00}$$

$$\omega_1^2 = \frac{1}{T^2}(1-D^2) = \omega_{00}^2(1-D^2)$$

$$s_{1,2} = -a \pm i\omega_1$$

(D damping coefficient; ω_{00} frequency of the undamped system) with no zero-order hold has the z-transfer function

$$G(z) = \frac{y(z)}{u(z)} = \frac{\left(\alpha K \dfrac{a^2 + \omega_1^2}{\omega_1}\sin\omega_1 T_0\right)z}{z^2 - (2\alpha\cos\omega_1 T_0)z + \alpha^2} = \frac{b_1 z}{(z-z_1)(z-z_2)}$$

Here $\alpha = e^{-aT_0}$.
The poles are

$$z_{1,2} = \alpha[\cos\omega_1 T_0 \pm i\sin\omega_1 T_0] = \alpha\, e^{\pm i\omega_1 T_0}$$

$$= e^{(-a \pm i\omega_1)T_0} \tag{3.5.13}$$

and the homogeneous equation becomes

$$y(k) - (2\alpha\cos\omega_1 T_0)y(k-1) + \alpha^2 y(k-2) = 0. \tag{3.5.14}$$

For arbitrary initial values $y(0)$ and the special initial value $y(1) = \alpha\cos\omega_1 T_0$, the solution of this equation is

$$y(k) = \alpha^k\cos\omega_1 kT_0 \cdot y(0). \tag{3.5.15}$$

The time behaviour is shown in Fig. 3.13 for positive values of α. Fig. 3.13a displays for constant $\omega_1 T_0 \triangleq 30°$ the transient behaviour for different α (hence different dampings).

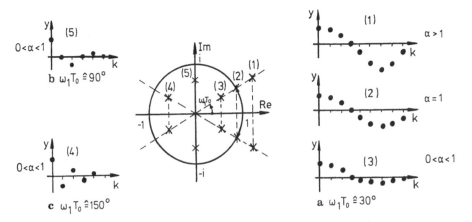

Fig. 3.13. Complex-conjugate pole-pair with corresponding modal behaviour (Shannon frequency: $\omega_{Sh} T_0 = \pi$)

If $\omega_1 T_0$ is increased, that means that e.g. ω_1 is increased for constant T_0, then the following results. For $\omega_1 T_0 \triangleq 90°$, the poles then are placed on the imaginary axis, $y(k)$ becomes 0 for $k = 1, 3, 5, \ldots$ and the sign changes for $k = 0, 2, 4, \ldots$. Alternating signs emerge for $90° < \omega_1 T_0 < 180°$. With $\omega_1 T_0 \triangleq 180°$ ω_1 transgresses Shannon's sampling frequency $\omega_{Sh} = \pi/T_0$, that means the oscillation with ω_1 cannot be further realized. If the complex-conjugate pole pairs are located in the z-plane within a circle of radius 1 (unit circle), then an asymptotically stable behaviour for $\alpha < 1$ results.

General Correspondence Between the Location of Poles in the s-Plane and z-Plane

For the real as well as for the complex-conjugate poles the transformation equation to map s-plane poles

$$s_j = \delta_j \pm i\omega_j \tag{3.5.16}$$

into the z-plane

$$z_j = e^{T_0 s_j} = e^{T_0(\delta_j \pm i\omega_j)} = e^{T_0 \delta_j} e^{\pm i\omega_j T_0}. \tag{3.5.17}$$

Consequently, the poles z_j have the magnitude and the phase angle

$$|z_j| = e^{T_0 \delta_j} \quad \text{und} \quad \varphi_j = \pm \omega_j T_0. \tag{3.5.18}$$

This is illustrated in Fig. 3.14. According to the transformation equation the following results:

a) The imaginary axis section $0 \leq i\omega_j \leq i\pi/T_0$ of the s-plane is mapped in the upper half circle with radius 1 of the z-plane.
b) The imaginary axis section $-i\pi/T_0 \leq i\omega_j \leq 0$ of the s-plane is mapped in the lower half circle with radius 1 of the z-plane.

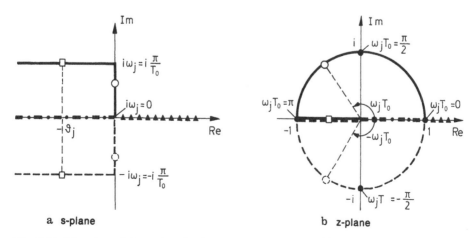

a s-plane b z-plane

Fig. 3.14. Mapping of the s-plane in the z-plane through $z = e^{T_0 s}$. **a** s-plane; **b** z-plane

c) The negative real axis $-\infty < \delta_j \leq 0$ of the s-plane ($i\omega_j = 0$) yields the positive axis section $0 < z_j \leq 1$ of the z-plane.

d) The positive real axis $\infty < \delta_j \leq 0$ of the s-plane results in the positive axis section $1 \leq z_j < \infty$ of the z-plane.

e) The parallels to the negative real axis for $i\omega_j = i\pi/T_0$ with $-\infty < \delta_j \leq 0$ of the s-plane are mapped in the negative axis section $-1 \leq z_j \leq 0$ of the z-plane.

f) Poles of the s-plane with $i\omega_j > i\pi/T_0$ (this means frequencies which do not satisfy Shannon's sampling theorem) are mapped inside the circle with radius 1 in the z-plane onto manifold Riemann areas.

g) The left-half s-plane is mapped in the inside of the circle with radius 1 (unit circle) of the z-plane. This is why *all asymptotically stable poles are inside the unit circle*.

h) The right-half s-plane are mapped outside of the unit circle which is where the *unstable poles* are located.

i) The *critically stable poles* on the imaginary axis of the s-plane are located on the unit circle.

Fig. 3.15 shows the pole positions in the s-plane and z-plane for constant characteristic values of a complex-conjugate pole-pair.

3.5.2 Stability Condition

According to section 3.5.1 a linear discrete-time system is *asymptotically stable*, if the poles of the transfer function or the roots of the *characteristic equation*

$$A(z) = (z-z_1)\ (z-z_2)\ \dots\ (z-z_m) = 0 \qquad (3.5.19)$$

are all located inside the unit circle. This then leads to

$$|z_j| < 1 \quad j = 1,2,\dots,m. \qquad (3.5.20)$$

If single poles are located on the unit circle, then the system is critically stable. For multiple poles on the unit circle, however, it becomes unstable.

3.5.3 Stability Analysis through Bilinear Transformation

The bilinear transformation

$$w = \frac{z - 1}{z + 1} \qquad (3.5.21)$$

maps the unit circle of the z-plane onto the imaginary axis of the w-plane. Therefore the inside of the unit circle is mapped into the left-half w-plane. Because the w-plane plays the same role as the s-plane for continuous-time systems the Hurwitz- or Routh-stability criterion can be applied [2.3]. For this purpose the inverse transformation

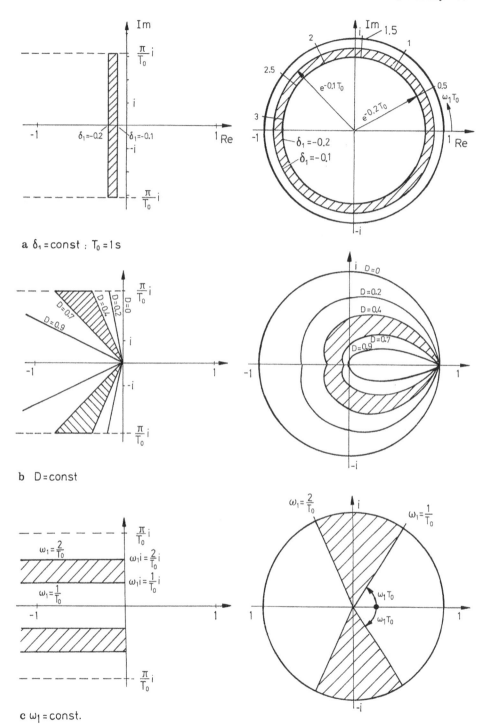

Fig. 3.15. Location of poles in the s-plane and z-plane for constant characteristic values, **a** δ_1 = const; T_0 = 1 s; **b** D = const; **c** ω_1 = const

$$z = \frac{1 + w}{1 - w} \tag{3.5.22}$$

is introduced into the denominator

$$A(z) = z^m + a_1 z^{m-1} + \dots + a_m \tag{3.5.23}$$

This leads to

$$\bar{A}(w) = \left(\frac{1 + w}{1 - w}\right)^m + a_1 \left(\frac{1 + w}{1 - w}\right)^{m-1} + \dots + a_m. \tag{3.5.24}$$

Multiplying by $(1 - w)^m$ yields

$$\bar{A}(w) = (1+w)^m + a_1(1+w)^{m-1}(1-w) + \dots + a_m(1-w)^m \tag{3.5.25}$$

Now the Hurwitz criterion can be applied for the polynomial

$$\bar{A}(w) = 0 \tag{3.5.26}$$

Hence the following is valid:

a) The system is not monotonic unstable, if all coefficients of the polynomial exist and carry the same algebraic sign.
b) The system is not oscillatory unstable, if the Hurwitz-determinants are positive for systems higher than second-order [2.21].

Example 3.10:
The second-order denominator is

$$A(z) = z^2 + a_1 z + a_2.$$

Then

$$\bar{A}(w) = \left(\frac{1+w}{1-w}\right)^2 + a_1 \left(\frac{1+w}{1-w}\right) + a_2$$

and

$$\bar{A}(w) = (1+w)^2 + a_1(1+w)(1-w) + a_2(1-w)^2$$

$$= (1 - a_1 + a_2)w^2 + 2(1 - a_2)w + (1 + a_1 + a_2).$$

If all coefficients of $\bar{A}(w)$ do exist and have positive signs it follows, in general, for second order systems that

$$A(z = -1) = 1 - a_1 + a_2 > 0$$
$$A(z = 1) \quad = 1 + a_1 + a_2 > 0$$
$$a_2 < 1$$

For higher order systems the application of this stability criterion is not so straightforward because of the calculation of the Hurwitz-determinants.

Example 3.11:

The stability analysis through bilinear transformation is to be applied to a sampled-data control loop consisting of the process

$$HG_p(z) = \frac{b_1'' z^{-1} + b_2'' z^{-2}}{1 + a_1'' z^{-1} + a_2'' z^{-2}}$$

and a proportional controller $G_R(z) = q_0$.

The characteristic equation

$$1 + HG_p(z) \cdot G_R(z) = 0$$

then becomes

$$z^2 + (a_1'' + b_1'' q_0) z + (a_2'' + b_2'' q_0) = 0$$

and the conditions are

$$q_0 < (1 - a_2'')/b_2''$$
$$q_0 > (a_1'' - a_2'' - 1)/(b_2'' - b_1'')$$
$$q_0 > - (1 + a_1'' + a_2'')/(b_1'' + b_2'')$$

With the values of the test process VIII (Appendix) follows

$$q_0 < 9{,}58; \; q_0 < 67{,}62; \; q_0 > -1$$

This indicates that the maximal permissible controller gain ($=$ loop gain) is $q_{0\,max} = 9.58$. Contrary to the continuous-time process which allows $q_0 \to \infty$, the gain of the sampled-data controller is bounded above by $q_{0\,max}$. This shows that in comparison with continuous control, sampled-data control has a smaller stability domain.

3.5.4 Schur–Cohn–Jury Criterion

By analogy to the Routh-Hurwitz stability criterion for systems with continuous signals, Schur, Cohn and Jury, compare e.g. [2.3, 2.15, 2.19, 2.21], developed conditions for the coefficients of a characteristic polynomial

$$A(z) = a_m' z^m + a_{m-1}' z^{m-1} + \dots + a_1' z + a_0' = 0 \tag{3.5.27}$$

provided the roots are located inside the unit circle. The following matrices are written

$$A_k = \begin{bmatrix} a_0' & a_1' & \cdots & a_{k-1}' \\ 0 & a_0' & \cdots & a_{k-2}' \\ & & & \vdots \\ 0 & \cdots & 0 & a_0' \end{bmatrix} \tag{3.5.28}$$

$$B_k = \begin{bmatrix} a_{m-(k-1)}' & \cdots & a_{m-1}' & a_m' \\ a_{m-(k-2)}' & \cdots & a_m' & 0 \\ & & & \\ a_m' & 0 & \cdots & 0 \end{bmatrix} \tag{3.5.29}$$

and the determinants

$$C_k = \det [A_k + B_k] \quad D_k = \det [A_k - B_k] \tag{3.5.30}$$

Necessary and sufficient conditions for the location of the roots of $A(z)$ inside of the unit circle are:

m even:	m odd:
$A(1)>0$	$(-1)^m A(-1)>0$
$C_2<0,\ D_2<0$	$C_1>0,\ D_1<0$
$C_4>0,\ D_4>0$	$C_3<0,\ D_3>0$
$C_6<0,\ D_6<0$	$C_5>0,\ D_5<0$
$\vdots$	$\vdots$

$$(3.5.31)$$

For higher order systems, however, the computational effort is comparatively high. Therefore another, more simple stability criterion based on the Schur–Cohn–Jury criterion is listed which is called "Jury's criterion" or "reduction method", [2.21, 2.25, 2.27].

The following scheme is written, assuming $a_m > 0$

row								
1	a'_0	a'_1	a'_2	$\ldots$	a'_{m-k}	$\ldots$	a'_{m-1}	a'_m
2	a'_m	a'_{m-1}	a'_{m-2}	$\ldots$	a'_k	$\ldots$	a'_1	a'_0
3	b_0	b_1	b_2	$\ldots$	b_{m-k}		b_{m-1}	
4	b_{m-1}	b_{m-2}	b_{m-3}	$\ldots$	b_k		b_0	
5	c_0	c_1	c_2	$\ldots$		c_{m-2}		
6	c_{m-2}	c_{m-3}	c_{m-4}			c_0		
$2m-5$	l_0	l_1	l_2	l_3				
$2m-4$	l_3	l_2	l_1	l_0				
$2m-3$	m_0	m_1	m_2					

$$(3.5.32)$$

The elements of the even rows are the elements of the preceeding row in inverse succession. They are followed by the elements of rows 3, 5, ... , $2m-5$, $2m-3$ from the determinants

$$b_k = \begin{vmatrix} a'_0 & a'_{m-k} \\ a'_m & a'_k \end{vmatrix} \qquad c_k = \begin{vmatrix} b_0 & b_{m-1-k} \\ b_{m-1} & b_k \end{vmatrix}$$

$$d_k = \begin{vmatrix} c_0 & c_{m-2-k} \\ c_{m-2} & c_k \end{vmatrix} \ldots \qquad\qquad (3.5.33)$$

$$\ldots m_0 = \begin{vmatrix} l_0 & l_3 \\ l_3 & l_0 \end{vmatrix} \qquad m_2 = \begin{vmatrix} l_0 & l_1 \\ l_3 & l_2 \end{vmatrix}$$

Necessary and sufficient conditions for the location of all roots of $A(z)$ inside the unit circle are

$$
\begin{aligned}
A(1) &> 0 \qquad\qquad (-1)^m A(-1) > 0 \\
|a_0'| &< a_m' \\
|b_0| &> |b_{m-1}| \\
|c_0| &> |c_{m-2}| \\
|d_0| &> |d_{m-3}| \\
&\ \,\vdots \qquad\quad \vdots \\
|m_0| &> |m_2|
\end{aligned}
\qquad\qquad\qquad (3.5.34)
$$

Hence, for an mth-order system $m+1$ conditions are obtained. Then for low order systems the following results hold

$$m=2: \quad A(1) > 0; \qquad A(-1) > 0; \qquad |a_0'| < a_2' \qquad\qquad (3.5.35)$$

$$m=3: \quad A(1) > 0; \qquad -A(-1) > 0; \qquad |a_0'| < a_3'; \qquad\qquad (3.5.36)$$
$$\qquad\qquad\qquad\qquad\qquad\qquad\qquad\quad |b_0| > |b_2|$$

If, in special cases, the first or the last element of a row is zero or complete rows are zero, then "singular cases" occur which have to be treated differently, [2.25].

For modified forms of the Jury-criterion one is referred e.g. to [2.21, 2.27].

Example 3.12:
A third order system has the characteristic equation

$$A(z) = z^3 + a_2'z^2 + a_1'z + a_0' = 0$$

The Jury-scheme then is

1	a_0'	a_1'	a_2'	1
2	1	a_2'	a_1'	a_0'
3	b_0	b_1	b_2	

and the stability conditions are

$$
\begin{aligned}
A(1) \quad &= 1 + a_2' + a_1' + a_0' > 0 \\
-A(-1) &= -1 + a_2' - a_1' + a_0' > 0 \\
|a_0'| &< 1 \\
|b_0| &> |b_2| \to |a_0'^2 - 1| > |a_0'a_2' - a_1'|
\end{aligned}
$$

It is because of the use of digital computers that characteristic equations are solved by *root calculation programs* instead of stability criterions. One then obtains not only a statement on the stability but also on the location of the poles.

3.5.5 Location of Zeros in the z-Plane

Unlike the poles, there is no simple relationship between the zeros of $G(s)$ and $G(z)$. The relations are rather complicated instead, as already shown in the transformation table. This is why general statements in closed form can be only made for special cases.

Using the continuous transfer function (3.5.1) and provided $n<m$, that is $G(s)\to0$ for $|s|\to\infty$ (passive, realizable transfer element and therefore without jump discontinuity, $G(s)$ has maximally $(n-1)$ zeros and $G(z)$ generally $(n-1)$ zeros ($b_0=0$ in (3.5.3)). For $G(z)$, however, zeros already occur, when $G(s)$ has no zeros, as can be seen by comparing the transformation tables. The zeros are further influenced by the holding element. This can be seen from Table 3.1 where the z-transfer functions for some simple transfer elements without and with zero-order hold are compared.

For transfer elements $G(s)=1/s^n$ without zeros Table 3.2 lists the zeros which result for $G(z)$ and $HG(z)$.

It can be seen that $G(z)$ for $n=3$ has a zero on the unit circle and for $n\geq4$ there is even a zero outside it. A first-order hold cancels the zero at $z_{01}=0$, however it introduces other zeros which for $n\geq2$ are located onto or outside the unit circle.

Also for lag elements $G(s)=1/\prod_{i=1}^{n}(s+a_i)$ without zeros, zeros occur at $G(z)$ for $n\geq1$ and $HG(z)\geq2$, compare Table 3.1. It is significant that, under certain conditions, the zeros for pure lag elements can be also placed outside the unit circle. This was examined in [3.16, 3.17]. General statements, however, are only possible for limit cases:

a) With $G(s)$ according (3.5.1) and $m<n$ the zeros of $HG(z)$ yield for the limit case $T_0\to0$:

$$- \ m \text{ zeros: } \lim_{T_0\to0} z_{0i} = \lim_{T_0\to0} e^{-T_0 s_{0i}} = 1$$

$-n-m-1$ zeros tend towards the zeros of $HG(z)$ of the element $1/s^{n-m}$

b) If $G(s)$ is asymptotically stable and $G(0)\neq0$ then for $T_0\to\infty$ all zeros of $HG(z)$ tend towards zero.

From this it follows that the zeros of transfer elements with a pole excess $n-m>2$ are located outside the unit circle provided the sampling time T_0 is sufficiently small.

Table 3.2. Location of the zeros of $G(z)$ and $HG(z)$ for $G(s)=1/s^n$

	$\dfrac{1}{s}$	$\dfrac{1}{s^2}$	$\dfrac{1}{s^3}$	$\dfrac{1}{s^4}$
$G(z)$				
$G(z)$	$z_{01}=0$	$z_{01}=0$	$z_{01}=0$ $z_{02}=-1$	$z_{01}=0$ $z_{02}=-0.268$ $z_{03}=-3.732$
$HG(z)$	$-$	$z_{01}=-1$	$z_{01}=-0.268$ $z_{02}=-3.732$	$z_{01}=-0.101$ $z_{02}=-1$ $z_{03}=-9.899$

Zeros outside the unit circle are called "unstable zeros" or one can speak about "unstable inverse transfer functions".

Example 3.13:
It is assumed that

$$G(s) = \frac{K}{(1+Ts)^3} = \frac{a^3 K}{(a+s)^3}; \quad T_0 = 10 \text{ s}$$

For the z-transfer function together with zero-order hold then

$$HG(z) = \frac{b_1 z^{-1} + b_2 z^{-2} + b_3 z^{-3}}{1 + a_1 z^{-1} + a_2 z^{-2} + a_3 z^{-3}} = \frac{(z-z_{01})(z-z_{02})}{(z-z_1)^3}$$

The poles and zeros dependent on the sampling time T_0 are

T_0/T	0.1	0.5	1.0	1.5	2.0	3.0	5.0
z_1	0.9048	0.6065	0.3679	0.2231	0.1353	0.0498	0.0067
z_{01}	−3.4631	−2.5786	−1.7990	−1.2654	−0.8958	−0.4537	−0.1152
z_{02}	−0.2485	−0.1832	−0.1238	−0.0827	−0.0547	−0.0232	−0.0038

As an example, for $T_0/T < 1.84$ it yields for the zero $z_{01} < -1.0$ indicating an outside of the unit circle location. The pure third order lag process with zero-order hold thus obtains a "inverse unstable" transfer function provided the sampling time is sufficiently small.

It may however also happen that $G(s)$ has a zero in the right half-plane, which makes it an *all-pass element*, and that $HG(z)$ does not possess zeros outside of the unit circle. For this case the sampling time has to be sufficiently large [3.16].
Deadtime elements represent another type of elements with non-minimal phase behaviour. For $G(s) = \exp(-T_t s)$ follows

$$HG(z) = z^{-d} = \frac{1}{z^d} \tag{3.5.37}$$

provided $d = T_t/T_0 = 0, 1, 2, \ldots$ is an integer. Hence, a deadtime element contains a d multiple pole at $z = 0$. Should, however, d be not an integer unstable zeros might occur.
This leads for $G(s) = (1/s) \exp(-T_t s)$ to

$$HG(z) = \frac{(T_0 - \tau) + \tau z^{-1}}{1 - z^{-1}} z^{-1} = \frac{(T_0 - \tau)z + \tau}{z(z-1)} \tag{3.5.38}$$

provided $0 < \tau < T_0$. Zeros outside the unit circle then result for $T_0/2 < \tau < T_0$ [3.16].
These examples show that $G(z)$ generally possesses $(n-1)$ zeros, provided $G(s)$ is without zeros and that these zeros may be located outside the unit circle provided the sampling time is small or the dead time is not an integer multiple of the sampling time. Hence, unstable zeros may always occur.

3.6 State Variable Representation

In modern control theory and especially in the design of control systems for processes with complicated behaviour and for multivariable processes, the state variable representation plays an important role. State variable representation not only allows the presentation of the input/output behaviour which, up to now has been exclusively considered, Fig. 3.16a, but also gives an insight into the inner relations through consideration of internal variables, the state variables x, Fig. 3.16b. These state variables describe the influence of the former input values and hence the state of the dynamic systems "memory".

There are various possible methods of determining the state variables. In the following, however, only two methods are presented: the solution of the continuous time vector differential equation for a linear system with zero-order hold and the introduction, by direct substitution, in the difference equation.

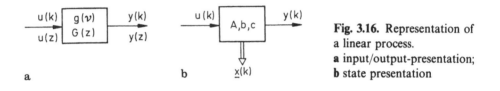

Fig. 3.16. Representation of a linear process.
a input/output-presentation;
b state presentation

3.6.1 The Vector Difference Equation Based on Vector Differential Equation

In the following it is assumed that the vector differential equation of the continuous-time system with single-input single-output and state variables $x(t)$ of dimension m is given. The state differential equation then is

$$\dot{x}(t) = Ax(t) + bu(t) \tag{3.6.1}$$

and for the output equation holds

$$y(t) = c^T x(t) + du(t) \tag{3.6.2}$$

There are various ways to define the state variables $x(t)$. With elementary state representation the definition of the state variables results directly from the equations of the theoretical modelling, by choosing the output variables of the storages as state variables. Other state variables can then be defined through mathematical transformations, as e.g. in the controllable canonical form and in the observable canonical form, the choice of which depends on the later application. The state differential equation can be solved with the Laplace-transform.

For (3.6.1)

$$sx(s) - x(0^+) = Ax(s) + bu(s) \tag{3.6.3}$$

becomes valid whereby $x(0^+)$ represents the initial state. Solving for $x(s)$ yields

$$x(s) = [sI - A]^{-1}x(0^+) + [sI - A]^{-1}bu(s). \tag{3.6.4}$$

Using the inverse Laplace transform one obtains

$$x(t) = \Phi(t) x(0^+) + \int_0^t \Phi(t-\tau) bu(\tau) d\tau \qquad (3.6.5)$$

where

$$\Phi(t) = \mathcal{L}^{-1}\{[sI-A]^{-1}\} = e^{At} \qquad (3.6.6)$$

This is analogous to the scalar case

$$\mathcal{L}^{-1}\{[s-a]^{-1}\} = e^{at} = \sum_{v=0}^{\infty} \frac{(at)^v}{v!}.$$

$\Phi(t)$ is called *transition matrix*, defined by the series expansion

$$\Phi(t) = e^{At} = \sum_{v=0}^{\infty} \frac{(At)^v}{v!}. \qquad (3.6.7)$$

Vector Difference Equation

For sampled input and output signals, the state representation can be derived simply from (3.6.5) and (3.6.7), provided the linear process is preceded by a zero-order hold as in Fig. 3.7. Then for the input signal:

$$u(t) = u(kT_0) \quad \text{for} \quad kT_0 \leq t < (k+1)T_0$$

and with the initial value $x(kT_0)$ for $kT_0 \leq t < (k+1)T_0$

$$x(t) = \Phi(t-kT_0)x(kT_0) + u(kT_0) \int_{kT_0}^t \Phi(t-\tau)bd\tau. \qquad (3.6.8)$$

becomes valid.

If the solution for only $t = (k+1)T_0$ is of interest, then

$$x((k+1)T_0) = \Phi(T_0)x(kT_0) + u(kT_0) \int_{kT_0}^{(k+1)T_0} \Phi((k+1)T_0-\tau)bd\tau \qquad (3.6.9)$$

and with the substitution $q = (k+1)T_0 - \tau$ and with $dq = -d\tau$ follows

$$x(k+1) = \Phi(T_0)x(k) + u(k) \int_0^{T_0} \Phi(q)bdq. \qquad (3.6.10)$$

By introducing the abbreviations

$$\left. \begin{array}{l} A = \Phi(T_0) = e^{AT_0} \\[2ex] b = \int_0^{T_0} \Phi(q)bdq \end{array} \right\} \qquad (3.6.11)$$

the *vector difference equation* is obtained

$$x(k+1) = A x(k) + bu(k) \qquad (3.6.12)$$

and using (3.6.2) the output equation

$$y(k) = c^T x(k) + du(k) \qquad (3.6.13)$$

with $c^T = c^T$ and $d = d$.

With series expansion the following calculations result for A and b [2.19]

$$
\left.
\begin{aligned}
A &= e^{AT_0} \approx I + \sum_{v=0}^{M} (AT_0)^{v+1} \frac{1}{(v+1)!} = I + AL \\[2ex]
L &= T_0 \sum_{v=0}^{M} (AT_0)^{v} \frac{1}{(v+1)!} \\[2ex]
b &\approx \int_0^{T_0} \sum_{v=0}^{M} A^v \frac{q^v}{v!} b \, dq = \sum_{v=0}^{M} A^v \int_0^{T_0} q^v dq \, b \frac{1}{v!} = Lb
\end{aligned}
\right\} \qquad (3.6.14)
$$

Hence, for finite element number $M+1$ of the series expansion one element more is used for A than for b. For large M this however is insignificant.

Example 3.14: Determination of the vector difference equation through continuous-time state representation. For a second-order oscillator the following differential equation is valid

$$\ddot{y}(t) + a_1 \dot{y}(t) + a_0 y(t) = b_0 u(t).$$

The corresponding state representation (in the controllable canonical form) with the state variables $x_1(t) = y(t)/b_0$ and $x_2(t) = \dot{y}(t)/b_0$ is written

$$
\dot{x}(t) = \begin{bmatrix} 0 & 1 \\ -a_0 & -a_1 \end{bmatrix} x(t) + \begin{bmatrix} 0 \\ 1 \end{bmatrix} u(t)
$$

$$y(t) = [b_0 \ 0] \, x(t).$$

From this A, b and c^T of the discrete-time representation can be calculated as follows

a) Inverse Laplace-transform

$$
[sI - A]^{-1} = \begin{bmatrix} s & -1 \\ a_0 & (s+a_1) \end{bmatrix}^{-1} = \frac{1}{s^2 + a_1 s + a_0} \begin{bmatrix} s+a_1 & 1 \\ -a_0 & s \end{bmatrix}
$$

The inverse Laplace transform yields for $a_1^2/4 - a_0 < 0$ and $\gamma = \sqrt{a_0 - a_1^2/4}$

$$
A = \Phi(T_0) = \begin{bmatrix}
\dfrac{1}{\gamma} e^{-\frac{a_1}{2}T_0} \left[\dfrac{a_1}{2} \sin\gamma T_0 + \gamma\cos\gamma T_0 \right] & \dfrac{1}{\gamma} e^{-\frac{a_1}{2}T_0} \sin\gamma T_0 \\[3ex]
-\dfrac{a_0}{\gamma} e^{-\frac{a_1}{2}T_0} \sin\gamma T_0 & \dfrac{1}{\gamma} e^{-\frac{a_1}{2}T_0} \left[-\dfrac{a_1}{2} \sin\gamma T_0 + \gamma\cos\gamma T_0 \right]
\end{bmatrix}
$$

$$
b = \begin{bmatrix}
\dfrac{1}{\gamma a_0} e^{-\frac{a_1}{2}T_0} \left[-\dfrac{a_1}{2} \sin\gamma T_0 - \gamma\cos\gamma T_0 \right] + \dfrac{1}{a_0} \\[3ex]
\dfrac{1}{\gamma} e^{-\frac{a_1}{2}T_0} \sin\gamma T_0
\end{bmatrix}
$$

b) *series expansion (break-off after the third term)*

$$A = \Phi(T_0) = e^{AT_0} \approx I + AT_0 + A^2 T_0^2/2 + \dots$$

$$= \begin{bmatrix} 1 & 0 \\ 0 & 1 \end{bmatrix} + T_0 \begin{bmatrix} 0 & 1 \\ -a_0 & -a_1 \end{bmatrix} + \frac{T_0^2}{2} \begin{bmatrix} -a_0 & -a_1 \\ a_0 a_1 & a_1^2 - a_0 \end{bmatrix}$$

$$= \begin{bmatrix} 1 - a_0 \dfrac{T_0^2}{2} & T_0\left(1 - a_1 \dfrac{T_0}{2}\right) \\ -a_0 T_0\left(1 - a_1 \dfrac{T_0}{2}\right) & 1 - a_1 T_0 + (a_1^2 - a_0)\dfrac{T_0^2}{2} \end{bmatrix}$$

$$b = \int_0^{T_0} \Phi(q) \begin{bmatrix} 0 \\ 1 \end{bmatrix} dq$$

$$= \begin{bmatrix} \left[\dfrac{1}{2} - \dfrac{a_1}{b} T_0\right] T_0^2 \\ \left[1 - \dfrac{a_1}{2} T_0 - \dfrac{a_1^2 - a_0}{b} T_0^2\right] \end{bmatrix}.$$

The results of both computation methods are now to be compared by a numerical example. For $D = 0.7$ and $\omega_{00} = 11/s$ and $a_1 = 2D\omega_{00} = 1,41/s$ and $a_0 = \omega_{00}^2 = 11/s^2$. This leads to

$T_0 = 1$ s:

a) $A = \begin{bmatrix} 0,6940 & 0,4554 \\ -0,4554 & 0,0563 \end{bmatrix}$ $b = \begin{bmatrix} 0,3059 \\ 0,4554 \end{bmatrix}$

b) $A = \begin{bmatrix} 0,5 & 0,3 \\ -0,3 & 0,08 \end{bmatrix}$ $b = \begin{bmatrix} 0,2667 \\ 0,46 \end{bmatrix}$

$T_0 = 0,1$ s:

a) $A = \begin{bmatrix} 0,9952 & 0,0932 \\ -0,0932 & 0,8648 \end{bmatrix}$ $b = \begin{bmatrix} 0,0048 \\ 0,0932 \end{bmatrix}$

b) $A = \begin{bmatrix} 0,9950 & 0,0930 \\ -0,093 & 0,8648 \end{bmatrix}$ $b = \begin{bmatrix} 0,0045 \\ 0,0932 \end{bmatrix}$

This example shows that calculation by hand can be only recommended for low-order systems and in the case of series expansion only for small sampling times. Therefore it is more expedient to work with computer programms using a larger number of elements of the series expansion.

3.6.2 The Vector Difference Equation Based on Difference Equation

If a difference equation for the system with discrete-time signals already exists, then its state representation can be derived by the introduction of state variables as shown in the following.

From the difference equation (3.4.20) and substituting of k with $k+n$, follows

$$y(k+n) + a_1 y(k+n-1) + \dots + a_n y(k)$$
$$= b_0 u(k+n) + b_1 u(k+n-1) + \dots + b_n u(k) \qquad (3.6.15)$$

To simplify matters let $m=n$. The case for $m \neq n$ can always be taken into account by setting the parameters to zero.

The corresponding transfer function is

$$G(z) = \frac{y(z)}{u(z)} = \frac{b_0+b_1z^{-1}+...+b_nz^{-n}}{1+a_1z^{-1}+...+a_nz^{-n}}$$

(3.6.16)

$$= \frac{b_0z^n+b_1z^{n-1}+...+b_n}{z^n+a_1z^{n-1}+...+a_n}.$$

The following state variables are introduced

$$y(k) = \qquad x_1(k) \qquad\qquad\qquad (3.6.17)$$

$$
\left.
\begin{aligned}
y(k+1) &= x_2(k) = x_1(k+1) \\
y(k+2) &= x_3(k) = x_2(k+1) \\
\vdots \qquad &\quad \vdots \qquad\quad \vdots \\
y(k+n-1) &= x_n(k) = x_{n-1}(k+1) \\
y(k+n) &\qquad = x_n(k+1)
\end{aligned}
\right\}
\qquad (3.6.18)
$$

This was based on the following considerations:

a) For the present time instant k the output signals $y(k+n-1), \ldots, y(k+1)$ are used as state variables. This corresponds with a "snapshot" of the present output signal $y(k)$ and the $n-1$ future values (or the output signal $y(k+n-1)$ and its $n-1$ past values).
b) Note, that with increasing time k to $k+1$, $k+1$ to $k+2$, e.g. $y(k)$ assumes the value of $y(k+1)$, $y(k+1)$ the value of $y(k+2)$, etc. thus leading to $x_1(k+1)$ $=x_2(k)$ and $x_2(k+1)=x_3(k)$. The difference equation hence expresses the "interval shifting" of the signals $y(k)$ for the current time k.

Substitution of (3.6.18) in (3.6.15) gives, for $b_n=1$ and $b_0, b_1, \ldots, b_{n-1}=0$

$$y(k+n) = x_n(k+1) = - a_1x_n(k) - a_2x_{n-1}(k) - ...$$
$$- a_nx_1(k) + 1u(k) \qquad\qquad (3.6.19)$$

(3.6.18) and (3.6.19) lead to the *vector difference equation* consisting of the *state difference equation*

$$
\begin{bmatrix} x_1(k+1) \\ x_2(k+1) \\ \vdots \\ x_n(k+1) \end{bmatrix}
=
\begin{bmatrix}
0 & 1 & 0 & \cdots & 0 \\
0 & 0 & 1 & \cdots & 0 \\
0 & 0 & 0 & & 1 \\
-a_n & -a_{n-1} & -a_{n-2} & \cdots & -a_1
\end{bmatrix}
\begin{bmatrix} x_1(k) \\ x_2(k) \\ \vdots \\ x_n(k) \end{bmatrix}
+
\begin{bmatrix} 0 \\ 0 \\ \vdots \\ 1 \end{bmatrix}
u(k) \quad (3.6.20)
$$

and the *output equation*

$$y(k) = [1 \ 0 \ ... \ 0] \quad \begin{bmatrix} x_1(k) \\ x_2(k) \\ \vdots \\ x_n(k) \end{bmatrix} \tag{3.6.21}$$

After introduction of a state vector x, a discrete-time system A matrix, a control vector b and an output vector c

$$x(k+1) = A\,x(k) + bu(k) \tag{3.6.22}$$

$$y(k) \quad = c^T x(k). \tag{3.6.23}$$

If the last two equations are solved one after another, it is expedient to write (3.6.23)

$$y(k+1) = c^T x(k+1) \tag{3.6.24}$$

On the one hand the system's matrix A expresses with the values 1 parallel to the diagonal the internal dependence of the state variables through the interval shifting. Whilst on the other hand the coefficients a_i indicate in the lowest row the dependence of the earliest state variable $x_n(k+1)$ on the remaining ones. This means that system-inherent internal feedbacks are considered. The control vector b indicates the effect of the input signal on the state variables and the output vector c how the output variables emerge from the state variables.

From (3.6.16) and (3.6.17) follows for $b_n = 1$ and $b_0, \ldots, b_{n-1} = 0$

$$y(z) \quad = \frac{1}{z^n + a_1 z^{n-1} + ... + a_n} u(z) = x_1(z). \tag{3.6.25}$$

If, however, $b_n \neq 1$ and $b_0, \ldots, b_{n-1} \neq 0$, (3.6.16) and (3.6.25) lead to

$$y(z) = b_n x_1(z) + b_{n-1} z x_1(z) + ... + b_0 z^n x_1(z)$$

or $\tag{3.6.26}$

$$y(k) = b_n x_1(k) + b_{n-1} x_1(k+1) + ... + b_0 x_1(k+n).$$

From (3.6.18)

$$y(k) = b_n x_1(k) + b_{n-1} x_2(k) + ... + b_1 x_n(k) + b_0 x_n(k+1). \tag{3.6.27}$$

is also valid. Here $x_n(k+1)$ comes from (3.6.19), leading finally to

$$y(k) = (b_n - b_0 a_n) x_1(k) + (b_{n-1} - b_0 a_{n-1}) x_2(k) +$$
$$... + (b_1 - b_0 a_1) x_n(k) + b_0 u(k). \tag{3.6.28}$$

In vector notation, the extended output equation becomes

$$y(k) = [(b_n - b_0 a_n) \dots (b_1 - b_0 a_1)] \begin{bmatrix} x_1(k) \\ \vdots \\ x_n(k) \end{bmatrix} + b_0 u(k) \qquad (3.6.29)$$

$$y(k) = c^T x(k) + d u(k).$$

For $b_0 = 0$ (3.6.29) gives

$$y(k) = [b_n \dots b_1] \begin{bmatrix} x_1(k) \\ \vdots \\ x_n(k) \end{bmatrix}. \qquad (3.6.30)$$

This choice of state variables is referred to as "controllable canonical form". Fig. 3.17 shows a block diagram of the state representation for a difference equation resulting from (3.6.18), (3.6.19) and (3.6.27). The following can be seen:

— the clockwise shifting of the state variable values through the blocks with the time shift operator z^{-1} (corresponding to $1/s$ for continuous-time signals)
— the internal feedback of the state variables with the weights a_i (homogeneous difference equation)
— forming of the output variable through summation of the state variables weighted by b_i.

Finally the vector difference equation becomes

$$x(k+1) = A x(k) + bu(k) \qquad (3.6.31)$$

$$y(k) = c^T x(k) + d u(k) \qquad (3.6.32)$$

Their block diagram is shown in Fig. 3.18.

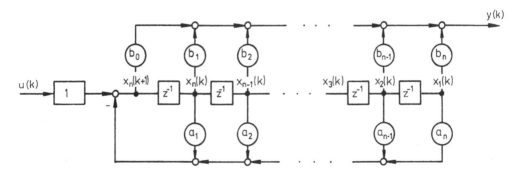

Fig. 3.17. Block diagram of the state representation of a difference equation

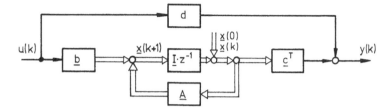

Fig. 3.18. Block diagram of a first order vector difference equation

3.6.3 Canonical Forms

The above state representation is only one of several possible realizations. By applying linear transformations

$$x_t = T\, x \tag{3.6.33}$$

an alternative state-space representation can be derived. T is necessarily a non-singular transformation matrix. The transformed representation then satisfies

$$x_t(k+1) = A_t x_t(k) + b_t u(k) \tag{3.6.34}$$

$$y(k) = c_t^T x_t(k) + d\, u(k) \tag{3.6.35}$$

with

$$\left. \begin{aligned} A_t &= T\, A\, T^{-1} \quad b_t = T\, b \\ c_t^T &= c^T T^{-1}. \end{aligned} \right\} \tag{3.6.36}$$

The characteristic equation (3.6.55) and hence the eigenvalues are not changed by the linear transformation. *Canonical forms* of state representation are specially structured forms of A_t, b_t and c_t. Some important canonical forms are given in Table 3.3 and their block diagrams in Fig. 3.19.

The *diagonal form*, corresponding to distinct eigenvalues, displays all the eigenvalues (modes) of A along the diagonal. Thus, for the discrete-time state variable model a system of decoupled first-order difference equations results. The superposition of the individual mode solutions then yields the output signals, as shown in the block diagram. Should A have multiple (non-distinct) eigenvalues, a diagonal form cannot be achieved. One then has to transform into a Jordan-form, (compare with e.g. [2.25]).

The *controllable canonical form* has an especially simple control vector b_t and, in the output vector c_t^T, the parameters b_i occur. With this form state feedback and state control systems can be easily developed as controllability is ensured for this representation [2.25].

Table 3.3. Canonical forms of state representation

	A_t	b_t	c_t	Remarks
diagonal form	$\begin{bmatrix} z_1 & 0 & \cdots & 0 \\ 0 & z_2 & \cdots & 0 \\ \vdots & \vdots & \ddots & \odot \\ 0 & 0 & \cdots & z_m \end{bmatrix}$	$=\begin{bmatrix} b_{1,D} \\ b_{2,D} \\ \vdots \\ b_{m,D} \end{bmatrix}$	$=\begin{bmatrix} c_{1,D} \\ c_{2,D} \\ \vdots \\ c_{m,D} \end{bmatrix}$	All eigenvalues $z_1, z_2, \ldots, z$ different. Correspond to part fraction expansion. A_t follo from: $T^{-1}A_t = AT^{-1}$. b_t or $c_t = [1\ 1 \ldots 1]^T$ if proce controllable and observable
column companion canonical form	$\begin{bmatrix} 0 & \cdots & 0 & -a_m \\ 1 & \cdots & 0 & -a_{m-1} \\ \vdots & \ddots & \vdots & \vdots \\ 0 & & 1 & -a_1 \end{bmatrix}$	$\begin{bmatrix} 1 \\ 0 \\ \vdots \\ 0 \end{bmatrix}$	$\begin{bmatrix} c^T b \\ c^T A b \\ \vdots \\ c^T A^{m-1} b \end{bmatrix}$	$c_t^T = [c_{1,s} c_{2,s} \cdots c_{m,s}]$ $c_t^T = [g(1), g(2), \ldots g(m)]$
controllable canonical form (regular form)	$\begin{bmatrix} 0 & 1 & \cdots & 0 \\ \vdots & & & \vdots \\ 0 & 0 & \cdots & 1 \\ -a_m & -a_{m-1} & \cdots & -a_1 \end{bmatrix}$	$\begin{bmatrix} 0 \\ \vdots \\ 0 \\ 1 \end{bmatrix}$	$\begin{bmatrix} b_m \\ \vdots \\ b_2 \\ b_1 \end{bmatrix}$	
row companion canonical form	$\begin{bmatrix} 0 & 1 & \cdots & 0 \\ \vdots & & \ddots & \vdots \\ 0 & 0 & & 1 \\ -a_m & -a_{m-1} & \cdots & -a_1 \end{bmatrix}$	$\begin{bmatrix} c^T b \\ c^T A b \\ \vdots \\ c^T A^{m-1} b \end{bmatrix}$	$\begin{bmatrix} 1 \\ 0 \\ \vdots \\ 0 \end{bmatrix}$	$b_t^T = [b_1{}_{,B} b_2{}_{,B} \cdots b_m{}_{,B}]$ $b_t^T = [g(1), g(2), \ldots, g(m)]$
observable canonical form	$\begin{bmatrix} 0 & \cdots & 0 & -a_m \\ 1 & \cdots & 0 & -a_{m-1} \\ \vdots & \ddots & \vdots & \vdots \\ 0 & \cdots & 1 & -a_1 \end{bmatrix}$	$\begin{bmatrix} b_m \\ b_{m-1} \\ \vdots \\ b_1 \end{bmatrix}$	$\begin{bmatrix} 0 \\ 0 \\ \vdots \\ 1 \end{bmatrix}$	

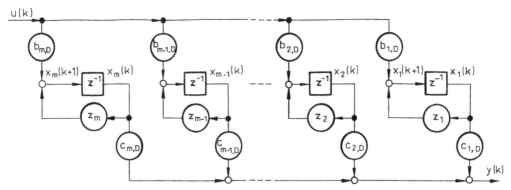

Diagonalform

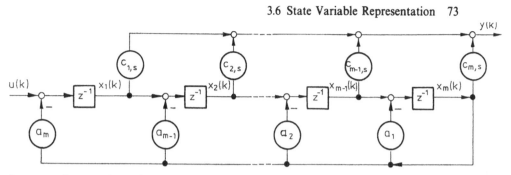

Column - Companion - Canonical - Form

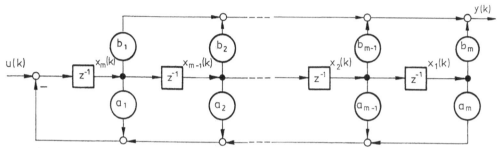

Controllable - Canonical - Form

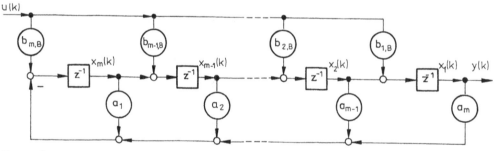

Row - Companion - Canonical - Form

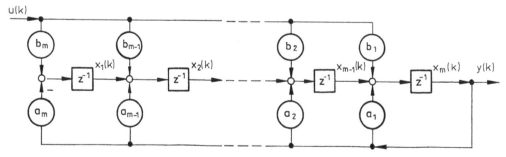

Observable - Canonical - Form

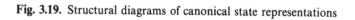

Fig. 3.19. Structural diagrams of canonical state representations

The *observable canonical form* is characterized by a particularly simple output vector c_t^T and contains, in the control vector, only the parameters b_i. This form is especially suited for generating state observers. Observability is ensured for this representation. The advantage of the two last-named forms is furthermore the use of the same a_i and b_i as for the input/output difference equation.

3.6.4 Processes with Deadtime

If a process with a *deadtime* $d = T_t/T_0 = 1, 2, \ldots$ is to be described by a state model which possesses a dynamic part with lumped parameters, one has to distinguish whether the deadtime exists at the input, at the output or between the state variables.

For *deadtime at the input* one has

$$x(k+1) \quad = A\,x(k) + bu(k-d) \tag{3.6.37}$$

$$y(k) \quad\quad = c^T x(k) \tag{3.6.38}$$

and for *deadtime at the output*

$$x(k+1) = A\,x(k) + bu(k) \tag{3.6.39}$$

$$y(k+d) = c^T x(k). \tag{3.6.40}$$

The deadtime can be represented as a series of d delay elements at the input or at the output, and the state variables of this series can be included in the state vector. For a deadtime at the input (c.f. Fig. 3.20) the following state representation holds

$$x_u(k+1) \; = \; A_u x_u(k) + b_u u(k) \tag{3.6.41}$$

$$u(k-d) \; = \; c_u^T x_u(k) \tag{3.6.42}$$

with

$$x_u^T(k) = [u(k-d)\,u(k-d-1)\ldots u(k-1)]$$

$$A_u = \begin{bmatrix} 0 & 1 & 0 & \cdots & 0 \\ 0 & 0 & 1 & \cdots & 0 \\ \vdots & & & \vdots & \vdots \\ 0 & 0 & 0 & \cdots & 1 \\ 0 & 0 & 0 & \cdots & 0 \end{bmatrix} \quad b_u = \begin{bmatrix} 0 \\ 0 \\ \vdots \\ \vdots \\ 1 \end{bmatrix} \tag{3.6.43}$$

$$c_u^T = [1 \;\; 0 \;\; 0 \;\; \cdots \;\; 0]$$

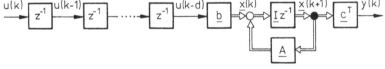

Fig. 3.20. State representation of a linear process with deadtime at the input

Note here that the matrix A contains one row with zeros. Combining (3.6.37) and (3.6.41) the state representation gives

$$\begin{bmatrix} x(k+1) \\ x_u(k+1) \end{bmatrix} = \begin{bmatrix} A & bc_u^T \\ 0 & A_u \end{bmatrix} \begin{bmatrix} x(k) \\ x_u(k) \end{bmatrix} + \begin{bmatrix} 0 \\ b_u \end{bmatrix} u(k) \tag{3.6.44}$$

$$y(k) = [c^T 0] \begin{bmatrix} x(k) \\ x_u(k) \end{bmatrix} \tag{3.6.45}$$

or more concisely as

$$x_d(k+1) = A_d x_d(k) + b_d u(k) \tag{3.6.46}$$

$$y(k) = c_d^T x_d(k). \tag{3.6.47}$$

If the controllable canonical form is chosen, one obtains

$$A_d = \begin{bmatrix} 0 & 1 & 0 & \cdots & 0 & 0 & 0 & \cdots & 0 \\ & \vdots & & & & \vdots & & & \\ 0 & 0 & 0 & \cdots & 1 & 0 & 0 & \cdots & 0 \\ -a_m & -a_{m-1} & & \cdots & -a_1 & 1 & 0 & \cdots & 0 \\ 0 & & & \cdots & 0 & 0 & 1 & \cdots & 0 \\ & \vdots & & & & \vdots & & & \\ & & & & 0 & 0 & \cdots & 1 \\ 0 & & & \cdots & 0 & 0 & 0 & \cdots & 0 \end{bmatrix} \quad b_d = \begin{bmatrix} 0 \\ \vdots \\ 0 \\ \vdots \\ 0 \\ 1 \end{bmatrix} \quad c_d = \begin{bmatrix} b_m \\ b_1 \\ 0 \\ \vdots \\ 0 \\ 0 \end{bmatrix} \tag{3.6.48}$$

The characteristic equation is then e.g. for $m=2$ and $d=1$

$$\det[zI - A_d] = z(z^2 + a_1 z - a_2)$$

that means, that the deadtime $d=1$ introduces a pole at $z=0$. Resulting from (3.6.43), a pure deadtime process d can be obtained from the general transfer element by choosing $b_1, b_2, \ldots, b_{m-1} = 0$ and $a_1, a_2, \ldots, a_m = 0$ with $m = d$.

The state representation of deadtime processes is treated for example in [3.1, 3.2]. Also compare section 9.1.

Example 3.15:
A state representation in the *controllable canonical form* is to be determined for the discrete process given by

$$G(z) = \frac{b_1 z^{-1}}{1 + a_1 z^{-1}} z^{-2}$$

a) According (3.6.48) it yields for a deadtime at the input

$$A_d = \begin{bmatrix} -a_1 & 1 & 0 \\ 0 & 0 & 1 \\ 0 & 0 & 0 \end{bmatrix} \quad b_d = \begin{bmatrix} 0 \\ 0 \\ 1 \end{bmatrix} \quad c_d = \begin{bmatrix} b_1 \\ 0 \\ 0 \end{bmatrix}$$

b) From

$$G(z) = \frac{b_1 z^{-1} + b_2 z^{-2} + b_3 z^{-3}}{1 + a_1 z^{-1} + a_2 z^{-2} + a_3 z^{-3}}$$

follows with $b_1 = b_2 = 0$; $b_3 = b_1$ and $a_2 = a_3 = 0$ according to the general controllable canonical form in Table 3.3

$$A_d = \begin{bmatrix} 0 & 1 & 0 \\ 0 & 0 & 1 \\ 0 & 0 & -a_1 \end{bmatrix} \quad b_d = \begin{bmatrix} 0 \\ 0 \\ 1 \end{bmatrix} \quad c_d = \begin{bmatrix} b_1 \\ 0 \\ 0 \end{bmatrix}$$

The deadtime is here realized at the output.

3.6.5 Solution of the Vector Difference Equation

Two possible solutions of the vector difference equation (3.6.12) and (3.6.13)

$$x(k+1) = A\,x(k) + b\,u(k)$$
$$y(k) \quad = c^T x(k) + d\,u(k)$$

are presented.

A first possibility which corresponds to the recursive solution of difference equations for a given input signal $u(k)$ and initial conditions $x(0)$ is

$$\begin{aligned} x(1) &= A\,x(0) + b\,u(0) \\ x(2) &= A\,x(1) + b\,u(1) \\ &= A^2 x(0) + Ab\,u(0) + b\,u(1) \end{aligned}$$

(3.6.49)

.
.
.

$$x(k) = A^k x(0) + \underbrace{\sum_{i=1}^{k} A^{i-1} bu(k-i)}.$$

$$\begin{matrix} \text{homo-} & \text{particular} \\ \text{geneous} & \text{solution} \\ \text{solution} & \text{(convolution sum)} \end{matrix}$$

where

$$A^k = \underbrace{A \cdot A \,\ldots\ldots\, A}_{k}.$$

$y(k)$ can finally be obtained from (3.6.13).

If $u(k)$ is given explicitly as a z-transform, a second possible solution can be used.

The z-transform furnishes:

$$3\{x(k)\} = x(z)$$
$$3\{x(k+1)\} = z[x(z) - x(0)]$$

(applying the shifting theorem to the left-hand side)

Then, it follows from (3.6.12)

$$z[x(z) - x(0)] = A\,x(z) + b\,u(z) \qquad (3.6.50)$$

or

$$x(z) = [zI-A]^{-1}zx(0) + [zI-A]^{-1}b\,u(z) \qquad (3.6.51)$$

and with (3.6.13)

$$y(z) = c^T[zI-A]^{-1}z\,x(0) + [c^T[zI-A]^{-1}b+d]\,u(z). \qquad (3.6.52)$$

Comparing (3.6.51) and (3.6.49) the following condition is obtained

$$A^k = 3^{-1}\{[zI-A]^{-1}z\}. \qquad (3.6.53)$$

3.6.6 Determination of the z-Transfer Function

With initial state $x(0)=0$ and (3.6.52) it follows that:

$$G(z) = \frac{y(z)}{u(z)} = c^T[zI-A]^{-1}b + d$$

$$= \frac{c^T \text{ adj } [zI-A]b + d \det [zI-A]}{\det [zI-A]} \qquad (3.6.54)$$

The uncancelled denominator of the z-transfer function results in the *characteristic equation*

$$\det [zI-A] = 0. \qquad (3.6.55)$$

This yields e.g. for a second-order system:

$$\det\left[z\begin{bmatrix}1 & 0\\0 & 1\end{bmatrix} - \begin{bmatrix}0 & 1\\-a_2 & -a_1\end{bmatrix}\right] = \det\begin{bmatrix}z & -1\\a_2 & z+a_1\end{bmatrix}$$

$$= z^2 + a_1 z + a_2 = 0$$

3.6.7 Determination of the Impulse Response

Applying (3.6.13) and (3.6.49) taking $x(0)=0$

$$y(k) = \sum_{i=1}^{k} c^T A^{i-1} bu(k-i) + d\,u(k). \tag{3.6.56}$$

Introducing

$$u(k) = \begin{cases} 1 & k=0 \\ 0 & k>0 \end{cases}$$

the impulse response is then given by

$$\begin{aligned} g(0) &= d \\ g(k) &= c^T A^{k-1} b \quad \text{für} \quad k>0. \end{aligned} \tag{3.6.57}$$

Here the following relationship between the impulse response and the z-transfer function is obtained, (c.f. (3.4.7))

$$G(z) = \sum_{k=0}^{\infty} g(k)\, z^{-k} = d + \sum_{k=1}^{\infty} c^T A^{k-1} b\, z^{-k}. \tag{3.6.58}$$

3.6.8 Controllability and Observability

It is of fundamental significance whether with dynamic systems in state representation it is possible to control from a given initial state to a given final state and to determine the inner state variables from the measurable inputs and outputs. This is performed by examining controllability and observability.

Controllability

The literature lists various stipulations of controllability. Here, the following is to be used:

A linear dynamic process is said to be controllable if there exists a realizable control sequence $u(k)$ which will drive the state for the finite time interval N from any initial state $x(0)$ to any final state $x(N)$.

(The term "reachability" is used instead of "controllability" for the special case when the system state is driven from $x(0)=0$.)

To obtain the input $u(k)$ one can start with (3.6.49). Then, for a process with one input

$$x(N) = A^N x(0) + [b,\, Ab\, ...\, A^{N-1} b]\, u_N \tag{3.6.59}$$

with

$$u_N^T = [u(N-1)\ u(N-2)\ ...u(0)]. \tag{3.6.60}$$

Here, the unknown input can be determined uniquely for $N=m$

$$u_m = Q_s^{-1} [x(m) - A^m x(0)] \tag{3.6.61}$$

$$Q_s = [b \; Ab \; ... \; A^{m-1}b] \tag{3.6.62}$$

if

$$\det Q_s \neq 0. \tag{3.6.63}$$

Q_s is called the controllability matrix. This matrix should not have linearly dependent columns of process rows. Hence, for a controllable system one obtains

$$\text{Rank } Q_s = m \tag{3.6.65}$$

with m as the order of the matrix A. For $N < m$ no solution exists for u, and for $N > m$ no unique solution exists.

Example 3.16:
The controllability matrix is to be determined for first-, second- and third-order processes by using the controllable canonical form.

a) $G(z) = \dfrac{b_1 z^{-1}}{1 + a_1 z^{-1}}$

Applying the controllable canonical form

$$A = -a_1; \; b = 1; \; c = b_1$$

the controllability matrix becomes

$$Q_s = b = 1$$

b) $G(z) = \dfrac{b_1 z^{-1} + b_2 z^{-2}}{1 + a_1 z^{-1} + a_2 z^{-2}}$

Written in controllable canonical form

$$A = \begin{bmatrix} 0 & 1 \\ -a_2 & -a_1 \end{bmatrix} \quad b = \begin{bmatrix} 0 \\ 1 \end{bmatrix} \quad c = \begin{bmatrix} b_2 \\ b_1 \end{bmatrix}.$$

and

$$\det Q_s = \det [b \; A \; b] = \det \begin{bmatrix} 0 & 1 \\ 1 & -a_1 \end{bmatrix} = -1$$

becomes valid.

c) $G(z) = \dfrac{b_1 z^{-1} + b_2 z^{-2} + b_3 z^{-3}}{1 + a_1 z^{-1} + a_2 z^{-2} + a_3 z^{-3}}$

On applying the controllable canonical form

$$A = \begin{bmatrix} 0 & 1 & 0 \\ 0 & 0 & 1 \\ -a_3 & -a_2 & -a_1 \end{bmatrix} \quad b = \begin{bmatrix} 0 \\ 0 \\ 1 \end{bmatrix} \quad c = \begin{bmatrix} b_3 \\ b_2 \\ b_1 \end{bmatrix}$$

and it holds

$$Q_s = [b \; Ab \; A^2b] = \begin{bmatrix} 0 & 0 & 1 \\ 0 & 1 & -a_1 \\ 1 & -a_1 & a_1^2-a_2 \end{bmatrix}.$$

This results in

$$\det Q_s = 1.$$

The controllability matrix is always nonsingular and independent of the parameters a_i and b_i, and hence on performing the transformation into the controllable canonical form it has been shown that the system is controllable.

Observability

A linear dynamic process is called *observable*, if any state $x(k)$ can be obtained from a finite number of output variables $y(k)$, $y(k+1)$, . . . , $y(k+N-1)$ and input variables $u(k)$, $u(k+1)$, . . . , $u(k+N-1)$. (If the state $x(k+N-1)$ can be determined, this can be called "reconstructable").

Conditions for the solution of this observability problem can be derived as follows.

With the output equation

$$y(k) = c^T x(k)$$

follows from (3.6.22)

$$\begin{aligned} y(k) &= c^T x(k) \\ y(k+1) &= c^T A \; x(k) + c^T b \; u(k) \\ y(k+2) &= c^T A^2 \; x(k) + c^T A \; b \; u(k) + c^T \; b \; u(k+1) \end{aligned}$$

.

.

.

$$y(k+N-1) = c^T A^{N-1} x(k) + [0, c^T b, c^T A \; b, ..., c^T A^{N-2} b] u_N \qquad (3.6.66)$$

where

$$u_N^T = [u(k+N-1) \; ... \; u(k+1) \; u(k)]. \qquad (3.6.67)$$

If the input vector u_N^T is completely known, m equations are required for a unique determination of the m unknowns of the state vector $x(k)$ in the system of equations (3.6.65). The system of equations (3.6.65) then becomes with $N = m$

$$y_m = Q_B x(k) + S \; u_m \qquad (3.6.68)$$

whence

$$y_m^T = [y(k)\,y(k+1)\dots y(k+m-1)]$$

$$u_m^T = [u(k+m-1)\dots u(k+1)\,u(k)]$$

$$Q_B = \begin{bmatrix} c^T \\ c^T A \\ \vdots \\ c^T A^{m-1} \end{bmatrix}$$

$$S = \begin{bmatrix} 0 & & \cdots & & 0 \\ \vdots & & & & c^T b \\ \vdots & & & & \vdots \\ 0 & c^T b & c^T\,Ab & \cdots & c^T A^{m-2} b \end{bmatrix}.$$

Then the required state vector is

$$x(k) = Q_B^{-1}\,[y_m - S\,u_m] \qquad (3.6.69)$$

if $\det Q_B \neq 0$. A dynamic process is, therefore, observable if the observability matrix Q has:

$$\text{Rank } Q_B = m$$

Q_B therefore has m linearly independent rows.

Example 3.17:
To be examined are the conditions on which first- and second-order processes are observable, using the controllable canonical form representation. The matrices and vectors are given in example 3.16:

a) $G(z) = \dfrac{b_1 z^{-1}}{1 + a_1 z^{-1}}$

$Q_B = c = b_1$

Hence, the process is observable if $b_1 \neq 0$.

b) $G(z) = \dfrac{b_1 z^{-1} + b_2 z^{-2}}{1 + a_1 z^{-1} + a_2 z^{-2}}$

The process is observable if:

$$\det Q_B = \det \begin{bmatrix} c^T \\ c^T A \end{bmatrix} = \det \begin{bmatrix} b_2 & b_1 \\ -b_1 a_2 & (b_2 - a_1 b_1) \end{bmatrix}$$

$$= b_2^2 + a_2 b_1^2 + a_1 b_1 b_2 \neq 0$$

Here, all parameters are important. The process is not observable if $b_1 = 0$ and $b_2 = 0$ or if, in the pulse transfer function

$$G(z) = \dfrac{b_1 z + b_2}{z^2 + a_1 z + a_2} = \dfrac{b_1(z - z_{01})}{(z - z_1)(z - z_2)}$$

the zero $z_{01} = -b_2/b_1$ is equal to one of the poles $z_{1,2} = -a_1/2 \pm \sqrt{a_1^2 - 4a_2}/2$ e.g. $z_{01} = z_1$ (pole-zero-cancellation) which follows from

$$b_2^2 + a_2 b_1^2 + a_1 b_1 b_2 = 0.$$

The process is observable for all other combinations of the parameters. The process, however, has to be either $b_1 \neq 0$ or $b_2 \neq 0$ and if $b_2 = 0$ then $a_2 \neq 0$.

This example shows that the parameters b_i play an essential role, since they generate the output variable as can be seen in Fig. 3.18.

If the observable canonical form is applied, the converse conditions are valid as demonstrated in examples 3.16 and 3.17, this requires a process which is observable and then controllable for the conditions of example 3.17.

These examples show that controllability or observability are lost if in the z-transfer function poles are cancelled (compensated) by zeros. This means that common poles and zeros should be avoided. With single-input/single-output systems the controllable and observable part forms a so-called *minimal realization* which is a representation with a minimal number of state variables.

For a more detailed study of the state representation of discrete-time systems it is referred to [2.18, 2.21, 2.25, 2.27].

3.7 Mathematical Models of Processes

Since the design of sophisticated and well-adjusted control algorithms presupposes the knowledge of mathematical process models, this section discusses some ways of obtaining discrete-time models for certain classes.

3.7.1 Basic Types of Technical Processes

Technical processes are characterized by the transformation and/or the transport of materials, energy and/or information (see e.g. DIN 66201); they can be classified according to

— the amplitude-time-behaviour of the signals
— the type of transport of materials, energy, information
— class of mathematical models.

Amplitude-time-behaviour of the Signals

From Fig. 3.21 one can distinguish

— continuous amplitude — continuous time
  ($\rightarrow$ processes with continuous signals)
— continuous amplitude — discrete time
  ($\rightarrow$ processes with discrete-time (sampled) signals. The pulses appearing can be amplitude modulated, pulse width modulated or frequency modulated)
— discrete amplitude — continuous time
  ($\rightarrow$ processes with stepwise signals)

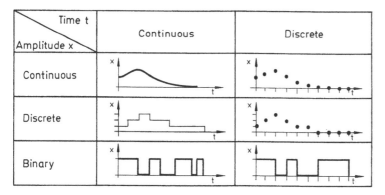

Fig. 3.21. Amplitude-time-behaviour of different signals

— discrete amplitude — discrete time
  (→ processes with sampled digital signals)
— binary amplitude — continuous or discrete time
— (→ processes with binary signals).

Type of Materials, Energy and Information Transport

According to the transport of materials, energy and/or information, technical processes can be divided as follows:

— Continuous Processes
  • Materials, energy, information flow in continuous streams
  • Once-through operation
  • Signals: many combinations as in Fig. 3.21 are possible
  • Mathematical models: linear and nonlinear, ordinary or partial differential equations or difference equations
  • Examples: pipeline, electrical power plant, electrical cable for analog signals. Many processes in power and chemical industries.

— Batch Processes
  • Materials, energy, information flow in "packets" or in interrupted streams
  • Process operation in a closed space
  • Signals: many combinations such as in Fig. 3.21 are possible
  • Mathematical models: mostly nonlinear, ordinary or partial differential equations or difference equations
  • Examples: many processes in chemical engineering: processes for chemical reactions, washing, drying, vulcanising.

— Piece-good Processes
  • Materials, energy, information are transported in "pieces" or "discrete samples"
  • Process operation: piecewise
  • Signals: mostly discrete (binary) amplitude. Continuous or discrete time.

- Mathematical models: flow schemes, digital simulation programs
- Examples: many processes in manufacturing technology. Processing of work pieces, transport of parts, transport in storages.

Class of Mathematical Models

The classes of mathematical models which describe the transient behaviour of continuous and batch processes can be divided as follows:

Ordinary differential equations (lumped parameters)	— Partial differential equations (distributed parameters)
linear	— nonlinear
linear in the parameters	— nonlinear in the parameters
time invariant	— time variant
parametric	— nonparametric
continuous signals	— discrete signals

The definitions of these ideas are found in many well-known books on system theory and control engineering.

In this volume, the design of control algorithms is treated with particular reference to continuous and batch processes. Since for digital control systems mainly mathematical models with discrete-time signals are of interest, some methods for obtaining these models are treated in the next sections.

3.7.2 Determination of the Process Model — Modelling and Identification

Mathematical process models can be obtained by theoretical or experimental process analysis [3.6–3.13].

In *theoretical analysis* (theoretical modelling) of continuous and batch processes the model is determined by the stating of balance equations, state equations and phenomenological laws. One then obtains (for continuous signals), in general, a system of ordinary and/or partial differential equations which lead to a theoretical model of the process with determined structure and parameters if the model can be solved explicitly. For the derivation of discrete-time models the following methods are recommended: approximation of continuous models by lumped-parameter models, simplification of the continuous models, then discretization or z-transformation according to section 3.7.3.

In the case of *experimental analysis* of the process (process identification) the mathematical model of the process is determined by using measured signals. Input and output signals of the process are evaluated by using identification methods such as their relationships are expressed by a mathematical model. This model can be nonparametric, as e.g. transient function of frequency response in tabular form, or parametric, for example a parametric differential or difference equation. Nonparametric models are obtained by evaluation of the measured signals using Fourier-analysis or correlation analysis, whilst parametric models are obtained by applying the methods of step response or frequency response fitting or by parameter estimation methods. For the design of control algorithms for digital processors, parametric models are especially suitable, since modern system theory is mainly based on these models as they contain the parameters explicitly, and as the synthesis of control algorithms can be performed discretely.

For identification of discrete-time parametric models, parameter estimation methods are especially suitable. Then for linear time invariant processes, models of the form:

$$y(z) = \underbrace{\frac{B(z^{-1})}{A(z^{-1})} z^{-d} u(z)}_{\text{process model}} + \underbrace{\frac{D(z^{-1})}{C(z^{-1})} v(z)}_{\text{disturbance model}} \qquad (3.7.1)$$

process model   disturbance model

c.f. (3.4.14) and (12.2.31) are assumed and the unknown parameters of the process and possibly also of the disturbance model are estimated based on the measured signals $u(k)$ and $y(k)$ [3.13]. For parameter estimation methods such as the following can be used: Least Squares, Instrumental Variables, Maximum Likelihood in non-recursive or recursive form. In recent years, using on-line and off-line computers, methods of process identification have been extensively developed and tested in practice. Many linear and nonlinear processes with and without perturbation signals in open and closed-loop, can be identified with sufficient accuracy. This is treated in chapter 24. There are program packages which are easy to operate and which contain methods for the determination of the model order and deadtime, c.f. chapter 30.

Unlike theoretical modelling, process models can be directly obtained in the form of discrete-time signals by using identification methods.

3.7.3 Calculation of z-Transfer Functions from s-Transfer Functions

The dynamic behaviour of linear, time-invariant processes with lumped parameters and continuous input- and output signal $u(t)$ and $y(t)$ is described by ordinary differential equations such as:

$$a_m y^{(m)}(t) + a_{m-1} y^{(m-1)}(t) + \dots + a_1 \dot{y}(t) + y(t)$$
$$= b_m u^{(m)}(t) + b_{m-1} u^{(m-1)}(t) + \dots + b_1 \dot{u}(t) + b_0 u(t) \qquad (3.7.2)$$

They are e.g. obtained through theoretical modelling. Applying the Laplace-transform the following s-transfer function emerges from (3.7.2)

$$G(s) = \frac{y(s)}{u(s)} = \frac{b_0 + b_1 s + \dots + b_{m-1} s^{m-1} + b_m s^m}{1 + a_1 s + \dots + a_{m-1} s^{m-1} + a_m s^m}$$
$$= \frac{B(s)}{A(s)}. \qquad (3.7.3)$$

For *small sample times* difference equations can be obtained by discretizing differential equations, as described in section 3.1.1.

For *larger sample times* the difference equations and the z-transfer functions are calculated most appropriately by the use of z-transform tables. For this, either the impulse response $g(t)=f(t)$ in analytical form or the s-transfer function $G(s)=f(s)$ in partial-fraction form is required. The corresponding transformed functions $f(z)=G(z)$ are taken from the z-transform tables. A partial-fraction expansion has to be performed for higher-order processes to obtain those terms of $G(z)$ which are to be found in the table. If there is e.g. a zero-order hold, (3.4.11) has to be used, and from the table $G(s)/s$ has to be taken. (In the following, $G(s)$ has to be replaced by $G(s)/s$, as in Example 3.18.) For example, the transfer function

$$G(s) = \frac{\sum\limits_{j=0}^{p+l} c_j s^j}{(s-s_0)^p \prod\limits_{i=1}^{l} (s-s_i)} \tag{3.7.4}$$

has to be expanded into

$$G(s) = \sum\limits_{q=1}^{p} \frac{A_{oq}}{(s-s_0)^q} + \sum\limits_{i=1}^{l} \frac{A_i}{(s-s_i)}$$

Here, the coefficients can be calculated using the residues

$$A_{oq} = \frac{1}{(p-q)!} \left[\frac{d^{p-q}}{ds^{p-q}} [(s-s_0)^p G(s)] \right]_{s=s_0} \tag{3.7.5}$$

$$A_i = [(s-s_i) G(s)]_{s=s_i}$$

According to (3.5.17), the poles of $G(s)$ and of $G(z)$ are directly mapped through $z=e^{T_0 s}$.

Example 3.18:
The z-transfer function of the process

$$G(s) = \frac{y(s)}{u(s)} = \frac{K}{(1+T_1 s)(1+T_2 s) \ldots (1+T_m s)} \qquad T_1 \neq T_2 \ldots \neq T_m$$

together with zero-order hold is to be calculated.
Solution:
1. Partial-fraction expansion of $G(s)/s$

$$\frac{G(s)}{s} = \frac{K \dfrac{1}{T_1} \dfrac{1}{T_2} \ldots \dfrac{1}{T_m}}{s \left(s+\dfrac{1}{T_1}\right)\left(s+\dfrac{1}{T_2}\right) \ldots \left(s+\dfrac{1}{T_m}\right)}$$

$$= \frac{A_0}{s} + \frac{A_1}{s+\dfrac{1}{T_1}} + \ldots \frac{A_m}{s+\dfrac{1}{T_m}}$$

$$A_i = \frac{-K \prod\limits_{\substack{j=1 \\ j\neq i}}^{m} \dfrac{1}{T_j}}{\prod\limits_{\substack{j=1 \\ j\neq i}}^{m} \left(-\dfrac{1}{T_i}+\dfrac{1}{T_j}\right)} \qquad i=1,\ldots,m$$

$$A_0 = K$$

2. Search for the corresponding z-transforms
From

$$\frac{G(s)}{s} = \frac{A_0}{s} + \sum_{i=1}^{m} \frac{A_i}{s+\dfrac{1}{T_i}}$$

it follows that:

$$\mathscr{L}\left\{\frac{G(s)}{s}\right\} = \frac{A_0}{1-z^{-1}} + \sum_{i=1}^{m} \frac{A_i}{1-e^{-\frac{T_0}{T_i}}z^{-1}}$$

3. From (3.4.11) one obtains

$$HG(z) = (1-z^{-1})\mathscr{L}\left\{\frac{G(s)}{s}\right\}$$

$$= \frac{A_0\prod\limits_{i=1}^{m}\left(1-e^{-\frac{T_0}{T_i}}z^{-1}\right) + \sum\limits_{i=1}^{m}(1-z^{-1})A_i\prod\limits_{j=1,j\neq i}^{m}\left(1-e^{-\frac{T_0}{T_j}}z^{-1}\right)}{\prod\limits_{i=1}^{m}\left(1-e^{-\frac{T_0}{T_i}}z^{-1}\right)}$$

For the parameters

$$m = 3;\ K = 1;\ T_1 = 10\ \text{s};\ T_2 = 7{,}5\ \text{s};\ T_3 = 5\ \text{s}$$

the parameters of $G(z)$ are given in Table 3.4 for different sample times T_0. With increasing sample time the following trends can be recognized:
a) The magnitudes of the parameters a_i decrease.
b) The magnitudes of the parameters b_i decrease.
c) The sum of the parameters, $\Sigma b_i = 1 + \Sigma a_i$, increases.
For larger sample times one gets $|a_3| \ll 1 + \Sigma a_i$ and $|b_3| \ll \Sigma b_i$, so that a_3 and b_3 can be neglected. In practice this means that a second-order model is obtained.

Another method for calculating $HG(z)$ from $G(s)$, which makes no use of z-transform tables, consists of the following approximation due to Tustin [3.3]

$$s \approx \frac{2}{T_0} \frac{z-1}{z+1}. \tag{3.7.6}$$

This approximation can be derived as follows:
The integral equation

$$y(t) = \frac{1}{T} \int_0^t u(t)\ dt \tag{3.7.7}$$

means that after taking the Laplace-transform operation

Table 3.4. Parameters of the z-transfer function $G(z)$ for the process
$$G(s) = \frac{1}{(1+10s)(1+7.5s)(1+5s)} \text{ with zero-order hold, for different sampling times } T_0.$$

T_0 in s	2	4	6	8	10	12
b_1	0.00269	0.0186	0.05108	0.09896	0.15867	0.22608
b_2	0.00926	0.0486	0.1086	0.17182	0.22570	0.26433
b_3	0.00186	0.0078	0.01391	0.01746	0.01813	0.01672
a_1	−2.25498	−1.7063	−1.2993	−0.99538	−0.76681	−0.59381
a_2	1.68932	0.958	0.54723	0.31484	0.18243	0.10645
a_3	−0.42035	−0.1767	−0.07427	−0.03122	−0.01312	−0.00552
$\Sigma b_i = 1 + \Sigma a_i$	0.01399	0.0750	0.17362	0.28824	0.40250	0.50712

$$\frac{y(s)}{u(s)} = \frac{1}{Ts}. \tag{3.7.8}$$

If the continuous integration is first replaced by *rectangular integration*, then for small T_0 the following equations are valid

$$y(k) \approx \frac{T_0}{T} \sum_{i=1}^{k} u(i-1)$$

$$y(k-1) \approx \frac{T_0}{T} \sum_{i=1}^{k-1} u(i-1)$$

$$y(k) - y(k-1) \approx \frac{T_0}{T} u(k-1)$$

$$y(z)[1-z^{-1}] \approx \frac{T_0}{T} u(z)z^{-1}$$

$$\frac{y(z)}{u(z)} \approx \frac{T_0 z^{-1}}{T(1-z^{-1})} = \frac{T_0}{T(z-1)} \tag{3.7.9}$$

Through correspondence of (3.7.8) and (3.7.9) for small sample times one gets

$$s \rightarrow \frac{1}{T_0}(z-1)$$

A better approximation of the continuous integration is obtained using *trapezoidal integration* as follows

$$y(k) \approx \frac{T_0}{T} \sum_{i=1}^{k} \frac{1}{2}[u(i) + u(i-1)]$$

$$y(k-1) \approx \frac{T_0}{T} \sum_{i=1}^{k-1} \frac{1}{2}[u(i) + u(i-1)]$$

$$y(k) - y(k-1) \approx \frac{T_0}{2T}[u(k) + u(k-1)]$$

$$\frac{y(z)}{u(z)} \approx \frac{T_0}{2T} \frac{z+1}{z-1}. \tag{3.7.10}$$

Hence, for small T_0 the correspondence

$$s \rightarrow \frac{2}{T_0} \frac{z-1}{z+1} \qquad (3.7.11)$$

results and is known as the Tustin approximation. This correspondence can also be obtained by the series expansion of $z = e^{T_0 s}$,

$$s = \frac{1}{T_0} \ln z \approx \frac{2}{T_0} \left[\frac{z-1}{z+1} + \frac{(z-1)^3}{3(z+1)^3} + \dots \right] \qquad (3.7.12)$$

obtained by terminating the sequence after the first term.

Example 3.19:
For the process

$$G(s) = \frac{1}{(1+10s)(1+5s)}$$

with zero-order hold, Table 3.5 gives the exact parameters of the z-transfer function $HG(z)$ together with the parameters resulting from the approximation (3.7.11)

$$H\tilde{G}(z) = G(s) \Big|_{s = \frac{2}{T_0} \frac{z-1}{z+1}}$$

for different sample times.
In this table, the maximum errors of the resulting transient function $\tilde{y}(k)$ are also given. Here

Table 3.5. Parameters of $HG(z)$ and $H\tilde{G}(z)$ for $\dfrac{2}{T_0} \dfrac{z-1}{z+1}$ and the resulting maximum error of the transient function of $H\tilde{G}(z)$. $G(s) = \dfrac{1}{(1+10s)(1+5s)}$.

T_0 in s	b_0	b_1	b_2	a_1	a_2	Σb_i	$(\Delta y/y_\infty)_{max}$ t in s
1		0.00906	0.00819	-1.72357	0.74082	0.01725	
	0.00433	0.00866	0.00433	-1.72294	0.74026	0.01732	+0,024 6
2		0.03286	0.02690	-1.48905	0.54881	0.05976	
	0.01515	0.03030	0.01515	-1.48485	0.54546	0.06061	+0.048 6
4		0.10869	0.07286	-1.11965	0.30119	0.18155	
	0.04762	0.09524	0.04762	-1.09524	0.28571	0.19048	+0.087 8
6		0.20357	0.11172	-0.85001	0.16530	0.31529	
	0.08654	0.17308	0.08654	-0.78846	0.13462	0.34615	+0.124 6
8		0.30324	0.13625	-0.65123	0.09072	0.43949	
	0.12698	0.25397	0.12698	-0.53968	0.04762	0.50794	+0.146 8
12		0.48833	0.15708	-0.39191	0.02732	0.63541	
	0.20455	0.40909	0.20455	-0.15909	-0.02273	0.81812	+0.20 0

$$\Delta y(k) = \bar{y}(k) - y(k)$$

$$y(\infty) = \lim_{k \to \infty} y(k).$$

By using this approximation of $HG(z)$, the parameter $b_0 \neq 0$ appears in $H\tilde{G}(z)$. Hence, a structural difference arises. For small sample times $T_0 \leq 2$ sec the parameters a_1 and a_2 agree relatively well and the maximum errors of the transient function are less than 5%. However, with an increasing sample time the errors become larger. If T_{95} can be defined as the settling time such that the output reaches 95% of the final value $y(\infty)$ of the transient function, for $T_{95} = 37$ sec Table 3.5 shows that:
If an error in the transient function

$$(\Delta y/y_\infty)_{max} = 0,05 \text{ bis } 0,1$$

is allowed, then the maximum sample time is

$$T_{95}/T_0 = 17,5 \text{ bis } 8.$$

Hence, the approximation (3.7.11) is only to be used for small sample times.

3.7.4 Simplification of Process Models for Discrete-time Signals

Process models, derivated by theoretical modelling can with a large number of storages with lumped parameters or with distributed parameter systems become quite extensive resulting with closed representation in a high model order of e.g. the transfer function. For many applications, however, approximation models with reduced extent or reduced model order are sufficient. Here, methods of model reduction are needed. They can be divided into methods for:

a) Model reduction through neglecting higher modes
b) Model reduction with correction of the rest model
c) Model reduction without consideration of the later application of the model
d) Model reduction with due consideration for the application specifications (e.g. closed-loop, simulation, forecasting)

Contributions for processes with continuous signals are e.g. described in [3.4, 3.5, 3.14, 3.15, 3.19, 3.20].

A simple method for model reduction with discrete-time signals is described as follows:

For linear process models with transfer functions of the form

$$G(s) = \frac{y(s)}{u(s)} = \frac{\prod\limits_{\beta=1}^{m} (1 + T_\beta s)}{(1 + 2DTs + T^2 s^2) \prod\limits_{\alpha=1}^{m-2} (1 + T_\alpha s)} e^{-T_t s}$$

$$= \frac{b_0 + b_1 s + \ldots + b_m s^m}{1 + a_1 s + \ldots + a_m s^m} e^{-T_t s} \tag{3.7.13}$$

the dynamic behaviour is approximately the same in both open- and closed-loop if the generalized sum of time constants

$$T_\Sigma = 2DT + \sum_{\alpha=1}^{m-2} T_\alpha - \sum_{\beta=1}^{m} T_\beta + T_t = (a_1 - b_1) + T_t \quad (3.7.14)$$

remains constant, [3.4, 3.5]. That means that the sum of energy, mass or momentum which is stored during a transient process has to remain constant if process models are simplified. It is to be supposed that this condition, at least for small sampling intervals, is also valid for discrete-time process models. If only the input/output behaviour is of interest, then the continuous model should be simplified as far as possible before it is converted to a discrete-time model. For example, small time constants T_i should be replaced by a deadtime

$$T_t = \sum_{i=1}^{l} T_i$$

or poles and zeros which are approximately equal should be cancelled considering (3.7.14), [3.4, 3.5].

If process models are obtained through identification methods, the deadtime has to be chosen such that the resulting model order m becomes as small as possible.

Now, some rules for the simplification of discrete-time models are given. A z-transfer function of the form

$$G(z) = \frac{y(z)}{u(z)} = \frac{b_0 + b_1 z^{-1} + \dots + b_m z^{-m}}{a_0 + a_1 z^{-1} + \dots + a_m z^{-m}} \quad (3.7.15)$$

is assumed. Changes in the parameters a_i and b_i result, in contrast to normalized continuous models, in changes in both the dynamic behaviour (normalized transient function) and the static behaviour (gain). Therefore, conditions for changes of the parameters a_i and b_i should be derived for which — during a transient process — the stored quantities and the gain remain constant.

The stored quantities for a dynamic process with proportional behaviour are, corresponding to the continuous case,

$$A' = T_0 \sum_{k=0}^{\infty} [u(k) - y(k)]. \quad (3.7.16)$$

If a step function with height u_0 is chosen as input signal and if

$$y(k) \approx K u_0 \text{ für } k \geq l$$

with K as the gain

$$K = \sum_{i=0}^{m} b_i / \sum_{i=0}^{m} a_i \quad (3.7.17)$$

and with $A'' = A'/T_0 u_0$ it follows that

$$A'' \approx \tilde{A} = \frac{1}{u_0} \sum_{k=0}^{l} [u(k) - y(k)] = \frac{1}{u_0} \left[(l+1)u_0 - \sum_{k=0}^{l} y(k) \right].$$

$$(3.7.18)$$

The difference equation of the process leads to the following system of equations

$$
\begin{array}{llll}
a_0 y(0) & = & & b_0 u(0) \\
a_0 y(1) & = -a_1 y(0) & & + b_0 u(1) & + b_1 u(0) \\
\vdots & & & & \\
a_0 y(m) & = -a_1 y(m-1) & \ldots -a_m y(0) & + b_0 u(m) & \ldots + b_m u(0) \\
\vdots & & & & \\
a_0 y(l) & = -a_1 y(l-1) & \ldots -a_m y(l-m) + b_0 u(l) & & \ldots + b_m u(l-m) \\
\vdots & & & & \\
a_0 y(l+m) & = -a_1 y(l+m-1) \ldots -a_m y(l) & & + b_0 u(l+m) \ldots + b_m u(l)
\end{array}
$$

$$a_0 \sum_{k=0}^{l+m} y(k) = -a_1 \sum_{k=0}^{l+m-1} y(k) \ldots -a_m \sum_{k=0}^{l} y(k) + b_0 \sum_{k=0}^{l+m} u(k) \ldots + b_m \sum_{k=0}^{l} u(k).$$

$$(3.7.19)$$

Hence, with $u(k) = u_0$ for $k \geq 0$ and (3.7.17)

$$\sum_{i=0}^{m} a_i \sum_{k=0}^{l} y(k) = (l+1)u_0 \sum_{i=0}^{m} b_i + u_0 [mb_0 + (m-1)b_1 + \ldots + b_{m-1}]$$
$$- Ku_0 [ma_0 + (m-1)a_1 + \ldots + a_{m-1}]$$

Substituting in Eq. (3.7.18) leads with $K = 1$

$$\tilde{A} = \frac{1}{\sum_{i=0}^{m} a_i} [m(a_0 - b_0) + (m-1)(a_1 - b_1) + \ldots + (a_{m-1} - b_{m-1})].$$

$$(3.7.20)$$

For small parameter changes one obtains with

$$\Delta \tilde{A} \approx \frac{\partial \tilde{A}}{\partial a_1} \Delta a_1 + \ldots + \frac{\partial \tilde{A}}{\partial a_m} \Delta a_m + \frac{\partial \tilde{A}}{\partial b_0} \Delta b_0 + \ldots + \frac{\partial \tilde{A}}{\partial b_m} \Delta b_m \qquad (3.7.21)$$

the approximation relation

$$\Delta \tilde{A} \approx \frac{1}{\sum_{i=0}^{m} a_i} \left[\sum_{i=0}^{m-1} (m-i)(\Delta a_i - \Delta b_i) - \tilde{A} \sum_{i=0}^{m} \Delta a_i \right] \qquad (3.7.22)$$

Considering changes of the stored quantity A it follows that:
a) The larger i, the smaller the effect of changes Δa_i or Δb_i

$$\left.\begin{aligned}
\frac{\partial \tilde{A}/\partial a_i}{\partial \tilde{A}/\partial a_0} &= \frac{(m-i)-\tilde{A}}{m-\tilde{A}} \\[2ex]
\frac{\partial \tilde{A}/\partial b_i}{\partial \tilde{A}/\partial b_0} &= \frac{m-i}{m}
\end{aligned}\right\} \qquad (3.7.23)$$

b) Using

$$\frac{\partial \tilde{A}/\partial b_i}{\partial \tilde{A}/\partial a_i} = \frac{(m-i)}{\tilde{A}-(m-i)} = \varkappa$$

leads to

$$\begin{aligned}
\varkappa \geq 1 \quad &\text{für} \quad (m-i) < \tilde{A} \leq 2(m-i) \\
\varkappa < 1 \quad &\text{für} \quad \tilde{A} > 2(m-i)
\end{aligned}$$

From (3.7.17) for small parameter changes it follows that, for the gain $K=1$,

$$K \approx \sum_{i=0}^{m} \frac{\partial K}{\partial a_i} \Delta a_i + \frac{\partial K}{\partial b_i} \Delta b_i$$

$$= \frac{1}{\sum\limits_{i=0}^{m} a_i} \sum_{i=0}^{m} (\Delta b_i - \Delta a_i). \qquad (3.7.24)$$

With $\Delta \tilde{A}=0$ and $\Delta K =0$ one obtains two equations for determining Δa_i and Δb_i. Since for $m \geq 1$ there are always more than two unknowns Δa_i and Δb_i, several solutions are possible.

A first solution is obtained directly from (3.7.22) and (3.7.24)

$$\Delta a_i = \Delta b_i \quad i = 0, 1, ..., m$$

and $\sum\limits_{i=0}^{m} \Delta a_i = 0.$ $\qquad (3.7.25)$

Hence, for $\tilde{A} \approx$ constant and $K \approx$ constant, small parameter changes are permitted if $\Delta a_1 = \Delta b_1$, $\Delta a_2 = \Delta b_2$ etc. and simultaneously $\Delta a_1 + \Delta a_2 + \ldots + \Delta a_m = 0$.

However, other solutions exist, as shown in the following example.

Example 3.20:
For the sample time $T_0 = 10$ sec one obtains from Table 3.4

$$HG(z) = \frac{0,1587\, z^{-1} + 0,2257\, z^{-2} + 0,0181\, z^{-3}}{1 - 0,7668\, z^{-1} + 0,1824\, z^{-2} - 0,0131\, z^{-3}}.$$

This process has the characteristic parameters $K=1$ and $\tilde{A}=2.75$. Since $a_3 \ll 1 + \Sigma a_i$ and $b_3 \ll \Sigma b_i$ (see Table 3.4) it is supposed that this process can also be described by a model of $m=2$.

It follows from (3.7.22) that:

$$\Delta \tilde{A} \approx \frac{1}{\Sigma a_i}[2(\Delta a_1 - \Delta b_1) + (\Delta a_2 - \Delta b_2) - \tilde{A}(\Delta a_1 + \Delta a_2 + \Delta a_3)] = 0$$

and from (3.7.24)

$$\Delta K \approx \frac{-1}{\Sigma a_i}[(\Delta a_1 - \Delta b_1) + (\Delta a_2 - \Delta b_2) + (\Delta a_3 - \Delta b_3)] = 0.$$

Now, $\tilde{a}_3 = 0$ and $\tilde{b}_3 = 0$ are set, i.e.

$$\Delta a_3 = -a_3 \text{ und } \Delta b_3 = -b_3.$$

In these two equations four unknowns remain, i.e. two variables can be chosen freely. It is assumed, for example that:

$$\Delta a_2 = -\Delta a_3 \text{ und } \Delta b_1 = 0.$$

It thus follows that:

$$2\Delta a_1 + (a_3 - \Delta b_2) - \tilde{A}(\Delta a_1 + a_3 - a_3) = 0$$

and
$$-0{,}75\Delta a_1 - \Delta b_2 = 0{,}0131$$

$$\Delta a_1 + (a_3 - \Delta b_2) + (-a_3 + b_3) = 0$$

$$\Delta a_1 - \Delta b_2 = -0{,}0181$$

and here

$$\Delta a_1 = -0{,}0178 \quad \Delta a_2 = -0{,}0131$$

$$\Delta b_2 = +0{,}0003 \quad \Delta b_1 = 0.$$

Finally, the following approximation is obtained

$$H\tilde{G}(z) \approx \frac{0{,}1587z^{-1} + 0{,}2260z^{-2}}{1 - 0{,}7847z^{-1} + 0{,}1693z^{-2}}.$$

This approximation has $\tilde{K} = 1.000$. $\tilde{A}$ has changed by 1‰.

For the simplification of discrete-time process models $\Delta \tilde{A} = 0$ and $\Delta K = 0$ can be assumed, based on the hypothesis that the stored physical quantities do not change during a transient. Based on (3.7.22) and (3.7.24) conditions are obtained for model simplification.

Part B
Control Systems for Deterministic Disturbances

4 Deterministic Control Systems

Deterministic control systems are control systems that are designed for external deterministic disturbances or deterministic initial values. Deterministic disturbances or initial values are variables which, unlike stochastic variables, can be described exactly in analytical form. Common control systems can be classified as *reference control systems* or *terminal control systems*. For their discussion a process with one manipulated variable $u(k)$, one controlled variable $y(k)$, the state variables $x(k)$ and the disturbances $v(k)$ are considered, as in Fig. 4.1. With reference control systems the controlled variable $y(k)$ has to follow a reference variable $w(k)$ as closely as possible, resulting in control errors $e(k) = w(k) - y(k)$ that are as small as possible, $e(k) \approx 0$. If the reference variable changes with time a *variable reference control system* or *tracking control system* is to be designed. If the controlled variable is a position, velocity or acceleration, this is also called a *servo control system*. If the reference variable is constant, this is called a *regulator*.

For *terminal control systems* a defined final state $x(N)$ of the process has to be reached and held at a prescribed or free final time point N. For both reference and terminal control systems, the influence of initial values $x(0)$ or disturbances $v(k)$ of the processes has to be compensated for as much as possible. The control problem, moreover, is such that for unstable processes a stable overall system is to be obtained through the feedback.

These problems can be solved, in general, by applying controllers which make use of the feedback of the process output $y(k)$ or the process states $x(k)$. The effect of feedback often can be improved by using additional feedforward control elements. In Fig. 4.1 block diagrams of control systems are presented for the case of a single feedback or feedforward controlled output variable y. The following notation is used: G_R for feedback controllers or feedback control algorithms; G_S for feedforward controllers or feedforward control algorithms.

Figure 4.1a shows a single-loop control. If the disturbance v is measurable, feedforward can be used as in Fig. 4.1b, in combination with a feedback loop for control of those disturbances which cannot be compensated by the feedforward control. If along the signal flow path between the manipulated variable and the controlled variable additional measured variables can be used for feedback, subsidiary controls or, as shown in Fig. 4.1c, cascaded control systems with major and minor control loops can be realized. A state variable feedback for stabilization or changing the modal behaviour of a process is shown in Fig. 4.1d.

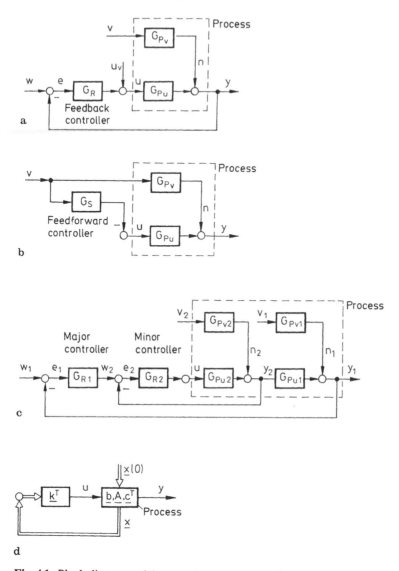

Fig. 4.1. Block diagrams of the most important control system structures for one controlled variable **a** single loop control; **b** feedforward control system; **c** cascaded control system; **d** state feedback

The design of control systems generally is made according to Fig. 4.2. Depending on the design method and the application, exact or approximate *mathematical models* of the *processes* and the *signals* (disturbances, reference variables, initial values) are used as the basis for design. Section 3.7 describes methods for obtaining process models. Frequently models of the signals can only be estimated approximately. For simplicity, step changes are often assumed, though they are rare in practice. However, by applying modern process computers it is possible to obtain more exact models of deterministic and stochastic signals without too much effort.

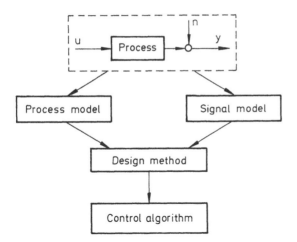

Fig. 4.2. To the design of control algorithms

In this book the design of linear control systems for linearizable time-invariant processes with sampled signals is considered. A schematic presentation of the most important control systems and their design principles is given in Fig. 4.3.

Two major groups are distinguished: parameter optimized control systems and structure optimized control systems. In the case of *parameter optimized control systems* the controller structure, i.e. the form and the order of the controller equation, is given and the free controller parameters are adapted to the controlled process, using an optimization criterion or using tuning rules. Control systems are called *structure optimal* if both the controller structure and the controller parameters are adapted optimally to the structure and the parameters of the process model. For both major groups subgroups can also be distinguished. In the case of parameter optimized controllers various lower-order controllers of PID-type, and in the case of structure optimal controllers cancellation and state controllers can be distinguished. For design, *tuning rules*, *performance criteria* and *pole placement* are commonly used. Fig. 4.3 also gives the names of the most important controllers and the suitability of the designs for *deterministic* or *stochastic disturbances*.

The choice of control performance plays a central role in controller design. For the cancellation controllers the time behaviour of the controlled variables is specified, either completely or after a final settling time. Little is specified of the time behaviour of the controlled variable in the case of pole placement. The poles describe just the single eigenfrequencies. However, their combined action, the resulting zeros and the behaviour for external disturbances are not included in this design.

A more comprehensive evaluation of control system behaviour and a more directed controller design is obtained by the introduction of performance criteria. In recent years, integral criteria have been used for the design of continuous control systems by integrating control errors, squares of control errors, absolute magnitudes of control errors etc., each of which could also be weighted with time. For discrete-time signals, these performance criteria are:

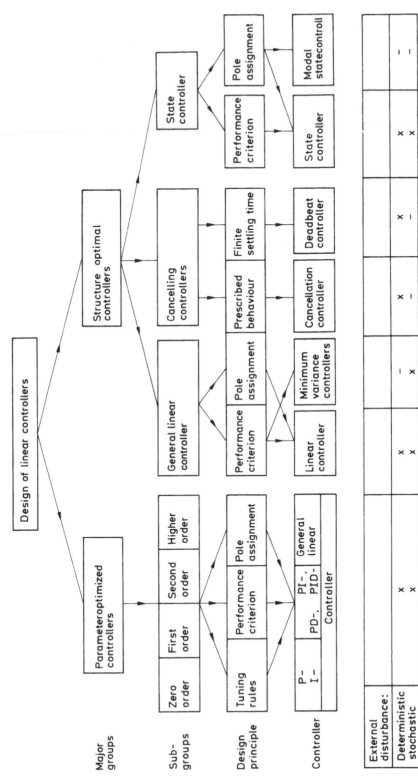

Fig. 4.3. Scheme for the design of linear controllers

$$I_1 = \sum_{k=0}^{\infty} e(k) \qquad \text{"sum of the error"}$$

$$I_2 = \sum_{k=0}^{\infty} e^2(k) \qquad \text{"sum of the square of the error"}$$

$$I_3 = \sum_{k=0}^{\infty} |e(k)| \qquad \text{"sum of the absolute magnitude of the error"}$$

$$I_4 = \sum_{k=0}^{\infty} k|e(k)| \qquad \text{"sum of time multiplied by the absolute value of error"}$$

Since I_1 cannot be used if the sign of $e(k)$ changes, I_2 is used more often. However, I_2 leads to heavily oscillating behaviour; better damped behaviour of the control variable is obtained by using I_3 or I_4.

In analytical design quadratic performance criteria are preferred as they have mathematical advantages. This is due to the fact that when searching for extremal values a single derivative results in relationships which are linear in $e(k)$. Additional degrees of freedom and the possibility of a more directed influence on the damping of the control systems behaviour result by the addition of quadratic deviations of the manipulated variable with a weighting factor r. Hence, a more general quadratic performance criterion is

$$I_5 = \sum_{k=0}^{\infty} [e^2(k) + r\, u^2(k)], \tag{4.1}$$

which for state control systems leads to the form

$$I_6 = \sum_{k=0}^{\infty} [x^T(k) Q\, x(k) + r\, u^2(k)] \tag{4.2}$$

These quadratic criteria are suited for both deterministic and stochastic signals, so are preferred in this book.

The steady-state behaviour of the system can be specified independently of the choice of the performance criterion. For step changes of the reference variable $w(k)$, the disturbances $n(k)$, $v(k)$ and $u_v(k)$ (Fig. 4.1a) no offset may result in general, i.e. $\lim_{k \to \infty} e(k) = 0$ and therefore, after using the final value theorem of the z-transformation

$$\lim_{z \to 1} (z-1)e(z) = 0$$

For the single control loop

$$e(z) = \frac{1}{1+G_R(z)G_P(z)} [(w(z)-n(z)] - \frac{G_P(z)}{1+G_R(z)G_P(z)} u_v(z). \tag{4.3}$$

and therefore, for a step change $1(z) = z/(z-1)$ different conditions result from various disturbances.

1. $w(k) = 1(k)$ and $n(k) = 1(k)$

$$\lim_{z \to 1} \frac{z}{1 + G_R(z) G_P(z)} = 0$$

$$\to \lim_{z \to 1} G_R(z) G_P(z) = \infty$$

2. $u_v(k) = 1(k)$ (disturbance at the process input)

$$\lim_{z \to 1} \frac{G_P(z) z}{1 + G_R(z) G_P(z)} = 0$$

a) $\lim_{z \to 1} G_P(z) = K_P = \text{const}$ (proportional behaviour)

$$\to \lim_{z \to 1} G_R(z) = \infty$$

b) $\lim_{z \to 1} G_P(z) = \infty$ (integral behaviour)

$$\to \lim_{z \to 1} G_R(z) = \infty$$

Therefore in all cases:

$$\lim_{z \to 1} G_R(z) = \infty \tag{4.4}$$

leads to a zero offset. This is given by a controller pole at $z = 1$

$$G_R(z) = \frac{Q(z)}{P'(z)(z-1)} \tag{4.5}$$

i.e. through integral action in the controller. A process pole at $z = 1$ with a proportional controller and for $w(k) = 1(k)$ and $n(k) = 1(k)$, leads to diminishing offsets. However, this is not the case for a constant disturbance $u_v(k) = 1(k)$

Similar requirements can be established for zero offsets in the case of linearly or quadratically changing reference signals. The controller then has to have double or triple poles at $z = 1$ (see [2.19]).

5 Parameter-optimized Controllers

5.1 Discretizing the Differential Equations of Continuous PID-Controllers

As parameter-optimized controllers have P-, PI- or PID-behaviour a number of attempts have been made to transfer their equations into sampled-data form by discretization. The established experience with continuous-time controllers could then be used, and, in principle, their well-known tuning rules could be applied. Furthermore, retraining of plant personnel is obviated [5.1 – 5.5].
The idealized equation of a PID-controller is

$$u(t) = K\left[e(t) + \frac{1}{T_I} \int_0^t e(\tau)\,d\tau + T_D \frac{de(t)}{dt} \right] \qquad (5.1.1)$$

with parameters (see e.g. DIN 19226)

K gain
T_I integration time
T_D derivative time (lead time)

For *small sample times* T_0 this equation can be turned into a difference equation by discretization. The derivative term is simply replaced by a first-order difference expression and the integral by a sum. The continuous integration may be approximated by either rectangular or trapezoidal integration, as in section 3.7.3.
Applying *rectangular integration* gives

$$u(k) = K\left[e(k) + \frac{T_0}{T_I} \sum_{i=0}^{k-1} e(i) + \frac{T_D}{T_0}(e(k) - e(k-1)) \right]. \qquad (5.1.2)$$

This is a nonrecursive control algorithm. For the formation of a sum all past errors $e(k)$ have to be stored. Since the value $u(k)$ of the manipulated variable is produced, this algorithm is called a "position algorithm", as in [5.1], [5.3].
However, recursive algorithms are more suitable for programing. These algorithms are characterized by the calculation of the current manipulated variable $u(k)$ based on the previous manipulated variable $u(k-1)$ and correction terms. To derive the recursive algorithm one has to subtract from (5.1.2)

$$u(k-1) = K\left[e(k-1) + \frac{T_0}{T_I}\sum_{i=0}^{k-2}e(i) + \frac{T_D}{T_0}(e(k-1)-e(k-2))\right]$$

and one obtains as PID-*control algorithm*

$$u(k)-u(k-1) = q_0e(k) + q_1e(k-1) + q_2e(k-2) \qquad (5.1.4)$$

with parameters

$$q_0 = K\left(1+\frac{T_D}{T_0}\right)$$

$$q_1 = -K\left(1+2\frac{T_D}{T_0}-\frac{T_0}{T_I}\right) \qquad (5.1.5)$$

$$q_2 = K\frac{T_D}{T_0}.$$

Now, only the current change in the manipulated variable

$$\Delta u(k) = u(k) - u(k-1)$$

is calculated and so this algorithm is also called a "velocity algorithm".

For a I-*control algorithm* it follows from (5.1.2) and (5.1.3) with missing P- and D-term that

$$u(k) - u(k-1) = K\frac{T_0}{T_I}e(k-1) \qquad (5.1.6)$$

Hence, the difference of the manipulated variable $\Delta u(k)$ is directly proportional to the previous control difference $e(k-1)$.

The distinction between the algorithms (5.1.2) and (5.1.4) is only significant for controllers with integral effect. Since for $T_0/T_I=0$ yields for the PD-*control algorithm* using (5.1.2)

$$u(k) = K\left[e(k) + \frac{T_D}{T_0}(e(k)-e(k-1))\right]$$

$$= K\left[\left(1+\frac{T_D}{T_0}\right)e(k) - \frac{T_D}{T_0}e(k-1)\right] \qquad (5.1.7)$$

and for the P-*control algorithm* with $T_D/T_0=0$

$$u(k) = Ke(k). \qquad (5.1.8)$$

It is one of the advantages of the PID-, PI- or I- control algorithm in recursive form that no special provisions are necessary for a smooth switching from manual to automatic operation, see section 5.8.

If the continuous integration is approximated by *trapezoidal integration* then one obtains from (5.1.1)

$$u(k) = K[e(k) + \frac{T_0}{T_I}\left(\frac{e(0)+e(k)}{2} + \sum_{i=1}^{k-1} e(i)\right)$$

$$+ \frac{T_D}{T_0}(e(k) - e(k-1))]. \tag{5.1.9}$$

After subtraction of the corresponding equation for $u(k-1)$, another recursive relation of the PID-control algorithm is obtained as follows

$$u(k) = u(k-1) + q_0 e(k) + q_1 e(k-1) + q_2 e(k-2)$$

with parameters

$$q_0 = K\left(1 + \frac{T_0}{2T_I} + \frac{T_D}{T_0}\right)$$

$$q_1 = -K\left(1 + 2\frac{T_D}{T_0} - \frac{T_0}{2T_I}\right) \tag{5.1.10}$$

$$q_2 = K\frac{T_D}{T_0}$$

For small sample times the parameters q_0, q_1 and q_2 can be calculated using the parameters K, T_I and T_D of the continuous-time PID-controllers by application of (5.1.5) or (5.1.7) provided they are already known.

Example 5.1
For a fourth-order process with time delay $T_u = 14$ sec and response time $T_G = 45$ sec the following continuous-time controller parameters were considered to be optimal

$$K = 2; T_D = 2,5 \text{ s}; T_I = 40 \text{ s}.$$

The sample time of the discrete-time PID-controller is assumed to be $T_0 = 1$ sec, which is relatively small. The parameters of the PID-control algorithm are then using (5.1.5) (forward rectangular integration)

$$q_0 = 7 \quad q_1 = -11,95 \quad q_2 = 5$$

and with (5.1.10) (trapezoidal integration)

$$q_0 = 7,025 \quad q_1 = -11,98 \quad q_2 = 5$$

Hence, the differences are considered negligibly small and either discretization method can be used.

5.2 Parameter-optimized Discrete Control Algorithms of Low-Order

For larger sample times the approximations of the continuous controller used in section 5.1 are no longer valid. Since, additionally a direct z-transformation of the continuous-time controller equation is not possible because of the derivative term, in this section the connections with continuous controllers are dropped.

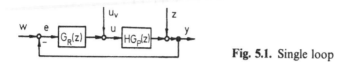

Fig. 5.1. Single loop

A simple control loop as shown in Fig. 5.1 is considered. The z-transfer function of the controlled process including a zero-order hold is

$$G_p(z) = \frac{y(z)}{u(z)} = \frac{B(z^{-1})}{A(z^{-1})} = \frac{b_0 + b_1 z^{-1} + \dots + b_m z^{-m}}{1 + a_1 z^{-1} + \dots + a_m z^{-m}} z^{-d}. \tag{5.2.1}$$

(In the sequel the order of the numerator and denominator polynominal are both assumed to be m. All other cases can be satisfied by setting single parameters to zero.)

The general transfer function of the linear controller is

$$G_R(z) = \frac{u(z)}{e(z)} = \frac{Q(z^{-1})}{P(z^{-1})} = \frac{q_0 + q_1 z^{-1} + \dots + q_\nu z^{-\nu}}{p_0 + p_1 z^{-1} + \dots + p_\mu z^{-\mu}}. \tag{5.2.2}$$

This algorithm can be realized if $p_0 \neq 0$. However, $\nu \leq \mu$ or $\nu > \mu$. Usually $q_0 \neq 0$, and $p_0 = 1$ is chosen.

For structure-optimized controllers μ and ν are functions of the orders of the process model. For example, in the deadbeat controllers $\nu = m$ and $\mu = m + d$. However, for parameter-optimized controllers the controller order can be smaller than the order of the process model, so $\nu \leq m$ and $\mu \leq m + d$. Parameter-optimized controllers, therefore, require less on-line computation.

When defining the structure of a parameter-optimized controller, one must generally ensure that changes of the reference variable $w(k)$ and of the disturbances $u_v(k)$ and $n(k)$, Fig. 5.1, do not lead to offsets in the control deviation $e(k)$. From the final value theorems of the z-transform it follows that the controller must have a pole at $z = 1$. The simplest control algorithms of νth therefore have the structure

$$G_R(z) = \frac{Q(z^{-1})}{P(z^{-1})} = \frac{q_0 + q_1 z^{-1} + \dots + q_\nu z^{-\nu}}{1 - z^{-1}}. \tag{5.2.3}$$

For $\nu = 1$ one obtains, by appropriate choice of parameters, a controller of PI-type, for $\nu = 2$ of PID-type, for $\nu = 3$ of PID_2-type, etc. The resulting difference equation means that

$$u(k) = u(k-1) + q_0 e(k) + q_1 e(k-1) + \ldots + q_v e(k-v). \qquad (5.2.4)$$

The parameters $q_0, q_1, q_2, \ldots, q_v$ must be matched to the process to obtain a good performance, and the following methods are available:

a) Based on a process model the controller parameters can be obtained by minimizing a performance criterion using *parameter optimization*. Analytical solutions are only possible for processes and controllers of very low order and numerical methods must, in general be used.

b) For a given process model the controller parameters are determined by *prescribed closed-loop poles* or by design methods *through the use of other forms of controllers*.

c) *Tuning rules* can be used which lead to approximately optimal controller parameters based on certain criteria. Here either the characteristic parameters of measured step responses are determined, or oscillation tests are made with proportional controllers set at the stability limit.

d) Starting with small values (giving a low-gain control), the controller parameters during closed-loop operation are systematically increased until the loop damping becomes too small. Then the parameters are decreased by some fraction (*trial-and-error method*).

If there is no specification for the control performance and if the process has a simple behaviour and a small settling time, then methods c) or d) may suffice. However, for stringent performance requirements or complicated or slow or changing process behaviour, one has to use a) or b). These methods are especially suitable for computer-aided design.

When synthesizing control systems by *numerical parameter optimization methods*, one is interested in a single characteristic value of the control performance. Therefore, for continuous-time signals integral criteria (which for discrete-time signals become sum criteria) are especially suited. The sum of quadratic control errors is preferred for mathematical reasons. It can also be interpreted as an averaged power and so is also suitable for other controller design methods. Hence in the following discussion *quadratic performance criteria* of the form

$$S_{eu}^2 = \sum_{k=0}^{M} [(e^2(k) + rK_p^2 \Delta u^2(k)] \qquad (5.2.6)$$

are used for parameter optimization (c.f. chapter 4). Here

$$e(k) = w(k) - y(k)$$

is the control error,

$$\qquad\qquad\qquad\qquad\qquad\qquad\qquad (5.2.7)$$

$$\Delta u(k) = u(k) - \bar{u}$$

the "manipulated variable deviation" from

— the final value $\bar{u} = u(\infty)$ for step disturbances
— the expectation $\bar{u} = E\{u(k)\}$ for stochastic disturbances

and r is a weighting factor on the manipulated variable. K_p is a scale factor, with proportional action processes the gain $K_p = y(\infty)/u(\infty)$ or with integral action processes the integration factor, so that the weighting factor r can be chosen independent from the factors K_p of the process.

In this quadratic performance criterion an *averaged quadratic control deviation*

$$S_e^2 = \overline{e^2(k)} = \frac{1}{M+1} \sum_{k=0}^{M} e^2(k) \tag{5.2.8}$$

and the averaged quadratic manipulated variable deviation or *averaged input power*

$$S_u^2 = \overline{\Delta u^2(k)} = \frac{1}{M+1} \sum_{k=0}^{M} \Delta u^2(k) \tag{5.2.9}$$

can be related by the appropriate choice of the weighting factor rK_p^2. If r is chosen to be small, then a small S_e^2 can be obtained using a large input power S_u^2. The more S_u^2 is weighted by r, the less the input changes and the larger the error becomes, so that the control loop has a more restrained behaviour.

Different methods for the design of parameter-optimized controllers will be described in the following sections.

5.2.1 Control Algorithms of First and Second Order

Control Algorithms of Second Order

Using (5.2.3) with $v = 2$

$$G_R(z) = \frac{q_0 + q_1 z^{-1} + q_2 z^{-2}}{1 - z^{-1}} \tag{5.2.10}$$

and according to (5.2.4)

$$u(k) = u(k-1) + q_0 e(k) + q_1 e(k-1) + q_2 e(k-2). \tag{5.2.11}$$

Assuming a step input

$$e(k) = 1(k) = \begin{cases} 1 & \text{für } k \geq 0 \\ 0 & \text{für } k < 0 \end{cases} \tag{5.2.12}$$

one obtains as controller step response

$$\begin{aligned}
u(0) &= q_0 \\
u(1) &= u(0) + q_0 + q_1 = 2q_0 + q_1 \\
u(2) &= u(1) + q_0 + q_1 + q_2 = 3q_0 + 2q_1 + q_2
\end{aligned}$$

.
.
.

$$u(k) = u(k-1) + q_0 + q_1 + q_2 = (k+1)q_0 + kq_1 + (k-1)q_2. \tag{5.2.13}$$

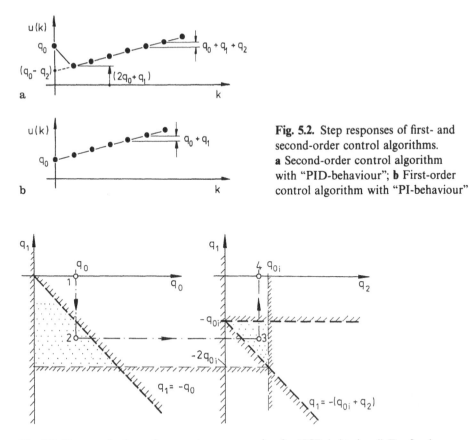

Fig. 5.2. Step responses of first- and second-order control algorithms. **a** Second-order control algorithm with "PID-behaviour"; **b** First-order control algorithm with "PI-behaviour"

Fig. 5.3. Ranges of values of parameters q_0, q_1 and q_2 for "PID-behaviour". For fixed q_{0i}, q_{1i} and q_{2i} must be placed between the dotted angles (according to line 1–2–3–4)

If $u(1) < u(0)$ and $u(k) > u(k-1)$ one obtains a discrete-time controller similar to the (generally used) continuous PID-controller with an additional first-order lag as shown in Fig. 5.2a. For the controller parameters with $q_0 > 0$ it follows that

From $u(1) < u(0)$: $\qquad\qquad\qquad q_0 + q_1 < 0 \qquad$ or $q_1 < -q_0$

From $u(k) > u(k-1)$ for $k \geqq 2$: $q_0 + q_1 + q_2 > 0$ or $q_2 > -(q_0 + q_1)$

For a positive controller gain $q_0 > q_2$, see (5.2.15). Summarizing the ranges of the various parameters

$$q_0 > 0; \quad q_1 < -q_0; \quad -(q_0 + q_1) < q_2 < q_0. \qquad\qquad (5.2.14)$$

The resulting parameter ranges are shown in Fig. 5.3. The parameter q_0 determines the manipulated variable $u(0)$ after the step input.

The following characteristic coefficients can be defined:

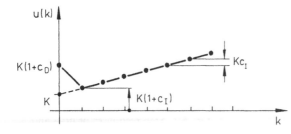

Fig. 5.4. Step response of a second-order control algorithm with gain K, lead coefficient c_D, integration coefficient c_I

$$
\begin{array}{ll}
K = q_0 - q_2 & \text{gain} \\
c_D = q_2/K & \text{lead coefficient} \\
c_I = (q_0 + q_1 + q_2)/K & \text{integration coefficient.}
\end{array}
\qquad (5.2.15)
$$

These characteristic coefficients are shown in the step response in Fig. 5.4. They have been defined so that for small sample times they are related to the parameters K', T'_D, T'_I of the continuous PID-control algorithms, following discretization through rectangular integration, (c.f. (5.1.5)) as shown by the following

$$
K = K'; \; c_D = \frac{T'_D}{T_0}; \; c_I = \frac{T_0}{T'_I}. \qquad (5.2.16)
$$

For small sample times the gains agree exactly; c_D is the ratio of lead time to sample time and c_I the ratio of sample time to integration time.

Applying (5.2.14) it follows that for usual PID-behaviour

$$
c_D > 0; \quad c_I > 0; \quad c_I < c_D \qquad (5.2.17)
$$

If these characteristic coefficients are substituted in (5.2.10) the z-transfer function becomes

$$
\begin{aligned}
G_R(z) &= \frac{K[(1+c_D) + (c_I - 2c_D - 1) \, z^{-1} + c_D z^{-2}]}{1 - z^{-1}} \\
&= K \left[1 + c_I \frac{z^{-1}}{1 - z^{-1}} + c_D(1 - z^{-1}) \right]
\end{aligned}
\qquad (5.2.18)
$$

Thus, in a similar way to the continuous-time PID-controller, separate channels for P-, I- and D-behaviour can be realized, as shown in Fig. 5.5. The single algorithms then are:

$$
\begin{array}{llr}
u_p(k) = Ke(k) & \text{(P-behaviour)} & \\
u_I(k) = u_I(k-1) + Kc_I e(k-1) & \text{(I-behaviour)} & \\
u_D(k) = Kc_D e(k) - Kc_D e(k-1) & \text{(D-behaviour)} & (5.2.19a) \\
u(k) = u_p(k) + u_I(k) + u_D(k) & &
\end{array}
$$

or in the original form

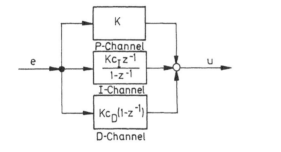

P-Channel

I-Channel

D-Channel

Fig. 5.5. Block diagram of the PID-control algorithm

$$G_R(z) = \frac{q_0 + q_1 z^{-1} + q_2 z^{-2}}{1 - z^{-1}}$$

with the characteristic values according to (5.2.15)

$$
\left.
\begin{aligned}
u_P(k) &= (q_0 - q_2)e(k) \\
u_I(k) &= u_I(k-1) + (q_0 + q_1 + q_2)e(k-1) \\
u_D(k) &= q_2(e(k) - e(k-1)).
\end{aligned}
\right\}
\qquad (5.2.19b)
$$

It should be noticed that the above second-order control algorithm is only similar to a continuous-time PID-controller with positive parameters if conditions (5.2.14) or (5.2.17) are satisfied. The parameters determined by the optimization can, depending on the process, the choice of optimization criterion and disturbance signal, fail to satisfy these conditions.

Control Algorithms of First-Order

Setting $q_2 = 0$ the z-transfer function becomes

$$G_R(z) = \frac{q_0 + q_1 z^{-1}}{1 - z^{-1}} \qquad (5.2.20)$$

and the difference equation is

$$u(k) = u(k-1) + q_0 e(k) + q_1 e(k-1).$$

The step response values then become

$$
\left.
\begin{aligned}
u(0) &= q_0 \\
u(1) &= u(0) + q_0 + q_1 = 2q_0 + q_1 \\
u(2) &= u(1) + q_0 + q_1 = 3q_0 + 2q_1 \\
&\quad . \\
&\quad . \\
&\quad . \\
u(k) &= u(k-1) + q_0 + q_1 = (k+1)q_0 + kq_1.
\end{aligned}
\right\}
\qquad (5.2.21)
$$

For $u(1) > u(0)$ the first-order control algorithm can be compared with a continuous PI-controller with no additional lag. With $q_0 > 0$ one obtains $q_0 + q_1 > 0$ or $q_1 > -q_0$.

In Fig. 5.2b the corresponding step response is shown. As in (5.2.15) the following characteristic coefficients can be defined:

$K = q_0$ gain

$c_I = (q_0 + q_1)/K$ integration coefficient

For PI-behaviour with positive characteristic coefficients

$$c_I > 0.$$

These factors, introduced in (5.2.20), lead to

$$G_R(z) = \frac{K[1 + (c_I - 1)z^{-1}]}{1 - z^{-1}}. \tag{5.2.23}$$

Choosing $q_0 = 0$ results in an *integral action controller* with transfer function

$$G_R(z) = \frac{q_1 z^{-1}}{1 - z^{-1}} \tag{5.2.24}$$

and difference equation

$$u(k) = u(k-1) + q_1 e(k-1). \tag{5.2.25}$$

This controller is equivalent to an integral action term with zero-order hold.

Other special cases are the *proportional plus derivative (PD) action controller*

$$G_R(z) = q_0 \quad \text{or} \quad u(k) = q_0 e(k) \tag{5.2.26}$$

and the *proportional derivative action controller*

$$G_R(z) = q_0 - q_2 z^{-1} \quad \text{or} \quad u(k) = q_0 e(k) - q_2 e(k-1) \tag{5.2.27}$$

obtained by putting $c_I = 0$ and thus $q_1 = -q_0 - q_2$ from (5.2.10).

The following example shows ranges of values in which the controller parameters may be chosen in order to obtain a stable closed-loop system.

Example 5.2

The stability limits of a sampled-data control loop consisting of a first-order process

$$G_P(s) = \frac{K_p}{1 + Ts}$$

together with a first-order hold and a discrete PI-controller:

$$G_R(z) = \frac{q_0 + q_1 z^{-1}}{1 - z^{-1}}$$

is to be examined and represented in a diagram for various sampling times.

For the dynamic process itself it follows that:

$$HG_P(z) = \frac{b_1 z^{-1}}{1 + a_1 z^{-1}} = \frac{b_1}{z + a_1}$$

with $a_1 = -e^{-\frac{T_0}{T}}$ and $b_1 = K_P\left(1 - e^{-\frac{T_0}{T}}\right)$.

The characteristic equation is

$$N(z) = 1 + G_R(z) HG_P(z) = z^2 + (a_1 + q_0 b_1 - 1)z + (q_1 b_1 - a_1) = 0$$

According to the stability conditions, example 3.10 (bilinear transformation) this results in

$$N(-1) = 2(1 - a_1) + b_1(q_1 - q_0) > 0$$
$$N(1) = b_1(q_0 + q_1) > 0$$
$$q_1 b_1 - a_1 < 1$$

and therefore

$$q_0 > -q_1$$

$$q_0 < q_1 + \frac{2(1 - a_1)}{b_1}$$

$$q_1 < \frac{1 + a_1}{b_1} = \frac{1}{K_P}$$

Fig. 5.6a depicts the resulting stability regions for different T_0/T. It can be seen that the larger the sampling time the smaller the stability region gets.

If the characteristic coefficients are introduced according to (5.2.22) the stability conditions are:

$$c_I = T_0/T_I > 0$$

$$c_I > 2\left(1 - \frac{(1 - a_1)}{b_1 K_R}\right) = 2\left(1 - \frac{1}{K_0} \frac{(1 + e^{-T_0/T})}{(1 - e^{-T_0/T})}\right)$$

$$c_I < 1 + \frac{1}{K_P K_R} = 1 + \frac{1}{K_0}$$

with $K_0 = K_R K_P$ as loop gain. In order to obtain one coordinate which is independent of the sampling time the characteristic coefficient T_I/T is considered instead of c_I. Then the following becomes valid

$$\frac{T_I}{T} > 0$$

$$\frac{T_I}{T} < \frac{T_0}{2T} \cdot \frac{K_0(1 - e^{-T_0/T})}{K_0 - 1 - (1 + K_0)e^{-T_0/T}} \quad \text{(denominator positive)}$$

$$\frac{T_I}{T} > \frac{T_0}{T} \cdot \frac{K_0}{1 + K_0}$$

This characteristic coefficient is represented in Fig. 5.6b. It thus follows that a continuous-time PI-controller has stable behaviour for all $K > 0$ and $T_I/T > 0$. With increasing sampling

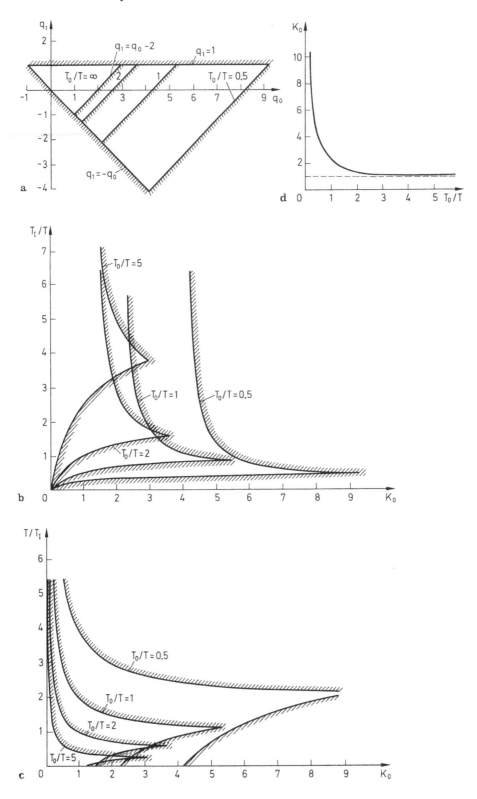

time the stabilization region becomes narrower and narrower. The larger the sampling time T_0, the smaller the values for the loop gain K_0 and the larger the values for the integration time T_I that have to be chosen. The basic course of the stability curves with the distinct peaks shows similarities to a PT_4-process with PID-controller with continuous-time signals, that is to a control system of apparently higher order.

Fig. 5.6c shows the reciprocal value T/T_I dependent on K_0. It can also be seen clearly that the stability region becomes smaller with an increasing T_0. Judging the stability regions, however, one has to note that good control performance can be only obtained for approximately $T_0/T < 0.6$.

Fig. 5.6d represents the critical loop gain for a PID-controller $(q_0 = -q_1)$. For $T_0 \to 0$ $K_0 \to \infty$ and for $T_0/T \to \infty K_0 \to 1$.

This example shows that, compared with the controller with continuous signals, the discrete controller leads to a smaller stability region. Hence the discrete PI-controller has to be set with a smaller gain and with larger integration time. The permissible control action becomes smaller with an increasing sampling time.

5.2.2 Control Algorithms with Prescribed Initial Manipulated Variable

The transfer function between the reference and the manipulated variable in closed-loop form is

$$\frac{u(z)}{w(z)} = \frac{G_R(z)}{1 + G_R(z) \, G_P(z)} = \frac{Q(z^{-1})A(z^{-1})}{P(z^{-1})A(z^{-1}) + Q(z^{-1})B(z^{-1})}$$

$$(5.2.28)$$

On introducing the process transfer function, (5.2.1), and the second-order controller transfer function, (5.2.10), and setting $b_0 = 0$, results in

$$[(1 - z^{-1})(1 + a_1 z^{-1} + \dots + a_m z^{-m})$$
$$+ (q_0 + q_1 z^{-1} + q_2 z^{-2})(b_1 z^{-1} + \dots + b_m z^{-m}) z^{-d}] u(z) \quad (5.2.29)$$
$$= (q_0 + q_1 z^{-1} + q_2 z^{-2})(1 + a_1 z^{-1} + \dots + a_m z^{-m}) w(z)$$

and for the manipulated variable

$$u(k) = (1 - a_1) \, u(k-1) + (a_1 - a_2) \, u(k-2) + \dots$$
$$- q_0 b_1 \, u(k-d-1) - (q_0 b_2 + q_1 b_1) \, u(k-d-2) + \dots$$
$$+ q_0 w(k) + (q_0 a_1 + q_1) \, w(k-1)$$
$$+ (q_0 a_2 + q_1 a_1 + q_2) \, w(k-2) + \dots \quad (5.2.30)$$

The first two manipulated variable values are, after a step change of the command variable $w(k) = w_0 1(k)$ given as:

Fig. 5.6a–d. Stability limits of a process with first-order hold and a PI-controller for different sampling times $\dfrac{T_0}{T_I} = 0.5$, 1; 2; 5.

1. Case $d = 0$

$$u(0) = q_0 w_0$$
$$u(1) = [q_0(2 - q_0 b_1) + q_1] w_0 \qquad (5.2.31)$$

2. Case $d \leq 1$

$$u(0) = q_0 w_0$$
$$u(1) = (2q_0 + q_1) w_0. \qquad (5.2.32)$$

Independently of the deadtime d, the value of $u(0)$ for a step change of the command variable depends only on the controller parameter q_0. Therefore by prescribing the manipulated variable $u(0)$, the parameter q_0 can be fixed, which, according to (5.1.15), is equivalent to $K(1 + c_D)$.

The correspondence between the first manipulated variable and the control parameter is useful during design when considering the allowable range of manipulated variable change. One only has to select a certain operating point of the control loop and the maximum process input change $u(0)$ for the (worst) case of a step change w_0 of the reference variable $w(k)$ (or the error $e(k)$) and then one simply sets $q_0 = u(0)/w_0$.

In order to avoid a larger manipulated variable $u(1)$ than $u(0)$ an inequality in the controller parameter q_1 must be considered. Hence from (5.2.31) and (5.2.32) for $u(1) \leq u(0)$

$$d = 0: q_1 \leq -q_0(1 - q_0 b_1)$$
$$d \geq 1: q_1 \leq -q_0. \qquad (5.2.33)$$

These inequalities are also valid for first-order controllers. If a small $u(0)$ is prescribed and results in a damped loop behaviour, $r = 0$ can be chosen in the optimization criterion of (5.2.6).

The determination of q_0 by prescribing $u(0)$ means that for a second-order controller only two parameters and for a first-order controller only one parameter must be optimized; this results in fewer calculations.

Of course, this design does not produce a controller which considers a hard restriction on the manipulated variable for all disturbances. The use of q_0 is only a simple design aid which regards the constraints on the manipulated variable for only one type of disturbance.

5.2.3. PID-Control Algorithm through z-Transformation

A continuous PID-controller with delayed differential part is followed by the transfer function

$$G_R(s) = K\left[1 + \frac{1}{T_I s} + \frac{T_D s}{1 + T_1 s}\right] \qquad (5.2.34)$$

If a z-transfer function with corresponding input/output behaviour is to be derived from (5.2.34), then a holding element has to be preconnected, c.f. example 3.8. Using (3.4.11) it holds for a zero-order hold

$$HG_R(z) = \frac{z-1}{z} \mathscr{L}\left\{\frac{G_R(s)}{s}\right\}$$

$$= \frac{z-1}{z} K \mathscr{L}\left\{\frac{1}{s} + \frac{1}{T_I s^2} + \frac{T_D}{1+T_1 s}\right\}$$

$$= K\left[1 + \frac{T_0}{T_I}\frac{1}{(z-1)} + \frac{T_D}{T_1}\frac{(z-1)}{(z-\gamma_1)}\right]$$

$$= K\left[1 + \frac{T_0}{T_I}\frac{z^{-1}}{1-z^{-1}} + \frac{T_D}{T_1}\frac{(1-z^{-1})}{(1-\gamma_1 z^{-1})}\right]$$

$$= \frac{q_0 + q_1 z^{-1} + q_2 z^{-2}}{(1-z^{-1})(1+\gamma_1 z^{-1})} = \frac{q_0 + q_1 z^{-1} + q_2 z^{-2}}{1 + p_1 z^{-1} + p_2 z^{-2}} \qquad (5.2.35)$$

with

$$\left.\begin{aligned}
q_0 &= K\left[1 + \frac{T_D}{T_1}\right] \\[2mm]
q_1 &= -K\left[1 - \gamma_1 + 2\frac{T_D}{T_1} - \frac{T_0}{T_I}\right] \\[2mm]
q_2 &= K\left[\frac{T_D}{T_1} + \left(\frac{T_0}{T_I} - 1\right)\gamma_1\right] \\[2mm]
\gamma_1 &= -e^{-\frac{T_0}{T_1}} \\[1mm]
p_1 &= \gamma_1 - 1 \\[1mm]
p_2 &= -\gamma_1
\end{aligned}\right\} \qquad (5.2.36)$$

(For analogue controllers $T_D/T_1 \approx 3 \ldots 10$ is used [5.25].)

Contrary to (5.1.5) these parameters are also valid for *large sample times*.

For the PI-controller ($T_D = 0$) the coefficients of the parameters which emerged through discretization of the difference equation (5.1.1) follow directly from (5.2.35)

$$q_0 = K \quad \text{and} \quad q_1 = -K\left(1 - \frac{T_0}{T_1}\right)$$

This, however, is not possible for the PID-controller because of the delayed differential element (also compare discussion of (3.4.19). Only q_0 agrees for $T_1 = T_0$. The other parameters differ because of $\gamma_1 = -1/e$. If, however, $T_1 = T_0$ and $\gamma_1 = 0$ are set in (5.2.36), then the values of (5.1.5) emerge and the controller structure transforms in (5.2.10). This again results in the block diagram of a PID-control algorithm as represented in Fig. 5.5. As for larger sample times the controller parameters of (5.2.36) cannot be taken for the continuous controller, one can immediately use the more simple controller of (5.2.10).

5.3 Modifications to Discrete PID-Control Algorithms

Many modifications to the discrete time PID algorithm have been published and these are based on different realizations of the discrete approximation to the continuous controller given by (5.1.1) to (5.1.4).

5.3.1 Different Evaluation of Control Variable and Reference Variable

Using (5.2.2) it follows that, for the controller,

$$P(z^{-1})u(z) = Q(z^{-1})e(z) = Q(z^{-1})[w(z)-y(z)]. \qquad (5.3.1)$$

Hence, the same polynominal $Q(z^{-1})$ is used for the reference variable w and the control variable y. A first modification to the basic PID controller algorithm is to treat the reference variable differently from the control variable. This leads to

$$
\left. \begin{aligned}
P(z^{-1})u(z) &= S(z^{-1})w(z) - Q(z^{-1})y(z). \\[4pt]
u(z) &= \frac{S(z^{-1})}{P(z^{-1})}w(z) - \frac{Q(z^{-1})}{P(z^{-1})}y(z) \\[4pt]
&= \frac{Q(z^{-1})}{P(z^{-1})}\left[\frac{S(z^{-1})}{Q(z^{-1})}w(z)-y(z)\right],
\end{aligned} \right\} \qquad (5.3.2)
$$

This is shown in Fig. 5.7a and corresponds to the introduction of a feed forward reference variable filter $G_{FW}(z)=S(z^{-1})/Q(z^{-1})$, Fig. 5.7b. This does not change the location of the poles. The following time domain specification is intended: in order to damp the relatively large manipulated variables which occur with *rapid changes of reference variables* the reference variable $w(k)$ is not considered in the derivative term [5.3]. Thus instead of the common PID-control algorithm

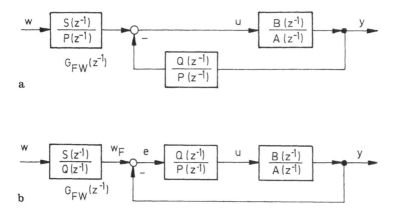

a

b

Fig. 5.7a, b. Different evaluation of control variable and reference variable

$$u(k) - u(k-1) = K\left[e(k) - e(k-1) \right.$$

$$\left. + \frac{T_0}{T_I} e(k-1) + \frac{T_D}{T_0}(e(k) - 2e(k-1) + e(k-2)) \right] \qquad (5.3.3)$$

c.f. (5.1.4), the modified algorithm then is

$$u(k) - u(k-1) = K\left[e(k) - e(k-1) \right.$$

$$\left. + \frac{T_0}{T_I} e(k-1) + \frac{T_D}{T_0}(y(k) + 2y(k-1) + y(k-2)) \right]. \qquad (5.3.4)$$

This corresponds in (5.3.2)

$$\left. \begin{array}{ll} S(z^{-1}) & = s_0 + s_1 z^{-1} \\[2mm] \text{with} \quad s_0 & = K \\[3mm] s_1 & = -K\left[1 - \dfrac{T_0}{T_I} \right] \end{array} \right\} \qquad (5.3.5)$$

$(T_D = 0$ in $(5.1.5))$

The amplitude changes of the manipulated variable are further reduced if the reference variable is only present in the integration term [1.12, 5.1, 5.3]

$$u(k) - u(k-1) = K\left[-y(k) + y(k-1) + \frac{T_0}{T_I} e(k-1) \right.$$

$$\left. + \frac{T_D}{T_0}(-y(k) + 2y(k-1) - y(k-2)) \right]. \qquad (5.3.6)$$

or

$$\left. \begin{array}{ll} S(z^{-1}) & = s_1 z^{-1} \\[3mm] \text{with} \quad s_1 & = K\dfrac{T_0}{T_I}. \end{array} \right\} \qquad (5.3.7)$$

With this algorithm it is then more appropriate to use $e(k)$ instead of $e(k-1)$ so that a change of the reference variable leads immediately to an input change.

These modified algorithms are less sensitive to the higher frequency signals of $w(k)$ than to those of $y(k)$. Therefore, for the same type of disturbance at e.g. the process input and the command variable, the differences between the controller parameters obtained by parameter optimization become smaller, (c.f. [5.8]). Large changes of manipulated variables can also be avoided by limiting the velocity of the command variable and/or the manipulated variable. Since these restrictions are effective for all disturbances, they should be used in preference to the modified control algorithms.

5.3.2 Different Discretizations of the Derivative Term

Other modifications are obtained by different realizations of the *derivative term*. If the controlled variable contains relatively high frequency noise which cannot be or is not supposed to be controlled, large unwanted manipulated variable changes can occur if the first-order difference

$$T_D \frac{\Delta e(k)}{T_0} = \frac{T_D}{T_0} (e(k) - e(k-1))$$

in the nonrecursive form, (5.1.2), or in the recursive form, (5.3.1)

$$\frac{T_D}{T_0} (e(k) - 2e(k-1) + e(k-2))$$

is used. The derivative term, however, can be necessary for improving control performance for medium frequency noise since, if it is not too large, a process pole can be approximately cancelled, the stability region can be increased and larger gains can be used. Therefore, one has to compromise.

One possibility is to choose T_D/T_0 smaller than the ideal value. One can also smooth the derivative action by using 4 values [5.2]. First an average value of the form

$$\bar{e}_k = \frac{1}{4} [e(k) + e(k-1) + e(k-2) + e(k-3)]$$

is taken and then all approximations to the first derivative are averaged in relation to $\bar{e}_k$. The derivative term for the nonrecursive form therefore becomes

$$T_D \frac{\overline{\Delta e_k}}{T_0} = \frac{T_D}{4} \left[\frac{e(k) - \bar{e}_k}{1,5\, T_0} + \frac{e(k-1) - \bar{e}_k}{0,5\, T_0} + \frac{\bar{e}_k - e(k-2)}{0,5\, T_0} + \frac{\bar{e}_k - e(k-3)}{1,5\, T_0} \right]$$

$$= \frac{T_D}{6T_0} [e(k) + 3e(k-1) - 3e(k-2) - e(k-3)]. \qquad (5.3.8)$$

For the recursive form of the *D*-part one obtains

$$\frac{T_D}{6T_0} [e(k) + 2e(k-1) - 6e(k-2) + 2e(k-3) + e(k-4)].$$
$$(5.3.9)$$

5.3.3 Delayed Differential Term

A further choice for small sampling times consists of using a differential term with a first-order lag as in the continuous transfer function

$$G_R(s) = K\left[1 + \frac{1}{T_I s} + \frac{T_D s}{1 + T_1 s}\right]$$

and by applying the bi-linear transformation $s \rightarrow 2(z-1)/T_0(z+1)$, as an approximation [2.19] (as in section 3.7). The resulting control algorithms becomes

$$u(k) = p_1 u(k-1) + p_2 u(k-2) + q_0 e(k) + q_1 e(k-1) + q_2 e(k-2)$$
$$(5.3.10)$$

with parameters

$$p_1 = -4c_1/(1+2c_1)$$

$$p_2 = (2c_1-1)/(1+2c_1)$$

$$q_0 = K\left[1 + 2(c_1 + c_D) + \frac{c_I}{2}(1+2c_1)\right]/[1+2c_1]$$

$$q_1 = K[c_I - 4(c_1 + c_D)]/[1+2c_1]$$

$$q_2 = K\left[c_1(2-c_I) + 2c_D + \frac{c_I}{2} - 1\right]/[1+2c_1]$$

with

$$c_1 = T_1/T_0; \ c_I = T_0/T_I; \ c_D = T_D/T_0.$$

The denominator polynomial is

$$P(z) = (1-z^{-1})\left(1 - \frac{2c_1-1}{2c_1+1}z^{-1}\right)$$

This shows a pole at $z_1 = 1$ (I-behaviour) and at $z_2 = (2T_1 - T_0)/(2T_1 + T_0)$. In order that the real pole satisfies $z_2 \geq 0$ (for non-oscillating behaviour according to Fig. 3.12), T_1 has to be chosen as $T_1 \geq T_0/2$.

If $T_1 = 0$ is assumed, a pole on the unit circle at $z_2 = -1$ results. The manipulated variable then would oscillate with a constant amplitude. This indicates that with discrete-time systems the differentiation without delay, possible for continuous-time systems, does not exist.

For a further discussion of these examples of modification to PID-control algorithms the reader is referred to [5.34]. Here, in order to *compensate the dead time caused by the zero-order hold* (approximately $T_0/2$), a predictor is suggested which predicts the control variable $y(k)$ by half a sampling time based on a known process model (e.g. first-order). Hence, $y(k+0.5)$ is taken. For the controller the deviation $e(k) = w(k) - y(k+0.5)$ is used. Advantages can only be expected if the sampling time is relatively large and the process model order (lag) small and no high frequency disturbance signals occur. Since the lags caused by the holding element are components of the process model, the structure optimal controllers based on process models consider these lags without further provisions.

5.4 Design Through Numerical Parameter Optimization

If a mathematical model of the process which is to be controlled already exists, numerical parameter optimization is the method which can be generally applied in order to determine the unknown parameters $q_0, q_1, \ldots, q_v$ of a parameter optimized controller. Here, a loss function V, which e.g. is a quadratic control performance criterion S_{eu}^2, Eq. (5.2.6), is by variation of the controller parameter

$$q^T = [q_0 \; q_1 \; \cdots \; q_v] \tag{5.4.1}$$

determined in such a way that V assumes a minimum, yielding

$$\frac{dV}{dq} = 0 \tag{5.4.2}$$

and

$$\frac{\partial^2 V}{\partial q \partial q^T} > 0. \tag{5.4.3}$$

Here, especially for higher order processes, numerical parameter optimization methods can be used.

5.4.1 Numerical Parameter Optimization

The general procedure is then as follows:

— Optimization step j:
   1. Assumption of initial values $q(j-1)$
   2. Calculation of the time behaviour of $e(k)$ and $u(k)$ for a given noise signal of the control loop and $k = 0 \ldots M$
   3. Calculation of the performance criterion $V(j)$
   4. Application of the numerical optimization method furnishes new controller parameters $q(j)$
— Optimization step $j+1$
   1. As initial values instead of $q(j-1) \rightarrow q(j)$ is assumed
   2. —4. same as above, etc.

The optimization is to be continued until a convergence criterion $\Delta V = V(j) - V(j-1) \leq \varepsilon$ is violated. The following shows some typical problems to be encountered when using this approach.

Initial Values

The computation time can be considerably reduced by prescribing suitable initial values which can be obtained through e.g.

$q(0) = 0$ if no better values are known

$q(0) = q_*$ whereby the parameters q_* originate from a "rapid" design method, see section 5.5.3.

$q(0) = q_0$ whereby q_0 are values based on former experience.

Calculation of the Performance Criterion

Usually $e(k)$ and $\Delta u(k)$ are calculated according the difference equations of the closed-loop control action in the time domain for $k = 0 \ldots M$ steps and an appropriate external excitation (stimulation). For a step change of the reference variable one obtains for the z-transform of the signals with $w(z) = 1/(1 - z^{-1})$ and consideration of $P(z^{-1}) = 1 - z^{-1}$ and $\Delta u(z) = (1 - z^{-1})u(z)$

$$
\left.
\begin{aligned}
e(z) &= \frac{A(z^{-1})}{P(z^{-1})A(z^{-1}) + Q(z^{-1})B(z^{-1})} \\[2em]
\Delta u(z) &= \frac{Q(z^{-1})A(z^{-1})}{P(z^{-1})A(z^{-1}) + Q(z^{-1})B(z^{-1})}
\end{aligned}
\right\}
\tag{5.4.4}
$$

Since only $Q(z^{-1})$ changes, the parameters of the terms $P(z^{-1})\, A(z^{-1})$ can be prescribed in the corresponding difference equations.

One can cut down on computation time considerably by using Parseval's theorem in order to calculate the summation in the z-domain [2.13, 2.18, 5.26]

$$
\left.
\begin{aligned}
S_e'^2 &= \sum_{k=0}^{\infty} e^2(k) = \frac{1}{2\pi i} \oint e(z)e(z^{-1})z^{-1}\,dz \\[2em]
S_u'^2 &= \sum_{k=0}^{\infty} \Delta u^2(k) = \frac{1}{2\pi i} \oint \Delta u(z)\Delta u(z^{-1})z^{-1}\,dz
\end{aligned}
\right\}
\tag{5.4.5}
$$

at which the integration on the unit circle is performed in positive direction. The equations can then be solved by using the residues [2.11, 2.13].

The corresponding denominator polynomials of the fractional rational terms, however, are not supposed to have roots on the unit circle. [12.4] gives a recursive algorithm to calculate (5.4.5). Here, poles may not be located outside the unit circle. Hence, the closed control loop has to be asymptotically stable. For orders of about $2 \leqq m \leqq 4$ the required calculation time is about 10–20% compared with calculation in the time domain [5.26]. Compare also [5.27].

Note, that this z-domain procedure is hardly worthwhile for multivariable systems because of their high-order polynomials.

Numerical Optimization Methods

With numerical optimization methods (hill-climbing method) one distinguishes between the simple *search methods* which only require the specific time function $V(j)$ and the *gradient methods* where the first derivations $V_q(j)$ or, as well, the second derivations $V_{qq}(j)$ (e.g. Newton-Raphson) have to be calculated. There also exist combinations of several methods (e.g. Fletcher-Powell). These methods are described e.g. in [5.6, 5.28–5.31, 5.33, 5.36].

The derivations of the performance criteria with respect to controller parameters are not known in analytical form. This is why especially search methods are used. Since their minima are often flat they, additionally, have numerical advantages compared with the gradient methods where numerical problems occur more distant from the optimum.

Because of the distinct convexity of the loss function, search methods, generally, cause no problems with single input/single output controllers in the range of the stable loop behaviour. Note, however, that usually the computational effort increases proportionally if v^3 if v is the number of the parameters to be optimized.

After extensive examinations a modified Hooke-Jeeves-search method has proven to be particularly suited, which not only allows to diminish the intervals but also to enlarge them depending on a successful proceeding [5.26].

5.4.2 Simulation Results for PID-Control Algorithms

During the design of control systems appropriate free parameters have to be selected. In the case of parameter-optimized discrete control algorithms these are the sampling time T_0 and the weighting factor r of the manipulated variable in the optimization criterion or the chosen initial manipulated variable $u(0)$. In order to assist in obtaining first estimates for their selection, this section shows some simulation results [5.7].

Free parameters cannot be chosen independently of the process under consideration and its technological properties. Therefore, very general rules cannot be given. However, results of two simulated test processes will show that qualitative results can be obtained which may be valid for similar processes.

Test Processes

For investigating control algorithms in closed loop operation, processes II and III which were proposed in [5.9] as test cases are used. The reader is referred to the Appendix for further details.

Process II: Process with nonminimal phase behaviour

Table 5.1. Parameters of process II

Sample time T_0 in s	1	4	8	16
b_1	−0.07289	−0.07357	0.13201	0.55333
b_2	0.09394	0.28197	0.34413	0.23016
a_1	−1.68364	−1.0382	−0.58466	−0.22021
a_2	0.70469	0.2466	0.06081	0.0037

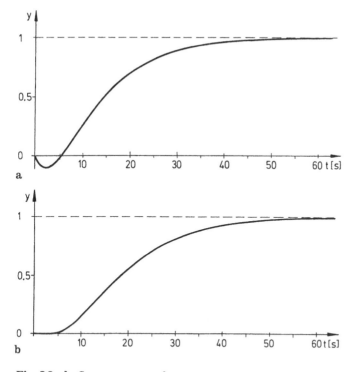

Fig. 5.8a, b. Step responses of test process II and III **a** Process II **b** Process III

$$G_{II}(s) = \frac{K(1-T_1s)}{(1+T_1s)(1+T_2s)} \qquad (5.4.6)$$

$K = 1; T_1 = 4 \text{ s}; T_2 = 10 \text{ s}$

$$G_{II}(z) = \frac{b_1z^{-1} + b_2z^{-2}}{1 + a_1z^{-1} + a_2z^{-2}} \qquad (5.4.7)$$

Figure 5.8a shows the step response of this process.

Process III: Process with low-pass behaviour and deadtime

$$G_{III}(s) = \frac{K(1+T_4s)}{(1+T_1s)(1+T_2s)(1+T_3s)} e^{-T_ts} \qquad (5.4.8)$$

$K = 1; T_1 = 10 \text{ s}; T_2 = 7 \text{ s}; T_3 = 3 \text{ s}; T_4 = 2 \text{ s}; T_t = 4 \text{ s}.$

$$G_{III}(z) = \frac{b_0 + b_1z^{-1} + b_2z^{-2} + b_3z^{-3}}{1 + a_1z^{-1} + a_2z^{-2} + a_3z^{-3}} z^{-d} \qquad (5.4.9)$$

Fig. 5.8b shows the step response of this process.

Control Algorithms with No Prescribed Initial Manipulated Variable

In this section, some results of digital computer simulations of the test processes II and III with the second-order control algorithms with no prescribed initial

Table 5.2. Parameters of process III

Sample time T_0 in s	1	4	8	16
d	4	1	1	1
b_0	0	0	0.06525	0.37590
b_1	0.00462	0.06525	0.25598	0.32992
b_2	0.00169	0.04793	0.02850	0.00767
b_3	−0.00273	−0.00750	−0.00074	−0.00001
a_1	−2.48824	−1.49863	−0.83771	−0.30842
a_2	2.05387	0.70409	0.19667	0.02200
a_3	−0.56203	−0.09978	−0.00995	−0.00010

manipulated variable are considered. All three controller parameters are optimized. The abbreviation for the controller given by (5.2.10) is: 3 PC-3 (3 Parameter Controllers with 3 optimized parameters). The criterion of (5.2.6) was used as optimization criterion. The controller parameters q_0, q_1 and q_2 were determined through numerical search using the method of Fletcher-Powell. The settling time taken was $M = 128$.

For a *step change of the reference variable*, the *control performance* expressed in the form of

$$S_e = \sqrt{\overline{e^2(k)}} = \sqrt{\frac{1}{N+1} \sum_{k=0}^{N} e^2(k)} \qquad \text{(Quadratic average of control deviation)}$$

$$(5.4.10)$$

$$y_m = y_{max}(t) - w(t) \qquad \text{(Max. overshooting)}$$

$$(5.4.11)$$

$$k_1 \quad \text{(Settling time for } |e(k)| \leq 0.01|w(k)|) \qquad (5.4.12)$$

and the corresponding *quadratic average of the manipulated variable deviation*

$$S_u = \sqrt{\overline{\Delta u^2(k)}} = \sqrt{\frac{1}{N+1} \sum_{k=0}^{N} \Delta u^2(k)}, \qquad (5.4.13)$$

the "manipulating effort" are functions of the sampling time T_0 and the weighting factor r of the manipulated variable in the optimization criterion (5.2.6). According to DIN 40110 the term "quadratic average" was chosen for E_e and S_u. Those values correspond to the "effective value" and the "root of the corresponding effective power" for an Ohm resistance, respectively.

Comment on the Choice of the Disturbances:

A step disturbance excites predominantly the lower frequencies and leads to a larger weighting of the integral action of the controller. In chapter 13 stochastic disturbances are used which contain higher frequency components and which constrain the proportional and derivative actions more severely.

Influence of the Sampling Time T_0:

Figure 5.9 shows the discrete values of the control and the manipulated variables for both processes after a step change of the command variable for the sampling times $T_0 = 1$; 4; 8 and 16 sec and for $r = 0$.

For the relatively small sampling time of $T_0 = 1$ sec one obtains an approximation to the control behaviour of a continuous-time PID-controller. With increasing sampling time the settling behaviour of the control variable of both the processes becomes less advantageous. With large sampling times $T_0 = 8$ sec and 16 sec the real course of the control variable cannot be sufficiently recorded. This means, that the value of S_e (5.4.10), which is defined for discrete signals should be used with caution as a measure of the control performance for $T_0 > 4$ sec. However, as the parameter optimization is based on the discrete-time signals (for computational reasons) S_e is used in the comparisons.

In Fig. 5.10 the control performance and the manipulating effort are shown as functions of the sample time. For process II the quadratic mean of the control deviation S_e, the overshoot y_m and the settling time k_1 increase with increasing sample time T_0, i.e. the control performance becomes worse. The manipulating effort S_u is at a minimum for $T_0 = 4$ sec and increases for $T_0 > 4$ sec and $T_0 < 4$ sec. For process III all three characteristic values deteriorate with increasing sample time. The manipulating effort is at a minimum for $T_0 = 8$ sec. The improvement of the control performance for $T_0 < 8$ sec is due to the fact that the manipulation effort increases considerably. Therefore the initial manipulated variable $u(0)$ also increases with decreasing sample time, Fig. 5.9b.

On the basis of these simulations it follows that for the given optimization criterion with $r = 0$, the appropriate sample interval for process II and III are $T_0 < 4$ sec and $T_0 < 8$ sec, respectively. The smaller sample interval enables a somewhat better control performance. On the other hand $T_0 = 16$ sec is unsuitable for either process because of a poor control performance.

To assist the choice of a suitable sample interval, the behaviour of the performance criterion S_{eu} due to (5.2.6) can also be used. This criterion takes into account both the control performance S_e and the manipulating effort S_u. For a weighting of the manipulated variable of $r = 0.25$ in Fig. 5.10 the value of S_{eu} is shown. This "mixed" criterion shows a flat minimum for process III for $T_0 = 5$ sec and for process II a flat behaviour for $T_0 < 8$ sec. Hence, a suitable region for the sample interval is: for process III approximately $T_0 = 3 \ldots 8$ sec and for process II approximately $1 \ldots 8$ sec. If T_{95} is the settling time of a step response defined as the time taken for the response to reach 95% of its final value (assuming zero initial conditions), the following rules can be used for choosing the sample time:

Process II : $\beta = T_{95}/T_0 = 4.4 \ldots 11.7$

$$(5.4.14)$$

Process III: $\beta = T_{95}/T_0 = 5.6 \ldots 15.0.$

Table 5.3 gives the controller parameters. With increasing sample time the parameters q_0, q_1 and q_2 become smaller. The controller gain K hardly changes for $T_0 \geqq 4$ sec, the lead factor c_D reduces and the integration factor c_I increases. The

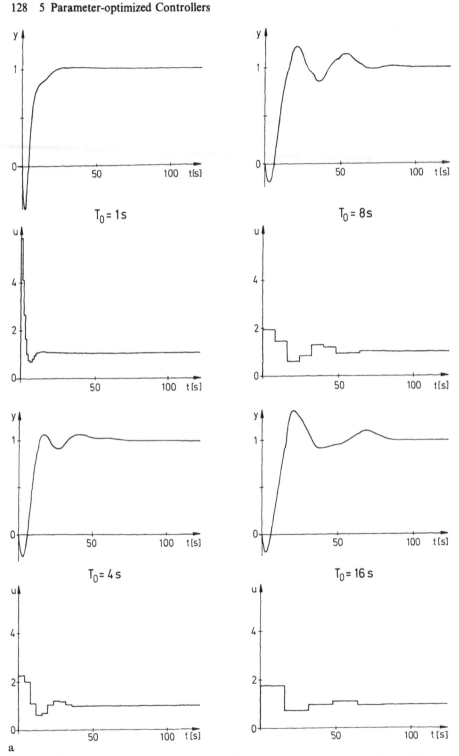

Fig. 5.9a. Step responses for changes of the reference variable for process II and different sample intervals T_0 and $r=0$

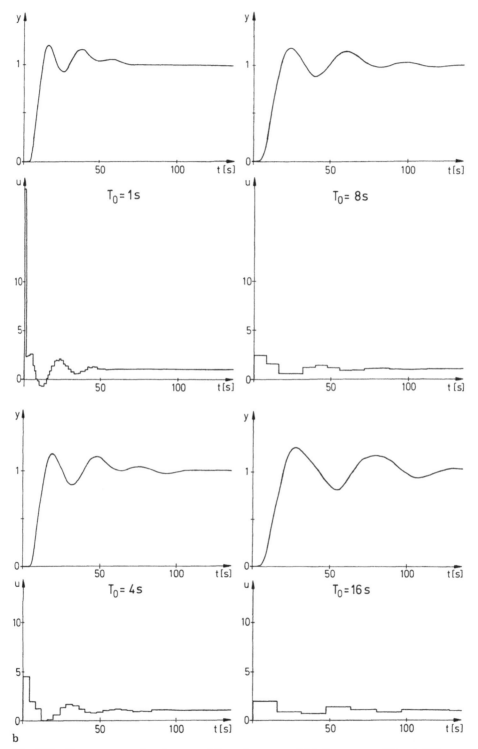

Fig. 5.9b. Step responses for changes of the reference variable for process III and different sample intervals T_0 and $r=0$

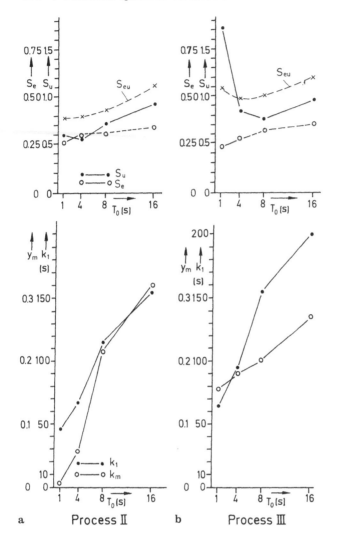

Fig. 5.10. Control performance and manipulating effort as functions of the sample time T_0 for $r=0$

a   Process II

b   Process III

Table 5.3. Controller parameters for different sample times T_0 and $r=0$

	Process II T_0 in s				Process III T_0 in s			
	1	4	8	16	1	4	8	16
q_0	5.958	2.332	2.000	1.779	19.408	4.549	2.437	1.957
q_1	−10.337	−3.074	−2.080	−1.089	−36.623	−7.160	−2.995	−1.660
q_2	4.492	1.105	0.748	0.361	17.370	3.030	1.158	0.667
K	1.466	1.227	1.252	1.418	2.038	1.519	1.279	1.290
c_D	3.065	0.901	0.597	0.255	8.524	1.994	0.905	0.517
c_I	0.077	0.297	0.534	0.742	0.076	0.275	0.469	0.748

inequalities (5.2.14) or (5.2.17) are satisfied for $T_0 = 1, 4$ and 8 sec, so that a control algorithm with normal PID-behaviour emerges.

Influence of the Weighting Factor r on the Manipulated Variable

For the sample time $T_0 = 1$ sec Fig. 5.11 shows step responses to changes of the reference variable as functions of the weighting factor r in the optimization criterion. A change from $r = 0$ to $r = 0.1$ leads to a larger increase of damped behaviour than the change from $r = 0.1$ to $r = 0.25$.

Fig. 5.12 shows the characteristic parameters of the control performance and the manipulating effort for $T_0 = 1; 4$ and 8 sec as functions of the weighting factor r. For both processes, with increasing r, the control performance S_e increases and the manipulating effort S_u decreases. For process III this effect is greater than for process II. The selection of the weighting factor r has greater influence on S_e and S_u for process III. Furthermore, r has smaller influence the greater the sample time.

The overshoot y_m also decreases with increasing r. The response time k_1 increases with r for $T_0 = 1$ sec. However, for $T_0 = 4$ sec and 8 sec, k_1 first decreases and then increases for greater r. The choice of r influences the characteristic values y_m and k_1 much more than S_e and S_u for all sample times. An increase of the weighting factor r of the manipulated variable in the optimization criterion (5.2.6) therefore results in a decrease of the manipulating effort S_u, an increase in S_e and a decrease in the overshoot y_m. The choice of r depends very much on the application. A suitable compromise between good control performance and small manipulation effort can be obtained in the region of $0.1 \leq r \leq 0.25$ if the process gain is unity.

Table 5.4 shows the controller parameters for the sample times $T_0 = 4$ sec and 8 sec. With increasing weighting r of the manipulated variable the parameters q_0, p_1 and q_2 decrease. K and c_D also decrease, whilst c_I hardly changes.

Table 5.4. Controller parameters for different weighting factors r and $T_0 = 4$ s and 8 s

	Process II r				Process III r		
	0	0.1	0.25		0	0.1	0.25
$T_0 = 4$ s				$T_0 = 4$ s			
q_0	2.332	1.933	1.663	q_0	4.549	2.688	2.049
q_1	−3.076	−2.432	−2.016	q_1	−7.160	−3.798	−2.723
q_2	1.117	0.816	0.637	q_1	3.030	1.398	0.916
K	1.215	1.117	1.026	K	1.519	1.290	1.133
c_D	0.919	0.730	0.621	c_D	1.994	1.083	0.808
c_I	0.307	0.284	0.277	c_I	0.275	0.223	0.213
$T_0 = 8$ s				$T_0 = 8$ s			
q_0	2.000	1.714	1.512	q_0	2.437	1.944	1.653
q_1	−2.080	−1.685	−1.423	q_1	−2.995	−2.222	−1.795
q_2	0.748	0.557	0.440	q_2	1.158	0.780	0.587
K	1.252	1.157	1.072	K	1.279	1.164	1.066
c_D	0.597	0.481	0.410	c_D	0.905	0.669	0.550
c_I	0.534	0.507	0.494	c_I	0.469	0.431	0.417

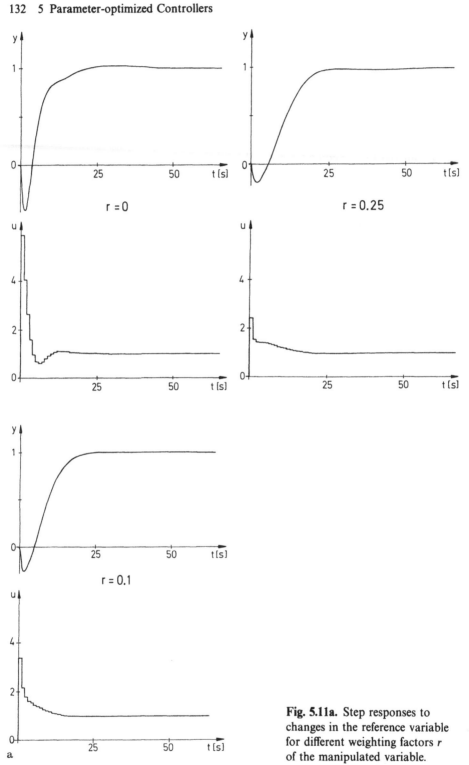

Fig. 5.11a. Step responses to changes in the reference variable for different weighting factors r of the manipulated variable. Process II. Sample time $T_0 = 1$ s

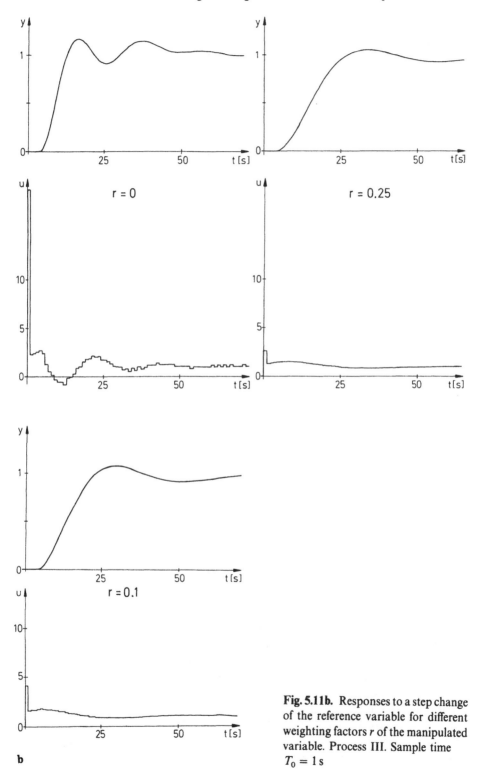

Fig. 5.11b. Responses to a step change of the reference variable for different weighting factors r of the manipulated variable. Process III. Sample time $T_0 = 1$ s

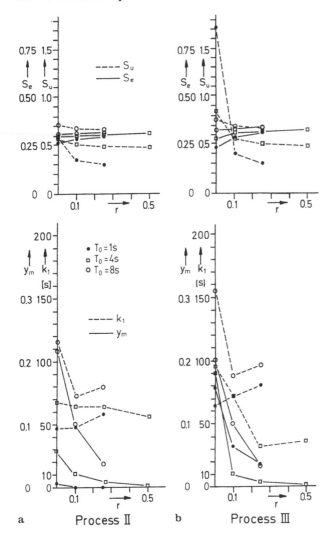

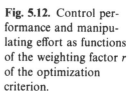

Fig. 5.12. Control performance and manipulating effort as functions of the weighting factor r of the optimization criterion.

a   Process II   b   Process III

Control Algorithms with Prescribed Initial Manipulated Variable u(0)

In section 5.2.2 it was shown that for a step change of the reference variable by 1 or w_0, the parameter q_0 of the control algorithm is equal to the manipulated variable $u(0)$ or $u(0)/w_0$ from (5.2.6). By properly choosing $u(0)$, taking into account an allowable region of the manipulated variable, the parameter q_0 can be readily determined. Then only two parameters q_1 and q_2 have to be optimized. The control algorithm is therefore called 3 PC-2.

Since the control behaviour can be constrained by assuming a relatively small $u(0)$, the weighting factor r in the optimization criterion (5.2.6) can be taken as zero when choosing the parameters.

Influence of the Prescribed Manipulated Variable

In Fig. 5.13 the responses to step changes in the command variable are shown for different values of the initial manipulated variable $u(0) = q_0$. Starting with a value $q_{0,opt}$, which results from optimization of all parameters for $r = 0$, a decrease of the chosen manipulated variable $u(0)$ results in a more restrained control behaviour. The overshoot y_m decreases. In Fig. 5.13b q_0 has been reduced such that both of the first two manipulated variables $u(0)$ and $u(1)$ are equal. However, the resulting overshoot increases again.

Fig. 5.14 shows all the characteristic values of the control performance and the manipulating effort for the practically significant sample times $T_0 = 4$ sec and $T_0 = 8$ sec. With a reduction in the chosen manipulated variable $u(0)$, i.e. decreasing q_0, the manipulating effort decreases and the control performance S_e increases slightly. The overshoot y_m and the response time k_1 also decrease for $T_0 = 8$ sec. For $T_0 = 4$ sec, the same trend occurs at first for both processes. If q_0 is chosen too small then both values increase again. A minimum occurs for y_m and k_1.

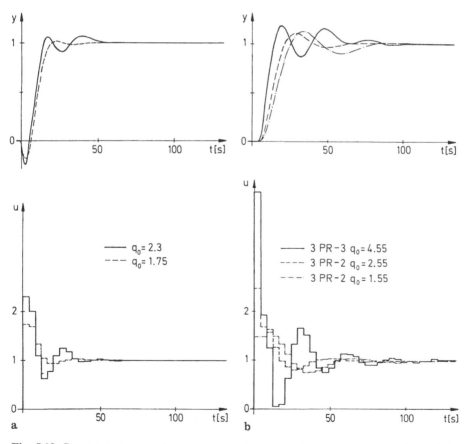

Fig. 5.13. Responses to step changes in the reference variable for different values of the chosen manipulated variable $u(0) = q_0$, $T_0 = 4$ s

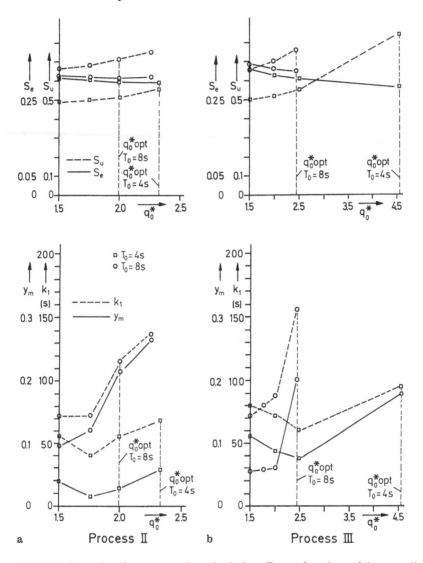

Fig. 5.14. Control performance and manipulating effort as functions of the prescribed initial manipulated variable $u(0)=q_0$.

If the manipulated variable $u(0) = q_0$ is prescribed not too small the control performance S_e deteriorates slightly. However, the manipulating effort S_u, the overshoot y_m and the response time k_1 decrease considerably.

A good choice of the initial manipulated variable $u(0)$ produces not only good control performance but also fewer computations in the parameter optimization. Table 5.5 shows the controller parameters for different values of $u(0)$. The parameters q_1 and q_2 follow the trend of q_0 but c_I hardly changes. For the same q_0 the other controller parameters hardly vary for both processes.

Table 5.5. Controller parameters for different chosen manipulated variable $u(0)=q_0$ for $T_0 = 4$ s and 8 s

	Process II					Process III			
	3 PC-2			3 PC-3		3 PC-2			3 PC-3
$T_0=4$ s					$T_0=4$ s				
q_0	1.500	1.750	2.000	2.332	q_0	1.500	2.000	2.500	4.549
q_1	−1.593	−2.039	−2.484	−3.076	q_1	−1.499	−2.406	−3.320	−7.160
q_2	0.375	0.591	0.810	1.105	q_2	0.223	0.656	1.097	3.030
K	1.125	1.159	1.190	1.227	K	1.277	1.344	1.403	1.519
c_D	0.333	0.511	0.681	0.901	c_D	0.175	0.488	0.783	1.994
c_I	0.251	0.261	0.274	0.295	c_I	0.176	0.186	0.198	0.275
$T_0=8$ s					$T_0=8$ s				
q_0	1.500	1.750	2.250	1.999	q_0	1.500	1.750	2.000	2.437
q_1	−1.338	−1.717	−2.405	−2.079	q_1	−1.451	−1.864	−2.280	−2.995
q_2	0.364	0.556	0.936	0.748	q_2	0.372	0.576	0.784	1.158
K	1.136	1.194	1.314	1.251	K	1.128	1.174	1.216	1.279
c_D	0.321	0.466	0.712	0.597	c_D	0.330	0.490	0.645	0.905
c_I	0.464	0.494	0.594	0.534	c_I	0.374	0.393	0.414	0.469

Conclusions Based on the Simulation Results

The simulation of a process of second-order with nonminimum phase behaviour and a low-pass third order process with deadtime lead to the following results for second-order control algorithms:

Choice of the Sample Time T_0

The smaller the sample time the better the control behaviour. However, if the sample time becomes very small, further improvement of the control behaviour can be only obtained by a considerable increase of the manipulating effort. Therefore, too small a sample time should not be chosen. For the selection of proper sample times the following rules can be used

$$T_0/T_{95} \approx 1/15 \ldots 1/4. \tag{5.4.15}$$

Here T_{95} is the settling time of the step response to 95% of the final value.

Choice of the Weighting Factor r

If all three parameters of the second-order control algorithm (3 PC-3) have to be optimized for a step change of the reference variable one obtains a good compromise between good control performance and small manipulating effort for

$$rK_p^2 \approx 0{,}05 \ldots 0{,}25 \tag{5.4.16}$$

where $K_p = G_p(1)$ is the process gain. The larger the sample time the smaller the influence of r.

Choice of the Initial Manipulated Variable $u(0) = q_0 w_0$

From the value of the parameter q_0, the initial manipulated variable $u(0)$ after a step change or control error can be determined. This has the advantage that one fewer parameter needs to be optimized by numerical search methods, so saving computing time. Furthermore, the control behaviour can simply be constrained and $r = 0$ can be set in the optimization criterion.

The magnitude of q_0 depends on the allowable range of the manipulated variable and of the process under consideration. Appropriate values are as follows: for process II with nonminimum phase behaviour $q_0 = 1.75$ and for the low-pass process III with deadtime $q_0 = 2.5$. However, q_0 can be chosen within a larger region depending on the allowable range of the manipulated variable. For the estimation of q_0 the maximum value of a step change in the reference variable must be assumed considering the inequality (5.2.33) (iterative procedure).

If the sample time is not too small, $u(0)$ can be chosen according to

$$q_0 \leqq 1/(1-a_1)(b_1 + b_2 + \ldots + b_m) \tag{5.4.17}$$

which is obtained for the modified deadbeat controller $DB(v+1)$ from (7.2.13).

5.5 PID-Controller Design Through Pole-Assignment, Compensation and Approximation

The parameter optimization methods treated in the last chapter are of general use, provided a definite minimum of the control performance criterion exists. However, they require a digital computer. It is therefore desirable to calculate the controller parameter directly from the process model. For unspecified process order and deadtime this is generally not achievable since the structure of the controller equation deviates from the structure of the process model equation. That is why simple relations exist for only a few special cases.

5.5.1 Pole-assignment Design

The characteristic equation of the closed-loop system is

$$1 + G_R(z)G_P(z) = 0 \tag{5.5.1}$$

and after inserting the particular polynomials is

$$P(z^{-1})A(z^{-1}) + Q(z^{-1})B(z^{-1})z^{-d} = 0 \tag{5.5.2}$$

The application of the PID-controller equation then yields

$$(1-z^{-1})(1+a_1z^{-1}+\ldots+a_mz^{-m})$$
$$+ (q_0+q_1z^{-1}+q_2z^{-2})(b_1z^{-1}+\ldots+b_mz^{-m})z^{-d} = 0 \tag{5.5.3}$$

or

$$\mathscr{A}(z^{-1}) = 1 + \alpha_1 z^{-1} + \ldots + \alpha_{m+2+d} z^{-(m+2+d)} = 0 \qquad (5.5.4)$$

so that

$$z^{(m+2+d)} + \alpha_1 z^{(m+2+d-1)} + \ldots + \alpha_{m+2+d} = 0 \qquad (5.5.5)$$

$$(z - z_{\alpha 1})(z - z_{\alpha 2}) \ldots (z - z_{\alpha(m+2+d)}) = 0 \qquad (5.5.6)$$

Hence, $(m+d+2)$ poles emerge.

The coefficients α_1 can be determined through pole-assignment (5.5.4) and by comparing the coefficients with (5.5.4) from $(m+d+2)$ equations the controller parameters can be obtained. Since the PID-controller is restricted to three controller parameters q_0, q_1 and q_2, a clear solution is only possible for

$$m+d+2 = 3 \text{ oder } m = 1-d \qquad (5.5.7)$$

with $m = 1$ for $d = 0$ or $m = 0$ for $d = 1$.

For $m > 1-d$ the poles $z_{\alpha i}$ respectively the coefficients cannot be assigned independent from each other, (c.f. section 11.1). Setting in (5.5.2) $P(z^{-1}) = (1-z^{-1})$ $\times (1 \times \gamma_1 z^{-1})$, one additional controller parameter is then obtained and this leads to

$$m+d+2 = 4 \text{ oder } m = 2-d \qquad (5.5.8)$$

e.g. $m = 2$ for $d = 0$, (c.f. (5.5.24)).

Example 5.3

The parameters of a PID-control algorithm for process VIII are to be determined in such a way that the poles are located as follows

$$z_1 = 0{,}125; \; z_2 = 0{,}375; \; z_{3,4} = 0{,}25 \pm 0{,}375i$$

(The poles of the process are $z_1 = 0.4493$; $z_2 = 0.6704$). Since $m = 2$, (5.5.8) is satisfied so that the controller

$$G_R(z) = \frac{q_0 + q_1 z^{-1} + q_2 z^{-2}}{(1-z^{-1})(1+\gamma_1 z^{-1})}$$

can be used. This leads to the following characteristic equation

$$(1-z^{-1})(1+\gamma z^{-1})(1+a_1 z^{-1}+a_2 z^{-2}) + (q_0+q_1 z^{-1}+q_2 z^{-2})(b_1 z^{-1}+b_2 z^{-2}) = 0.$$

Following multiplying and comparing the coefficients with the characteristic equation (5.5.5) one obtains

$$\gamma + q_0 b_1 = -(z_1+z_2+z_3+z_4) + 1 - a_1$$
$$\gamma(a_1-1) + q_0 b_2 + q_1 b_1 = z_1 z_2 + z_3 z_4 + (z_1+z_2)(z_3+z_4) - a_2 + a_1$$
$$\gamma(a_2-a_1) + q_1 b_2 + q_2 b_1 = -z_1 z_2(z_3+z_4) - z_3 z_4(z_1+z_2) + a_2$$
$$q_2 b_2 - \gamma a_2 = z_1 z_2 z_3 z_4$$

This equation system can be first solved for γ and then for the q_i:

$$\gamma = 0,3429; \quad q_0 = 7,1455; \quad q_1 = -6,5758; \quad q_2 = 1,5477$$

The step response of the controller becomes

$$u(0) = 7,15; u(1) = 5,26; u(2) = 8,03; u(3) = 9,19$$
$$u(4) = 10,91; u(5) = 12,44.$$

The expected PID behaviour is then achieved.

5.5.2 Design as a Cancellation Controller

Through prescribing the command variable behaviour $G_w(z) = y(z)/w(z)$ for closed-loop with given process model the cancellation controller can be calculated (c.f. section 6.2) according

$$G_R(z) = \frac{1}{G_P(z)} \cdot \frac{G_w(z)}{1 - G_w(z)} \tag{5.5.9}$$

If for the controller

$$G_R(z^{-1}) = \frac{Q(z^{-1})}{P(z^{-1})} = \frac{Q(z^{-1})}{(1-z^{-1})P'(z^{-1})} \tag{5.5.10}$$

is assumed, then

$$G_w(z) = \frac{y(z)}{w(z)} = \frac{B(z^{-1})z^{-d}Q(z^{-1})}{A(z^{-1})P'(z^{-1})(1-z^{-1}) + B(z^{-1})z^{-d}Q(z^{-1})}$$
$$= \frac{\mathscr{B}_w(z^{-1})}{\mathscr{A}(z^{-1})}. \tag{5.5.11}$$

In consequence, different assumptions on $G_p(z^{-1})$ and $G_w(z^{-1})$ can be made, resulting in a PID-controller.

Design According to Dahlin and Smith

In [5.20, 6.6] a first-order lag is suggested for the command variable behaviour

$$G_w(z) = \frac{\beta_1 z^{-1}}{1 + \alpha_1 z^{-1}} z^{-d} \quad (\beta_1 = 1 + \alpha_1) \tag{5.5.12}$$

From this

$$G_R(z) = \frac{A(z^{-1})}{B(z^{-1})} \frac{\beta_1 z^{-1}}{1 + \alpha_1 z^{-1} - \beta_1 z^{-(d+1)}} \tag{5.5.13}$$

results and for a special process with $b_2 = 0$

$$G_P(z^{-1}) = \frac{b_1 z^{-1}}{1 + a_1 z^{-1} + a_2 z^{-2}} \tag{5.5.14}$$

one obtains [5.21] the common PID-controller with

$$G_R(z) = \frac{\beta_1(1+a_1z^{-1}+a_2z^{-2})}{b_1(1-z^{-1})} \qquad (5.5.15)$$

and hence

$$q_0 = \beta_1/b_1; \; q_1 = q_0a_1; \; q_2 = q_0a_2. \qquad (5.5.16)$$

Design According to Wittenmark and Åström

Another possibility [5.23] is to assume a process model of the form

$$G_P(z) = \frac{b_1z^{-1}+b_2z^{-2}}{1+a_1z^{-1}+a_2z^{-2}}. \qquad (5.5.17)$$

For the case of a controller with command variable prefilter according (5.3.2) the command variable behaviour is given by

$$G_w(z) = \frac{y(z)}{w(z)} = \frac{S(z^{-1})B(z^{-1})}{P(z^{-1})A(z^{-1}) + Q(z^{-1})B(z^{-1})} \qquad (5.5.18)$$

According to (5.2.35), for the PID-controller with first-order lag it follows that

$$G_R(z) = \frac{Q(z^{-1})}{P(z^{-1})} = \frac{q_0+q_1z^{-1}+q_2z^{-2}}{(1-z^{-1})(1+\gamma_1z^{-1})} \qquad (5.5.19)$$

It also follows that for the special choice

$$S(z^{-1}) = q_0 + q_1 + q_2 = S_0 \qquad (5.5.20)$$

for the command variable control behaviour then

$$G_w(1) = 1. \qquad (5.5.21)$$

The command variable control behaviour can now be expressed by the following second-order lag

$$G_w(z) = \frac{S_0(b_1z^{-1}+b_2z^{-2})}{1+\alpha_1z^{-1}+\alpha_2z^{-2}} \qquad (5.5.22)$$

where the coefficients of the denominator respectively the poles can be chosen freely, e.g. with reference to the continuous-time oscillator, (c.f. (3.5.11)), with

$$\left.\begin{aligned}\alpha_1 &= -2e^{-D\omega_{00}T_0}\cos\omega_{00}T_0\sqrt{1-D^2}\\[1mm]\alpha_2 &= e^{-2D\omega_{00}T_0}\end{aligned}\right\} \qquad (5.5.23)$$

(D damping constant; ω_{00} natural frequency).

On comparing the coefficients the controller parameters q_0, q_1, q_2 and γ_1 are then obtained as follows

$$(1-z^{-1})(1+\gamma_1 z^{-1})A(z^{-1}) + (q_0+q_1 z^{-1}+q_2 z^{-2})B(z^{-1})$$

$$= 1 + \alpha_1 z^{-1} + \alpha_2 z^{-2} = \mathscr{A}(z^{-1}) \qquad (5.5.24)$$

by solving

$$\begin{bmatrix} b_1 & 0 & 0 & 1 \\ b_2 & b_1 & 0 & (a_1-1) \\ 0 & b_2 & b_1 & (a_2-a_1) \\ 0 & 0 & b_2 & -a_2 \end{bmatrix} \begin{bmatrix} q_0 \\ q_1 \\ q_2 \\ \gamma_1 \end{bmatrix} = \begin{bmatrix} \alpha_1-a_1-1 \\ \alpha_2-a_2+a_1 \\ a_2 \\ 0 \end{bmatrix}$$

$$\boldsymbol{D}\cdot\boldsymbol{q} = \boldsymbol{\alpha} \qquad (5.5.25)$$

for

$$\boldsymbol{q} = \boldsymbol{D}^{-1}\cdot\boldsymbol{\alpha}. \qquad (5.5.26)$$

From (5.5.24) it follows that

$$G_R(z)\cdot G_P(z) = \frac{Q(z^{-1})}{P(z^{-1})}\frac{B(z^{-1})}{A(z^{-1})}$$

$$= \frac{1}{P(z^{-1})A(z^{-1})}[\mathscr{A}(z^{-1}) - P(z^{-1})A(z^{-1})] \qquad (5.5.27)$$

and the numerator polynomial of the process is compensated. Therefore this controller can only be applied for processes whose zeros are located within the unit circle (i.e. a minimum phase system).

Design According to Banyasz and Keviczky

In [5.35] it is proposed to assume

$$G_P(z) = \frac{b_1 z^{-1}}{1+a_1 z^{-1}+a_2 z^{-2}} z^{-d} \qquad (5.5.28)$$

as process model and as in (5.5.15)

$$G_R(z) = \frac{q_0(1+a_1 z^{-1}+a_2 z^{-2})}{1-z^{-1}}. \qquad (5.5.29)$$

as controller. This yields for the command variable control behaviour

$$G_w(z) = \frac{b_1 q_0 z^{-1}}{1-z^{-1}+b_1 q_0 z^{-1} z^{-d}} z^{-d} \qquad (5.5.30)$$

The denominator is identical with the characteristic equation of an I-controller $q_0 b_1 z^{-1}/(1-z^{-1})$ for a deadtime process z^{-d}. For a phase margin of this control loop of $60°$ it then follows that

$$q_0 = \frac{1}{b_1(2d+1)} \tag{5.5.31}$$

For $d=0$ this yields the same command variable control behaviour as given by (5.5.12). A corresponding design method for the extended numerator of the process model $B(z^{-1}) = b_0(1+\gamma z^{-1})$ is given in [26.50].

This shows that these "direct" design methods for discrete-time PID-controllers are restricted to second-order processes. Additionally they presume special types of process models. Therefore these methods can only be applied in special cases and are not suited for a larger class of processes ($m > 2$, $d > 0$, zeros outside of the unit circle, etc.).

5.5.3 Design of PID-Controllers Through Approximation of Other Controllers

Structurally optimal controllers often have the advantage of a direct design, since the process model forms the basis of an analytical solution. Hence one can try to design a structurally optimal controller for a given process model followed by an approximation through a PID-controller. Since this, however, requires the design of structurally optimal controllers, this procedure will be treated later.

5.6 Tuning Rules for Parameter-optimized Control Algorithms

In order to obtain approximately optimal settings of parameters for continuous-time controllers with PID-behaviour, so-called "tuning rules" are often applied. These rules are mostly given for low-pass processes, and are based on experiments with a P-controller at the stability limit, or on time constants of processes. A survey of these rules is e.g. given in [5.14]. Well-known rules are for example those by Ziegler and Nichols [5.14].

The application of these rules in modified form for discrete-time PID-control algorithms has been attempted. [5.15] gives the controller parameters for processes which can be approximated by the transfer function

$$G_P(s) = \frac{1}{1+Ts} e^{-T_t s} \tag{5.6.1}$$

However, the resulting controller parameters can also be obtained by applying the rules for continuous-time controllers if the modified deadtime $(T_t + T_0/2)$ is used instead of the original deadtime of the sample and hold procedure.

5.6.1 Tuning Rules for Modified PID-controllers

Tuning rules which are based on the characteristics of the process step response and on experiments at the stability limit, have been treated in [5.16] for the case of modified control algorithms according to (5.3.6). These are given in Table 5.6.

5.6.2 Tuning Rules Based on Measured Step Functions

To obtain a more detailed view of the dependence of the parameters of the control algorithm:

$$u(k) = u(k-1) + q_0 e(k) + q_1 e(k-1) + q_2 e(k-2) \qquad (5.6.2)$$

on the process parameters for low-pass processes, on the control performance criterion and on the sample time, a digital computer simulation study [5.18] has been made. Processes with the transfer function

$$G_P(s) = \frac{1}{(1+Ts)^n} \qquad (5.6.3)$$

with zero-order hold, orders $n = 2, 3, 4$ and 6, and sample times $T_0/T = 0.1; 0.5$ and 1.0 were assumed and transformed into z-transfer functions. The controller parameters q_0, q_1 and q_2 were optimized using the Fletcher–Powell method with the quadratic control performance criterion, (c.f. (5.2.6))

$$S_{eu}^2 = \sum_{k=0}^{M} [e^2(k) + r\, \Delta u^2(k)] \qquad (5.6.4)$$

The controller parameters K, c_D and c_I corresponding to control weightings of $r = 0, 0.1$ and 0.25 and unit step changes in the command variable $e(k)$ are shown in Figures 5.15, 5.16 and 5.17 (tuning diagrams). The controller parameters are shown as functions of the ratio T_u/T_G of the process transient functions in Table 5.6. The relationship between the quantities T_u/T or T_G/T and T_u/T_G can be taken from Fig. 5.18.

These figures show that:

a) With increasing T_u/T_G (increasing order n)
   — the gain K decreases
   — the lead factor c_D increases
   — the integration factor c_I decreases
b) With increasing sample time T_0
   — K decreases
   — c_D decreases
   — c_I increases
c) With increasing weighting factor r in the performance criterion
   — K decreases
   — c_D decreases
   — c_I decreases

Table 5.6. Tuning rules for controller parameters according to Takahashi [5.16] based on the rules of the Ziegler–Nichols control algorithm tuning rules:

$$u(k)-u(k-1)=K\left[y(k-1)-y(k)+\frac{T_0}{T_I}[w(k)-y(k)]+\frac{T_D}{T_0}[2y(k-1)-y(k-2)-y(k)]\right]$$

	Step response measurement			Oscillation measurement		
	K	$\dfrac{T_0}{T_I}$	$\dfrac{T_D}{T_0}$	K	$\dfrac{T_0}{T_I}$	$\dfrac{T_D}{T_0}$
P	$\dfrac{T_G}{T_u+T_0}$	—	—	$\dfrac{K_{crit}}{2}$	—	—
PI	$\dfrac{0.9\,T_G}{T_u+T_0/2}-\dfrac{0.135\,T_G T_0}{(T_u+T_0/2)^2}$	$\dfrac{0.27\,T_G T_0}{K(T_u+T_0/2)^2}$	—	$0.45\,K_{crit}-0.27\,K_{crit}\dfrac{T_0}{T_P}$ smaller values if $T_0\approx 4\,T_u$	$0.54\dfrac{K_{crit}}{K}\dfrac{T_0}{T_P}$	—
PID	$\dfrac{1.2\,T_G}{T_u+T_0}-\dfrac{0.3\,T_G T_0}{(T_u+T_0/2)^2}$	$\dfrac{0.6\,T_G T_0}{K(T_u+T_0/2)^2}$	$\dfrac{0.5\,T_G}{K\,T_0}$	$0.6\,K_{crit}-0.6\,K_{crit}\dfrac{T_0}{T_P}$	$1.2\dfrac{K_{crit}}{K}\dfrac{T_0}{T_P}$	$\dfrac{3}{40}\dfrac{K_{crit}}{K}\dfrac{T_P}{T_0}$
	Not applicable for $T_u/T_0\to 0$			Range of validity: $T_0\leqq 2\,T_u$.	Not recommended for $T_0\approx 4\,T_u$.	

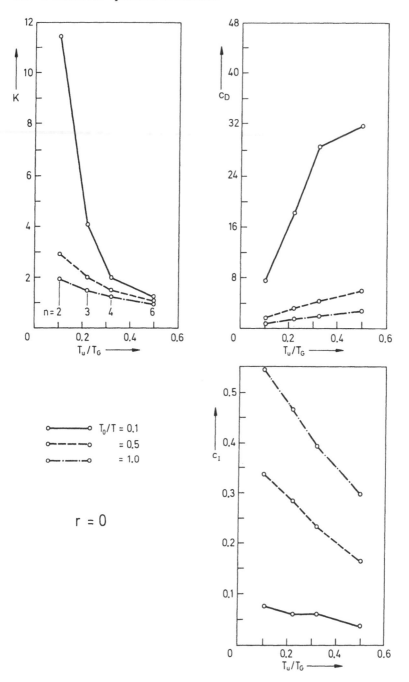

Fig. 5.15. Optimal controller parameters of the control algorithm 3 PC-3 (PID-behaviour) due to the performance criterion (5.6.4) with $r=0$ for processes $G_P(s)=\dfrac{1}{(1+Ts)^n}$. The controller parameters are given, according to (5.2.15), by: $K=q_0-q_2$; $c_D=q_2/K$; $c_I=(q_0+q_1+q_2)/K$

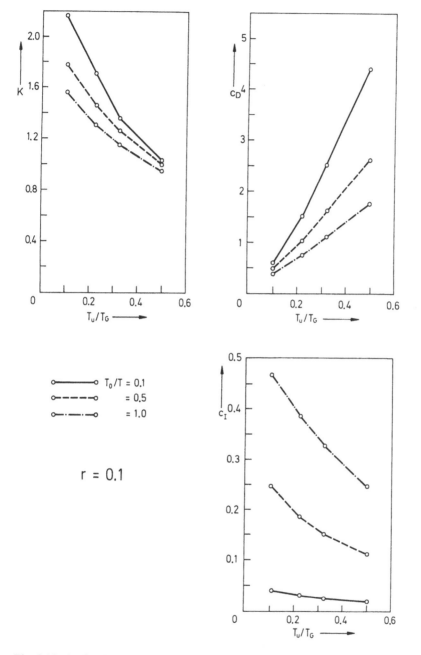

Fig. 5.16. Optimal controller parameters of the control algorithm 3 PC-3 (PID-behaviour) due to the performance criterion (5.6.4) with $r=0.1$ for processes of the form:
$G_P(s)=\dfrac{1}{(1+Ts)^n}$. The controller parameters are given once again, according to (5.2.15), by:
$K=q_0-q_2$; $c_D=q_2/K$; $c_I=(q_0+q_1+q_2)/K$

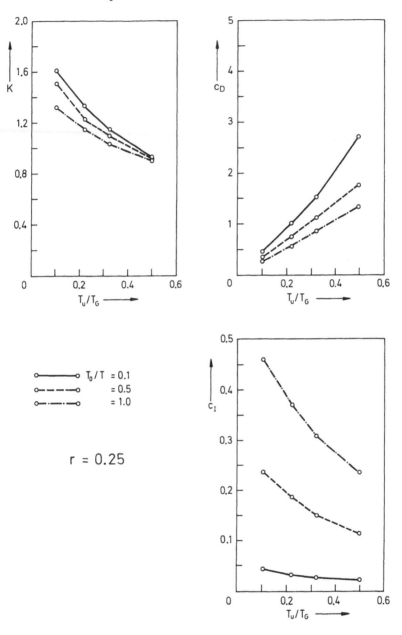

Fig. 5.17. Optimal controller parameters of the control algorithm 3 PC-3 (PID-behaviour) due to the performance criterion (5.6.4) with $r=0.25$ for processes of the form: $G_P(s)=\dfrac{1}{(1+Ts)^n}$. The controller parameters are given by: $K=q_0-q_2$; $c_D=q_2/K$; $c_I=(q_0+q_1+q_2)/K$

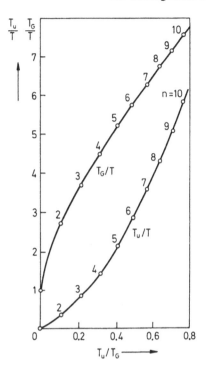

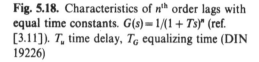

Fig. 5.18. Characteristics of nth order lags with equal time constants. $G(s) = 1/(1 + Ts)^n$ (ref. [3.11]). T_u time delay, T_G equalizing time (DIN 19226)

With the aid of these tuning diagrams the controller parameters of a 3-parameter-control algorithm can be determined as follows, on the basis of a measured process step response.

1. The tangent to the point of inflexion is drawn and the characteristic values T_u, T_G and T_u/T_G, and the gain $K_p = y(\infty)/u(\infty)$ are determined.
2. From Fig. 5.18 it follows that T_u/T' and T_G/T''. Thus $T = \frac{1}{2}(T' + T'')$.
3. After selecting the sample time T_0, the ratio T_0/T is determined.
4. The Figures 5.15 to 5.17 show, after suitable choice of the control weighting factor r, the characteristic coefficients K_0, c_D and c_I which depend on T_u/T_G and T_0/T. Here, K_0 is the loop gain $K_0 = KK_p$. (Because of $K_p = 1$ in the Fig.: $K_0 = K$)
5. From $K = K_0/K_p$, c_D and c_I and (5.2.15), the controller parameters follow:

$$q_0 = K(1 + c_D)$$
$$q_1 = K(c_I - 2c_D - 1)$$
$$q_2 = Kc_D.$$

Though the tuning diagrams in Fig. 5.15 to 5.17 depend on equal time constants, this procedure for determining the controller parameters can also be used for low-pass processes with widely differing time constants, as simulations have shown (c.f. section 3.7.3).

Table 5.7. Comparison of the results of tuning rules for the controller parameters based on step response characteristics. Process III. $T_0 = 4$ s; $K = 1$; $T_u/T_G = 6.6$ s/25 s = 0.264

Process III $T_0 = 4$ s	K	$\dfrac{T_0}{T_I} = c_I$	$\dfrac{T_D}{T_0} = c_D$	Process model
Parameter optimization for $r = 0 \dots 0.25$ (Table 5.4)	1.52 ... 1.13	0.28 ... 0.21	1.99 ... 0.81	Process III
According Takahashi [5.16] (Table 5.6)	2.34	0.33	1.3	$\dfrac{1}{T_G s} e^{-T_u s}$
According Fig. 5.15 to 5.17 $r = 0 \dots 0.25$	1.7 ... 1.2	0.27 ... 0.17	3.8 ... 0.85	$\dfrac{1}{(1 + Ts)^n}$

This can also be recognized in Table 5.7 where a comparison is made for process III, showing the optimized controller parameters and the parameters based on the tuning rules of Table 5.6 and based on the tuning diagrams in Figures 5.15 to 5.17. The tuning diagrams yield controller parameters which compare well with the optimal values. Applying the tuning rules according to Table 5.6 (left part) the gain K is too large. c_D and c_I, however, compare quite well.

Example 5.4:
In a low-pass process $T_u = 14$ sec, $T_G = 45$ sec and $K_p = 2$ were obtained from the measured step function. Sample time $T_0 = 10$ sec.

1. $T_u/T_G = 0.31$
2. $T = 10$ s
3. $T_0/T = 1$
4. $r = 0.1$; $K_0 = 1.15$; $c_I = 0.33$; $c_D = 1.1$
5. $K = 1.15/2 = 0.575$; $q_0 = 1.2075$; $q_1 = -1.5928$; $q_2 = 0.6325$.

5.6.3 Tuning Rules with Oscillation Tests

With reference to the Ziegler-Nichols tuning rules, the steps used in tuning different processes (test process II (all-pass), test process VI, test process VIII (low-pass) with deadtime $d = 0, 1, 2$) were determined according Table 5.8 [5.23, 5.24].

First, the controller is switched as discrete-time P-controller with sample time T_0. Then the gain factor is being increased until reaching K_{crit} and steady oscillations with period duration T_p appear. The parameters q_0, q_1 and q_2 of the PID- controller then follow using Table 5.8 and (5.1.5).

Example 5.5:
In a low-pass process (test process VI) the following critical quantities were determined after an oscillation test

$$K_{crit} = 5.03 \quad T_p = 34 \text{ s}$$

Table 5.8. Tuning rules for discrete-time PID-controllers from an oscillation test. K_{crit} controller gain of the P-controller at stability limit, oscillation duration at stability limit

Controller	K	T_I	T_D
PI	$<0.45\,K_{crit}$	$0.4\,T_p$	–
PID	$<0.6\;K_{crit}$	$0.5\,T_p$	$0.12\,T_p$

According to Table 5.8 the following controller parameters can be derived

$$K = 3; \; T_I = 17 \text{ s}; \; T_D = 4 \text{ s}$$

With sample time $T_0 = 4$ sec

$$q_0 = 6; \; q_1 = -9,705; \; q_2 = 3.$$

For a given process model, tuning rules can be also used for a purely mathematical determination of the PID-controller parameters [5.23, 5.24], (c.f. section 26.5.3).

5.7 Choice of Sample Time for Parameter-optimized Control Algorithms

As is well known, sampled-data controllers have generally inferior performance over continuous control systems. This is sometimes explained by the fact that sampled signals contain less information than continuous signals. However, the information and also the use of this information is of interest. As the class and the frequency spectrum of the disturbance signals also play an important role, general remarks on the control performance of sampled-data systems are difficult to make. However, for parameter-optimized controllers one can assume, in general, that the control performance deteriorates with increasing sample time [5.34]. Therefore, the sample time should be as small as possible if only the control performance is of importance.

The choice of sample time depends not only on the achievable control performance but also on:

— the desired control performance
— the process dynamics
— the spectrum of disturbances
— the actuator and its motor
— the measurement equipment
— the requirements on the operator
— the computational load or costs per control loop
— the identified process model.

These factors will be discussed.

When considering the *control performance*, it can be seen from Figures 5.9 and 5.10 that a sample time $T_0 = 4$ sec, compared with $T_0 = 1$ sec which is a good approximation to the continuous case, leads to only a small deterioration in control performance. If only the control performance is of interest the sample time can usually be greater than required by the approximation of the continuous control loop. Some rules of thumb for determining the sample time based on the approximation of the continuous control loop behaviour are given in Table 5.9.

The *process dynamics* have great influence on the sample time in terms of both the transfer function structure and its time constants. Rules for the sample time, therefore, are given in Table 5.9 as functions of the time delay, deadtime, sum of the constants etc. In general, the larger the time constant the larger the sample time.

Now the dependence of the sample time on the *disturbance signal spectrum* or its bandwidth is considered. As is well known, for control loops three frequency ranges can be distinguished [5.15] (c.f. section 11.4):

The low frequency range ($0 \leq \omega \leq \omega_1$): disturbances of the control variable are reduced.

The medium frequency range ($\omega_1 < \omega \leq \omega_2$): disturbances are amplified.

The high frequency range ($\omega_2 < \omega < \infty$): disturbances are not affected by the loop.

Control loops in general have to be designed such that the medium frequency range comes within that range of the disturbance signal spectrum where the magnitude of the spectrum is small. In addition, disturbances with high and medium frequency components must be filtered in order to avoid unnecessary variations in the manipulated variable. If disturbances up to the frequency $\omega_{max} = \omega_1$ have to be controlled approximately as in a continuous loop, the sample time has to be chosen in accordance with *Shannon's sampling theorem* as follows

$$T_0 < \pi/\omega_{max}.$$

Particularly with an *actuator* having a long rise-time it is inappropriate in general to take too small a sample time, since it can happen that the previous manipulated variable has not been acted upon when a new one arrives. If the *measurement equipment* furnishes discrete-time signals, as in chemical analysers or rotating radar antennae, the sample time is already determined. An *operator* generally wants a quick response of the manipulated and control variable after a change of the reference variable at an arbitrary time. Therefore, the sample time should not be larger than a few seconds. Moreover, in a dangerous situation such as after an alarm, one is basically interested in a small sample time. To minimize the *computational* load or the *costs* for each control loop, the sample time should be as large as possible.

If the control design is based on identified *process models*, and if parameter estimation methods are used for the identification, then the sample time should not be too small in order to avoid the numerical difficulties which result from the approximate linear dependence in the system equations for small sample times [3.13].

This discussion shows that the sample time has to be chosen according to many requirements which partially are contradictory. Therefore, suitable compromises must be found in each case. In addition, to simplify the software organization one must use the same sample time for several control loops. In Table 5.9 rules for

Table 5.9. Summary of rules for determining the sample time for low-pass processes. f: eigen-frequency of the closed-loop in cycles/s; T_t: deadtime; T_{95}: 95% settling time of the step response; T_u: delay time, c.f. Table 5.6

Criteria to determine the sample time	Literature	Determination of the sample time	Sample time process III [s]	Remarks
	[5.10], [5.3]	$T_0 \approx (1/8 \ldots 1/16)\dfrac{1}{f}$	3 ... 1	
	[5.10], [5.3]	$T_0 \approx (1/4 \ldots 1/8)T_t$	—	Processes with dominant deadtime
Larger settling time as with continuous PI-controller: 15%	[5.11], [5.17]	$T_0 \approx (1.2 \ldots 0.35)T_u$ $T_0 \approx (0.35 \ldots 0.22)T_u$	4.5	$0.1 \leqq T_u/T \leqq 1.0$ $1.0 \leqq T_u/T \leqq 10$
Compensation of disturbances until ω_{max} as contin. loop		$T_0 = \pi/\omega_{max}$ (Shannon, Nyquist)	8 ... 2	ω_{max} chosen such that for the process $|G(\omega_{max})| = 0.01 \ldots 0.1$
Simulation, section 5.4	[5.7]	$T_0 \approx (1/6 \ldots 1/15)T_{95}$	8 ... 3	
Identification of the process model	[3.13]	$T_0 \approx (1/6 \ldots 1/12)T_{95}$	8 ... 4	

choosing the sample time are summarized, based on current literature. Note, that rules which are based on approximating the continuous control performance frequently predict too small a sampling time. Considering only the control performance, about 6 to 15 samples per settling T_{95} are sufficient, at least for low-pass processes. For some processes in the power and chemical industries the sample times given in Table 5.10 have often been proposed [5.5, 5.12, 5.13].

Table 5.10. Recommended sample times for processes in the power and chemical industry

Control variable	Sample interval T_0 [s]
Flow	1
Pressure	5
Level	10
Temperature	20

5.8 Supplementary Functions of Digital PID-Controllers

Up to now different versions and properties of PID-algorithms for normal
operation around a fixed operation point were considered. For application,
however, several supplementary functions are required. They will be briefly viewed
in the following.

a) *Prevention of integral wind-up*
For large control deviations $e(k)$ the controlling element generally drives to a
restriction and the integral acting term of the control algorithm

$$u(k) = u(k-1) + q_0 e(k) + q_1 e(k-1) + q_2 e(k-2) \qquad (5.8.1)$$

produces continually increasing values of the manipulated variable. When the
control variable then reaches the reference value and the control deviation sign is
reverse the integral acting algorithms takes long to integrate backwards the
manipulated variable and to quit again the restriction. This, therefore results in
control deviations which last too long [5.14, 5.32].
   Hence, e.g. undesired long and large overshoots occur when starting-up a
process. The measures taken against these undesired properties depend on the
programming of the control algorithm and the type of motor element.
   For proportionally acting motor elements there are the following possibilities
to prevent antireset wind-up:

1. *Zeroing the integrator*
The control algorithm is programmed according (5.2.19a) and the integral term
set to zero by $c_I = 0$ or $e(k-1) = 0$, if $u(k) = u_{max}$ or $u(k) = u_{min}$
2. *Conditioned integration*
The integration is only executed if $|e(k)| < e_{max}$

Possibility (1) presupposes that the calculated manipulated variable $u(k)$ agrees
well in the computer with the manipulated variable of the controlling element.
Disturbances, however, can destroy this correspondence. This possibility, therefore, is only recommended if an actuator position feedback $u_A(k)$ takes place, (c.f.
chapter 29).

If the motor element possesses integral acting behaviour then the control algorithm can be programmed as follows

$$\Delta u(k) = q_0 e(k) + q_1 e(k-1) + q_2 e(k-2) \tag{5.8.2}$$

Note, that the motor element is part of the controller transfer function, (c.f. chapter 29). The motor element thus represents the integrator and automatically takes into account its restriction, or the control step $\Delta u(k)$ required by the control algorithm (5.8.2) is not carried out since the controlling element is automatically switched off by its limit value switch. Hence, no special measures have to be taken to prevent antireset wind-up.

b) *Bumpless transfer from manual to automatic automation*
During commisioning the manipulated variable, in general, is controlled manually in order to obtain a certain value of the control variable. A bumpless transfer from manual to automatic automation now requires that the manually controlled manipulated variable u_M and the manipulated variable generated by the controller agree well. During transformation one can adapt the command variable to the momentary control variable. Then, the command variable can be left there or, after a preprogramed ramp function, can be brought to the desired value. In order to obtain during the transfer manipulated variables which agree well, $u_M(k) = u_R(k)$, supplementary measures have to be taken. First, one has to provide that during manual automation the integrator does not wind up the always existing control deviations and produces much too large $u_R(k)$. Therefore, measures as described in a) have to be taken, e.g. for proportional motor elements to switch off the integral term in (5.2.19) or in Fig. 5.5 through zeroing the input or through $c_I = 0$. The command variable for the remaining P- or PD-controller can then be tuned in such a way that $u_M(k) = u_R(k)$. This takes into account a possibly remaining control deviation. The internal realization of these measures also depend on whether the manual operation of the control element is executed by conventional analog methods in the command device or through a digital controller.

c) *Deadzone for the control difference*
For certain processes one is not interested in controlling small deviations (e.g. to increase the lifetime of the actuator). For this a *deadzone* is introduced into the controller, which can be written as:

$$u(k) - u(k-1) = 0 \, f\ddot{u}r \, |e(k)| \leq e_{min}.$$

d) *One-sided acting control difference*
If a process is operating close to a critical operating point it may be desired to generate manipulated variables which act fast if the control variable moves towards the limit and act slowly in the opposite direction [5.25]. Here, with (5.2.19b) it follows that:

$$u_D(k) = q_2((e(k) - e(k-1)) \quad \text{if} \quad e(k) < 0$$
$$u_D(k) = 0 \quad \text{if} \quad e(k) \geq 0$$

e) *Structural change of the controller*

If PID-control algorithms are programed in the form of (5.2.18), then it is easily possible by zeroing different parameters to switch from PID-controllers to PD-controllers ($c_I = 0$) or PI-controllers ($c_D = 0$) or P-controllers ($c_I = 0$). This e.g. is required for the start-up of processes where first of all one closes the control loop with a P-controller without changing the command variable $w(k)$ followed by switching to a PI-controller if $y(k) \approx w(k)$. This avoids a too strong over-shoot caused by a too strong wind-up through the integral part (see above).

For more details on supplementary functions of digital PID-controllers the reader is referred to e.g. [5.25]. Altogether one can state that it is much easier to realize these functions through suitable programing than through supplementary switching in analogous technique which is described e.g. in [5.32].

6 General Linear Controllers and Cancellation Controllers

As first forms of structurally optimal input/output controllers this chapter considers the general linear controller and the cancellation controller the orders of which are directly connected with the process model order.

6.1 General Linear Controller

The general linear controller with transfer function

$$G_R(z) = \frac{u(z)}{e(z)} = \frac{Q(z^{-1})}{P(z^{-1})} = \frac{q_0 + q_1 z^{-1} + \dots + q_\nu z^{-\nu}}{1 + p_1 z^{-1} + \dots + p_\mu z^{-\mu}} \tag{6.1.1}$$

represents in this form the *general input/output controller*. From this controller emerge as special cases the parameter-optimized controllers of low-order and the cancellation and deadbeat controllers which will be treated later.

If the controller, according (6.1.1) is connected with the process

$$G_P(z) = \frac{y(z)}{u(z)} = \frac{B(z^{-1})}{A(z^{-1})} z^{-d} = \frac{b_1 z^{-1} + \dots + b_m z^{-m}}{1 + a_1 z^{-1} + \dots + a_m z^{-m}} z^{-d} \tag{6.1.2}$$

in a control loop, then the command transfer function

$$
\begin{aligned}
G_w(z) = \frac{y(z)}{w(z)} &= \frac{G_R(z) G_P(z)}{1 + G_R(z) G_P(z)} \\
&= \frac{Q(z^{-1}) B(z^{-1}) z^{-d}}{P(z^{-1}) A(z^{-1}) + Q(z^{-1}) B(z^{-1}) z^{-d}}
\end{aligned} \tag{6.1.3}
$$

and the closed-loop transfer functions result, c.f. Fig. 5.1

$$
\begin{aligned}
G_n(z) = \frac{y(z)}{n(z)} &= \frac{1}{1 + G_R(z) G_P(z)} \\
&= \frac{P(z^{-1}) A(z^{-1})}{P(z^{-1}) A(z^{-1}) + Q(z^{-1}) B(z^{-1}) z^{-d}}
\end{aligned} \tag{6.1.4}
$$

$$G_u(z) = \frac{y(z)}{u_v(z)} = \frac{G_p(z)}{1 + G_R(z) G_p(z)}$$

$$= \frac{P(z^{-1}) B(z^{-1}) z^{-d}}{P(z^{-1}) A(z^{-1}) + Q(z^{-1}) B(z^{-1}) z^{-d}} \qquad (6.1.5)$$

In general these forms can be written as

$$G_*(z) = \frac{\mathcal{B}_*(z^{-1})}{\mathcal{A}(z^{-1})} = \frac{[\beta_0 + \beta_1 z^{-1} + ... + \beta_r z^{-r}]_*}{1 + \alpha_1 z^{-1} + ... + \alpha_\ell z^{-\ell}} \qquad (6.1.6)$$

In this * means w, n or u. The order ℓ is

$$\ell = max \; [m + \mu; \; m + d + v]. \qquad (6.1.7)$$

These different transfer functions show that even though the denominator polynomial $\mathcal{A}(z^{-1})$ of the closed loop is independent from the injection point of the external signal, the numerator polynomial is not.

6.1.1 A General Linear Controller for Specified Poles

If the poles z_{α_i} in

$$\mathcal{A}(z) = (z - z_{\alpha 1}) \; ... \; (z - z_{\alpha \ell}) = 0 \qquad (6.1.8)$$

or the resulting characteristic equation

$$\mathcal{A}(z^{-1}) = 1 + \alpha_1 z^{-1} + ... + \alpha_\ell z^{-\ell} = 0 \qquad (6.1.9)$$

are specified, the controller parameters can be determined by comparing the coefficients in

$$\mathcal{A}(z^{-1}) = (1 + p_1 z^{-1} + ... + p_\mu z^{-\mu})(1 + a_1 z^{-1} + ... + a_m z^{-m})$$

$$+ (q_0 + q_1 z^{-1} + ... + q_v z^{-v})(b_1 z^{-1} + ... + b_m z^{-m}) z^{-d} = 0$$

To avoid steady state offsets, $G_w(1)$ has to be set to one. From (6.1.3) it follows that $P(1)A(1) = 0$ and this is generally fulfilled for

$$\sum_{i=1}^{\mu} p_i = -1. \qquad (6.1.11)$$

There are $(\ell + 1)$ equations giving unique determination of the $(\mu + v + 1)$ unknown controller parameters. Hence

$$\mu + v + 1 = \ell + 1. \qquad (6.1.12)$$

In (6.1.7) two cases must be distinguished

a) $\mu \geq v+d \rightarrow \ell = m+\mu$.

(6.1.12) gives $v=m$. Hence $\mu \geq m+d$.

b) $\mu \leq v+d \rightarrow \ell = m+d+v$.

(6.1.12) gives $\mu = m+d$. Hence $v \geq m$.

If the smallest possible order numbers are chosen

$$v = m \quad \text{und} \quad \mu = m + d \tag{6.1.13}$$

then in all cases it is possible to determine uniquely the controller parameters. (6.1.10) and (6.1.11) lead to the system of equations

$$
\underbrace{
\begin{bmatrix}
1 & 0 & & 0 & 0 & & & 0 \\
a_1 & 1 & 0 & & & d & & \\
& a_1 & 1 & 0 & 0 & 0 & & 0 \\
\vdots & & \ddots & & b_1 & 0 & \cdots & 0 \\
a_m & \vdots & \ddots & 0 & \vdots & b_1 & & \\
0 & a_m & & 1 & & & & \\
& & & a_1 & b_m & & & 0 \\
\vdots & \vdots & \ddots & \vdots & \vdots & & & b_1 \\
& & & & & & \ddots & \\
0 & 0 & \cdots & a_m & 0 & \cdots & & b_m \\
1 & 1 & \cdots & 1 & 0 & & \cdots & 0
\end{bmatrix}
}_{\substack{m+d \qquad\qquad m+1 \\ R}}
\underbrace{
\begin{bmatrix}
p_1 \\ \vdots \\ \vdots \\ \vdots \\ \vdots \\ p_{m+d} \\ q_0 \\ q_1 \\ \vdots \\ \vdots \\ q_m
\end{bmatrix}
}_{\theta_R}
=
\underbrace{
\begin{bmatrix}
\alpha_1 - a_1 \\ \vdots \\ \vdots \\ \vdots \\ \alpha_m - a_m \\ \alpha_{m+1} \\ \vdots \\ \vdots \\ \alpha_{2m+d} \\ -1
\end{bmatrix}
}_{\alpha}
$$

$$\tag{6.1.14}$$

and the unknown parameters are obtained from

$$\theta_R = R^{-1}\alpha \tag{6.1.15}$$

if $\det R \neq 0$.

As already remarked in chapter 4 there is more freedom to place the poles within the stability region. Therefore moderate trial and error should lead to suitable time responses of the controlled and manipulated variables. Note, that for a given characteristic equation or for given poles, the zeros $\mathcal{B}_*(z)=0$ of the transfer functions (6.1.3) to (6.1.5) are also determined if the system of equations (6.1.14) is unique, i.e. (6.1.12) is valid. If, however, $v>m$ then the zeros of $G_w(z)$, and if $\mu>m+d$, the zeros of $G_n(z)$ can be influenced as well. For $v=m$ and $\mu=m+d$ the zeros of the process appear in $G_w(z)$ and $G_u(z)$ and the process poles appear in the zeros of $G_n(z)$. This means that the process itself "dictates" some of the zeros of the closed-loop transfer functions.

6.1.2 General Linear Controller Design Through Parameter Optimization

The parameters $q_0, \ldots, q_v$ and $p_1, \ldots, p_\mu$ of the general linear controller can also be determined through numerical optimization of a control performance criterion, as already described in section 5.4. The computational effort, however, augments considerably with increasing number of parameters. Low-order control algorithms, treated in chapter 5, are therefore preferred for this design method.

Yet, simple direct design methods for linear controllers are possible through specifications for the closed loop behaviour. Cancellation and deadbeat linear emerge as special cases of the general linear controller. These will be considered in the next two sections.

6.2 Cancellation Controllers

The main objective in tracking control system design is to make the controlled variable y follow the command input w as closely as possible. If the model G_p of a stable process is known exactly and if there is no other disturbance, this problem could be solved by using a feedforward control system as in Fig. 6.1. Ideally, one could require that the output y follows the input w exactly. This would be the case if

$$G_S(z) = G_S^0(z) = \frac{1}{G_P(z)}. \tag{6.2.1}$$

If $G_S^0(z)$ is realizable, this feedforward element would "compensate" the process completely since it has the reciprocal transfer function behaviour.

For processes with time lags, however, the feedforward element is not realizable and one has to add a "realizability term"

$$G_S(z) = G_S^0(z) G_S^R(z) = \frac{1}{G_P(z)} G_S^R(z) \tag{6.2.2}$$

which leads to deviations between w and y. In considering the cancellation of poles and zeros of the feedforward element and the process, the effects discussed on page 162 have to be taken into account.

If the assumptions made for the design of this feedforward control element do not hold, e.g. the process model is not known exactly and disturbances arise, one has to change the system to a feedback control system as in Fig. 6.2.

Unlike the feedforward control system, one cannot require $e(t) = w(t) - y(t) = 0$ for $t \geq 0$ for a feedback control system. The simple reason is that a manipulated variable u can only be produced by a control deviation that is non-zero at least in a transitory state. Therefore, the closed-loop transfer function

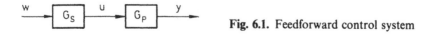

Fig. 6.1. Feedforward control system

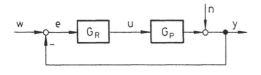

Fig. 6.2. Feedback Control System

$$G_w(z) = \frac{y(z)}{w(z)} = \frac{G_R(z)G_P(z)}{1+G_R(z)G_P(z)} = \frac{\mathscr{B}_w(z)}{\mathscr{A}(z)} \qquad (6.2.3)$$

is specified $(G_w(z) \neq 1)$ and the resulting controller

$$G_R(z) = \frac{1}{G_P(z)} \cdot \frac{G_w(z)}{1-G_w(z)} = \frac{A(z^{-1})}{B(z^{-1})z^{-d}} \cdot \frac{\mathscr{B}_w(z^{-1})}{\mathscr{A}(z^{-1})-\mathscr{B}_w(z^{-1})} \qquad (6.2.4)$$

is determined. The controller transfer function consists of the inverse process transfer function and an additional term which depends on the given closed-loop transfer function. Therefore, a part of the controller cancels the poles and zeros of the process.

The design of these "cancellation" controllers is not only restricted to the reference variable as input but can also be made for specified disturbances. For a given $G_n(z) = y(z)/n(z)$ it is e.g.

$$G_R(z) = \frac{1}{G_P(z)} \frac{1-G_n(z)}{G_n(z)}. \qquad (6.2.5)$$

For the design of these cancellation controllers, many papers have been published, especially for continuous signals. Discrete cancellation controllers have been described in [2.4, 2.14, 6.1–6.3].

For prescribing the required closed-loop transfer function $G_w(z)$ or $G_n(z)$, the following restrictions have to be noted:

Realizability

If a z-transfer function of the following form:

$$G(z) = \frac{\beta_0 + \beta_1 z + \dots + \beta_n z^n}{1 + \alpha_1 z + \dots + \alpha_m z^m} \qquad (6.2.6)$$

is given, then the realizability condition is $n \leq m$ is $\alpha_m \neq 0$. In the transfer functions of the form

$$G_R(z) = \frac{Q_\nu(z)}{P_\mu(z)} \quad \text{und} \quad G_P(z) = \frac{B_n(z)}{A_m(z)}$$

the indices describe the orders of the single polynomials. With (6.2.3) it follows that the closed-loop behaviour is given by

$$G_w(z) = \frac{Q_v B_n}{P_\mu A_m + Q_v B_n}.$$

If $G_R(z)$ and $G_P(z)$ are realizable, i.e. if $v \leq \mu$ and $n \leq m$, it follows that the order of the polynomials of $G_w(z)$ are given by

$$G_w(z): \frac{\text{order } (v+n)}{\text{order } (\mu+m)}.$$

The pole excess of $G_w(z)$ is therefore

$$Pe = (\mu - v) + (m - n) \tag{6.2.7}$$

To make this as small as possible, $v = \mu$ is chosen. The pole excess of the closed-loop transfer function $G_w(z)$ is therefore

$$Pe(G_w) = (m - n)$$

and is also equal to the pole excess of the process $G_P(z)$

$$Pe(G_P) = (m - n) = Pe(G_w). \tag{6.2.8}$$

This means that because of the realizability condition, the pole excess of the command transfer function $G_w(z)$ for the controller order $\mu \geq v$ has to be greater or equal than the pole excess of the process [2.19].

In general, the z-transfer function of the process given by (6.1.2) has $b_0 = 0$, since for the processes without jumping properties at least the sensor of the actuator has a time lag. Then it follows in (6.2.6) $n = m - 1$, i.e. a pole excess of one, so that e.g.

$$G_w(z) = z^{-1}$$

could be assumed.

Cancellation of Poles and Zeros

If the cancellation controller $G_R(z)$ given by (6.2) and the process $G_{P0}(z)$ are in a closed-loop, the poles and zeros of the processes are cancelled by the zeros and poles of the controller if the process model $G_P(z)$ matches the process exactly. Since the process models $G_P(z) = B(z)/A(z)$ used for the design practically never describe the process behaviour exactly, the corresponding poles and zeros will not be cancelled exactly but only approximately. For poles $A^+(z)$ and zeros $B^+(z)$ which are "sufficiently spread" in the inner of the unit disc of the z-plane, this leads to only small deviations of the assumed behaviour $G_w(z)$ in general. However, if the process has poles $A^-(z)$ or zeros $B^-(z)$ near of the outside the unit circle one has to be careful. The *real process* will be described by

$$G_{PO}(z) = \frac{B_0(z)}{A_0(z)} = \frac{B_0^+(z) B_0^-(z)}{A_0^+(z) A_0^-(z)} \tag{6.2.9}$$

and the *process model* by

$$G_P(z) = \frac{B(z)}{A(z)} = \frac{B^+(z) B^-(z)}{A^+(z) A^-(z)} \tag{6.2.10}$$

This concept can easily be explained by using the characteristic equation. From (6.2.4) it follows that

$$1 + G_R(z) G_P(z) = 1 + \frac{A(z) z^d}{B(z)} \cdot \frac{\mathcal{B}_w(z)}{\mathcal{A}(z) - \mathcal{B}_w(z)} \cdot \frac{B_0(z)}{A_0(z) z^d} = 0$$

or

$$B(z) [\mathcal{A}(z) - \mathcal{B}_w(z)] A_0(z) z^d + A(z) z^d \mathcal{B}_w(z) B_0(z) = 0$$

$$\mathcal{A}(z) A_0(z) z^d B(z) + \mathcal{B}_w(z) z^d [A(z) B_0(z) - A_0(z) B(z)] = 0 \tag{6.2.12}$$

If the process model and the process have only small differences then the following approximation holds

$$A(z) B_0(z) - A_0(z) B(z) \approx 0 \tag{6.2.13}$$

Consequently

$$\mathcal{A}(z) A_0(z) z^d B(z) = \mathcal{A}(z) A_0^+(z) A_0^-(z) z^d B^+(z) B^-(z) \approx 0. \tag{6.2.14}$$

This means that because of the roots of $A_0^-(z)$ and $B^-(z)$ a weakly damped or even unstable control behaviour occurs. For a process model which exactly matches the process, it follows from (6.2.11) that for the corresponding order reduction, $\mathcal{A}(z) = 0$. This means that stability would not depend on the process and controller but only on the prescribed command control behaviour which, of course, is not possible. This indicates that stating the characteristic equation cancellations are not allowed.

A second possibility to deal with the problem is discussing the command transfer function [2.4]. If now the cancellation controller compensates ideally for the uncritical poles and zeros inside the unit circle

$$G_R(z) = \frac{A_0^+(z) A^-(z)}{B_0^+(z) B^-(z)} \frac{G_w(z)}{1 - G_w(z)} \tag{6.2.15}$$

one obtains for the resulting command behaviour

$$G_{w,res}(z) = \frac{A^-(z)B_0^-(z)G_w(z)}{A_0^-(z)B^-(z)[1-G_w(z)]+A^-(z)B_0^-(z)G_w(z)}$$

(6.2.16)

and with $A^-(z)=A_0^-(z)+\Delta A^-(z)$

$$B^-(z)=B_0^-(z)+\Delta B^-(z)$$

it follows that

$$G_{w,res} = \frac{A_0^- B_0^- G_w + \Delta A^- B_0^- G_w}{A_0^- B_0^- + A_0^- \Delta B^-[1-G_w]+\Delta A^- B_0^- G_w}.$$

(6.2.17)

For $\Delta A^-(z)=0$ and $\Delta B^-(z)=0$, the poles of this transfer function are near or outside the unit circle. They are, however, exactly cancelled by the zeros. For small differences $\Delta A^-(z)$ and $\Delta B^-(z)$ the poles change by correspondingly small amounts and are therefore no longer cancelled. Then a weakly damped control behaviour or, if the poles are outside the unit circle, an unstable behaviour results. Therefore, one should not design cancellation controllers for processes with poles or zeros outside or near the unit circle in the z-plane. One always has to take into account that differences $\Delta A^-(z)$ and $\Delta B^-(z)$ occur in practice.

Therefore the design of cancellation controllers according to (6.2.4) has to be restricted to processes whose poles and zeros are located well inside the unit circle. Therefore application of these cancellation controllers is very restricted, compare section 3.5.

Behaviour Between the Sampling Points

Unlike cancellation controllers for continuous signals, with discrete-time signals the behaviour at only the sampling points is given through prescribing a certain $G_w(z)$. If $G_w(z)$ is not chosen properly, the desired behaviour at these sampling points is obtained, but between the sampling points deviations such as oscillations or "ripples" can occur. These ripples are mostly weakly damped and result in large changes of the manipulated variables for so called "minimal prototype responses"

$$G_w(z) = z^{-1} \text{ or } 2z^{-1}-z^{-2} \text{ or } 3z^{-1}-3z^{-2}+z^{-3}$$

(6.2.18)

[2.4, 2.19].

For processes with deadtime d Dahlin [6.3] suggested the choice of command behaviour based on a first-order lag delayed by the deadtime.

$$G_w(z) = \frac{\beta_1 z^{-1}}{1+\alpha_1 z^{-1}}z^{-d}; \ \beta_1 = 1 + \alpha_1$$

(6.2.19)

With this the following controller results

$$G_R(z) = \frac{A(z^{-1})}{B(z^{-1})} \cdot \frac{\beta_1 z^{-1}}{1 + \alpha_1 z^{-1} - \beta_1 z^{-(1+d)}} \qquad (6.2.20)$$

Hence all process poles and zeros are cancelled. In order to obtain an oscillation-free controller it is further suggested that for poles $(z - \alpha_i)$ located close to or outside the unit circle these can be replaced with $z = 1$ by $(1 - \alpha_i)$.

However, the problem with oscillations can be bypassed through requiring that

$$G_w(z) = \frac{1}{K_P} G_P(z) \qquad (6.2.21)$$

where K_P is the process gain. Then one obtains the so-called *predictor controller* which can be of advantage for processes with large deadtimes, (see chapter 9 for further discussion).

Though the design of cancellation controllers has the advantage of simplicity, it is not recommended in general because of the above restrictions. In particular, for processes of higher order it becomes difficult to prescribe the closed-loop behaviour, so that other design methods become more appropriate.

7 Controllers for Finite Settling Time (Deadbeat)

The ripples between the sampling points that can appear with the cancellation controllers can be avoided if a finite settling time is required for both the controlled variable and the manipulated variable. Jury [2.3, 7.1] has called this behaviour "deadbeat-response". For a step change of the reference variable the input and the output signal of the process have to be in a new steady state after a definite finite settling time.

In order to determine a deadbeat controller a process model with discrete signals is assumed in [2.3]. A method for process models with continuous signals is described e.g. in [2.21].

In the following discussion, methods for the design of deadbeat controllers are described which are characterized by an especially simple derivation and for which the resulting synthesis requires little calculation.

7.1 Deadbeat Controller Without Prescribed Manipulated Variable

Process Without Deadtime

A step change of the reference variable at the instant $k=0$ is assumed, giving

$$w(k) = 1 \text{ für } k = 0, 1, 2, \ldots \tag{7.1.1}$$

For deadtime $d=0$ the requirement for minimal settling time is

$$
\begin{aligned}
y(k) &= w(k) = 1 \quad \text{für} \quad k \geq m \\
u(k) &= u(m) \qquad \text{für} \quad k \geq m.
\end{aligned}
\tag{7.1.2}
$$

The z-transforms of the reference, controlled and manipulated variables become [7.2], for the case where $b_0 = 0$,

$$w(z) = \frac{1}{(1-z^{-1})} \tag{7.1.3}$$

$$y(z) = y(1)z^{-1} + y(2)z^{-2} + \ldots + 1[z^{-m} + z^{-(m+1)} + \ldots] \tag{7.1.4}$$

$$u(z) = u(0) + u(1)z^{-1} + \ldots + u(m)[z^{-m} + z^{-(m+1)} + \ldots]. \tag{7.1.5}$$

Dividing (7.1.4) and (7.1.5) by (7.1.3), one obtains

$$\frac{y(z)}{w(z)} = p_1 z^{-1} + p_2 z^{-2} + \dots + p_m z^{-m} = P(z) \qquad (7.1.6)$$

$$p_1 = y(1)$$
$$p_2 = y(2) - y(1)$$
.
.
.
$$p_m = 1 - y(m-1)$$

$$\frac{u(z)}{w(z)} = q_0 + q_1 z^{-1} + \dots + q_m z^{-m} = Q(z) \qquad (7.1.7)$$

$$q_0 = u(0)$$
$$q_1 = u(1) - u(0)$$
.
.
.
$$q_m = u(m) - u(m-1).$$

It should be notified that

$$p_1 + p_2 + \dots + p_m = 1 \qquad (7.1.8)$$

$$q_0 + q_1 + \dots + q_m = u(m) = \frac{1}{K_P} = \frac{1}{G_P(1)} . \qquad (y(m)=1) \quad (7.1.9)$$

The closed-loop transfer function is

$$G_w(z) = \frac{y(z)}{w(z)} = \frac{G_R(z) G_P(z)}{1 + G_R(z) G_P(z)} . \qquad (7.1.10)$$

The controller $G_R(z)$ is now to be determined in such a way that the deadbeat behaviour, described in (7.1.6) and (7.1.7) results. Solving (7.1.10) for $G_R(z)$ one obtains the equation of the cancellation controllers, see (6.2.4)

$$G_R(z) = \frac{1}{G_P(z)} \frac{G_w(z)}{1 - G_w(z)} . \qquad (7.1.11)$$

Comparison of (7.1.6) and (7.1.10) leads to

$$G_w(z) = P(z). \qquad (7.1.12)$$

Moreover, it follows from (7.1.6) and (7.1.7) that

$$G_P(z) = \frac{y(z)}{u(z)} = \frac{P(z)}{Q(z)} \qquad\qquad (7.1.13)$$

and the controller given by (7.1.11) then becomes

$$G_R(z) = \frac{Q(z)}{1 - P(z)} = \frac{q_0 + q_1 z^{-1} + \dots + q_m z^{-m}}{1 - p_1 z^{-1} - \dots - p_m z^{-m}}. \qquad (7.1.14)$$

The parameters of this controller are obtained through comparison of the coefficients in (7.1.13) as follows

$$q_1 = a_1 q_0 \qquad p_1 = b_1 q_0$$
$$q_2 = a_2 q_0 \qquad p_2 = b_2 q_0$$

$$\begin{matrix} \cdot & & \cdot \\ \cdot & & \cdot \\ \cdot & & \cdot \end{matrix} \qquad\qquad (7.1.15a)$$

$$q_m = a_m q_0 \qquad p_m = b_m q_0.$$

and from (7.1.8) it also follows that

$$q_0 = \frac{1}{b_1 + b_2 + \dots + b_m} = u(0). \qquad\qquad (7.1.15b)$$

The controller parameters, therefore, can be calculated in a very simple way. It can be seen, that the initial value of the manipulated variable $u(0)$ depends only on the sum of the b-parameters of the process. Since this sum decreases with decreasing sampling time, the manipulated variable $u(0)$ increases the smaller the sampling time becomes.

Introducing (7.1.15) into (7.1.14), it follows for the deadbeat controller (without deadtime) that

$$G_R(z) = \frac{u(z)}{e_w(z)} = \frac{q_0 A(z^{-1})}{1 - q_0 B(z^{-1})}. \qquad\qquad (7.1.16)$$

Hence the denominator $A(z^{-1})$ of the process transfer function is cancelled.

This deadbeat controller can be regarded as a *cancellation controller* because of (7.1.11). However, the closed-loop transfer function, (7.1.12) and (7.1.6) is only determined as a result of the design and is not prespecified as described in section 6.2. Distinguishing between the real process $G_{P0}(z)$ and the process model used for calculating the controller one obtains the following closed-loop transfer function

$$G_w(z) = \frac{y(z)}{w(z)} = \frac{q_0 A(z^{-1}) B_0(z^{-1})}{A_0(z^{-1}) - q_0 A_0(z^{-1}) B(z^{-1}) + q_0 A(z^{-1}) B_0(z^{-1})}$$

The characteristic equation is therefore

$$A_0(z^{-1}) + q_0[-A_0(z^{-1}) B(z^{-1}) + A(z^{-1}) B_0(z^{-1})] = 0 \qquad (7.1.18)$$

Assuming that process and process model agree approximately well, it then follows that

$$A_0(z^{-1}) \approx 0 \quad \text{or} \quad z^m A_0(z^{-1}) = A_0(z) \approx 0 \tag{7.1.19}$$

If the poles of the process are located sufficiently inside the unit circle, then the closed control loop is asymptotically stable. Therefore, the deadbeat controller can be only used for asymptotically stable processes. If $G_{P0}(z) = G_P(z)$ agree precisely, then

$$G_w(z) = P(z) = p_1 z^{-1} + ... + p_m z^{-m}$$

$$= \frac{p_1 z^{m-1} + ... + p_m}{z^m} = \frac{q_0 B(z)}{z^m} \tag{7.1.20}$$

becomes valid. The characteristic equation is therefore:

$$1 + G_R(z) G_P(z) = z^m = 0 \tag{7.1.21}$$

Hence, the control loop has an mth-order pole in the origin of the z-plane. After an initial deviation the control loop settles after m steps into the steady state. For unprecise agreement the steady state will be only reached for $k \to \infty$, according (7.1.19).

Process with Deadtime

For a process with deadtime one uses the process model [5.7]

$$G_P(z) = \frac{b_1 z^{-(1+d)} + ... + b_m z^{-(m+d)}}{1 + a_1 z^{-1} + ... + a_m z^{-m}}$$

$$= \frac{\bar{b}_1 z^{-1} + ... + \bar{b}_{d+1} z^{-(1+d)} + ... + \bar{b}_\nu z^{-\nu}}{1 + a_1 z^{-1} + ... + a_m z^{-m} + ... + a_\nu z^{-\nu}} \tag{7.1.22}$$

On considering the following

$$\left.
\begin{aligned}
&\bar{b}_1 = \bar{b}_2 = ... = \bar{b}_d = 0 \quad && a_{m+1} = 0 \\
&\bar{b}_{1+d} = b_1 \quad && \vdots \\
&\bar{b}_{2+d} = b_2 \quad && a_\nu = 0. \\
&\vdots \\
&\bar{b}_\nu = b_m
\end{aligned}
\right\} \tag{7.1.23}$$

As the command behaviour it is now required that

$$\left.
\begin{aligned}
y(k) &= w(k) = 1 \quad && \text{für} \quad k \geq \nu = m+d \\
u(k) &= u(m) \quad && \text{für} \quad k \geq m
\end{aligned}
\right\} \tag{7.1.24}$$

(7.1.3) to (7.1.15) can be applied using (7.1.22). Then it follows from (7.1.22) and (7.1.13) that

$$
\left.
\begin{aligned}
q_0 &= \frac{1}{b_1 + b_2 + \dots + b_m} = u(0)\text{(accord. to 7.1.15b)} \\[2mm]
q_1 &= a_1 q_0 \\
q_2 &= a_2 q_0 \qquad\qquad\qquad p_1 = \bar{b}_1 q_0 = 0 \\
&\ \ \vdots \qquad\qquad\qquad\qquad\qquad \vdots \\
q_m &= a_m q_0 \qquad\qquad\qquad p_d = \bar{b}_d q_0 = 0 \\
q_{m+1} &= a_{m+1} q_0 = 0 \qquad p_{1+d} = \bar{b}_{1+d} q_0 = b_1 q_0 \\
&\ \ \vdots \qquad\qquad\qquad\qquad\qquad \vdots \\
q_v &= a_v q_0 = 0 \qquad\qquad p_v = \bar{b}_v q_0 = b_m q_0
\end{aligned}
\right\}
\qquad (7.1.25)
$$

and the controller transfer function is

$$
G_R(z) = \frac{q_0 + q_1 z^{-1} + \dots + q_m z^{-m}}{1 - p_{1+d} z^{-(1+d)} - \dots - p_{m+d} z^{-(m+d)}}
\qquad (7.1.26)
$$

From (7.1.25) and (7.1.26) the transfer function of the deadbeat controller DB(v) for processes with deadtime becomes

$$
G_R(z) = \frac{u(z)}{e_w(z)} = \frac{q_0 A(z^{-1})}{1 - q_0 B(z^{-1}) z^{-d}}.
\qquad (7.1.27)
$$

The characteristic equation which corresponds with (7.1.19) is

$$
z^{m+d} A_0(z^{-1}) = z^d A_0(z) \approx 0.
\qquad (7.1.28)
$$

Therefore, the command transfer function for an exact match of process model and process is

$$
G_w(z) = \frac{q_0 B(z^{-1}) z^{-d}}{1} = \frac{q_0 B(z)}{z^{(m+d)}}
\qquad (7.1.29)
$$

and the characteristic equation becomes

$$
z^{(m+d)} = 0.
\qquad (7.1.30)
$$

Example 7.1. Deadbeat Controller DB(v). ($v = m$).
For the low-pass process III, described in section 5.4.2, one obtains for $T_0 = 4$ s the following coefficients of the deadbeat controller by using (7.1.25)

$$
q_0 = 9{,}523 \quad q_1 = -14{,}285 \quad q_2 = 6{,}714 \quad q_3 = -0{,}952
$$

$$
p_1 = 0 \quad p_2 = 0{,}619 \quad p_3 = 0{,}457 \quad p_4 = -0{,}0762
$$

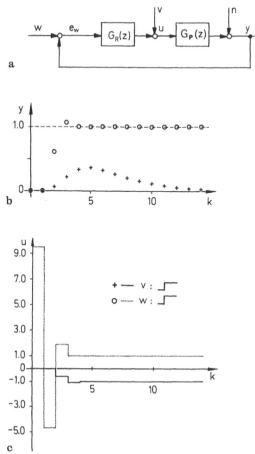

Fig. 7.1. Responses of the control loop
with a deadbeat controller of order v
(normal order) (without prescribed
manipulated variable) and process III
for a step change of w and a step
change of v.
a block diagram of the control loop;
b behaviour of the controlled variable;
c behaviour of the manipulated variable

In Fig. 7.1 the responses to step changes in the command variable w and the disturbance
variable v can be seen. The required deadbeat behaviour for the change of w is produced.

7.2 Deadbeat Controller with Prescribed Manipulated Variable

If the finite settling time is increased from m to $m+1$ then one value of the
manipulated variable can be prescribed. Since the first manipulated variable $u(0)$
generally is the largest one, this value should be reduced by prescribing it [5.7].
   In (7.1.4) and (7.1.5) one more step is admitted. Then (7.1.6) and (7.1.7) become

$$P(z) = p_1 z^{-1} + p_2 z^{-2} + \ldots + p_{m+1} z^{-(m+1)} \tag{7.2.1}$$

$$Q(z) = q_0 + q_1 z^{-1} + \ldots + q_{m+1} z^{-(m+1)} \tag{7.2.2}$$

Comparison of the coefficients in (7.1.13) then leads to

$$\frac{b_1 z^{-1} + \dots + b_m z^{-m}}{1 + a_1 z^{-1} + \dots + a_m z^{-m}} = \frac{p_1 z^{-1} + \dots + p_{m+1} z^{-(m+1)}}{q_0 + q_1 z^{-1} + \dots + q_{m+1} z^{-(m+1)}}.$$

(7.2.3)

This equation can only be satisfied if the right hand term has the same root in both the denominator and the numerator. Hence

$$\frac{P(z)}{Q(z)} = \frac{(p_1' z^{-1} + \dots + p_m' z^{-m})(\alpha - z^{-1})}{(q_0' + \dots + q_m' z^{-m})(\alpha - z^{-1})}.$$

(7.2.4)

If the coefficients in (7.2.3) are compared one obtains, after dividing by q_0'

$$\begin{aligned}
q_1' &= a_1 q_0' & p_1' &= b_1 q_0' \\
q_2' &= a_2 q_0' & p_2' &= b_2 q_0' \\
&\ \ \vdots & &\ \ \vdots \\
q_m' &= a_m q_0' & p_m' &= b_m q_0'.
\end{aligned}$$

(7.2.5)

Now, the numerator and denominator in (7.2.4) are written fully, and from the comparison of the coefficients with the right hand side of (7.2.3) and (7.2.4), the following equations are obtained

$$\begin{aligned}
q_0 &= \alpha q_0' & p_1 &= \alpha p_1' \\
q_1 &= (\alpha q_1' - q_0') & p_2 &= (\alpha p_2' - p_1') \\
&\ \ \vdots & &\ \ \vdots \\
q_m &= (\alpha q_m' - q_{m-1}') & p_m &= (\alpha p_m' - p_{m-1}') \\
q_{m+1} &= -q_m' & p_{m+1} &= -p_m'.
\end{aligned}$$

(7.2.6)

From (7.1.7) one gets

$$q_0 = \alpha q_0' = u(0)$$

(7.2.7)

and with (7.2.1) or (7.2.6)

$$p_1 + \dots + p_{m+1} = 1$$

Now, it follows from (7.2.6) and (7.2.5)

$$q_0' = q_0 - \frac{1}{b_1 + b_2 + \dots + b_m} = q_0 - \frac{1}{\Sigma b_i}.$$

(7.2.8)

The parameters of the controller using (7.2.7) and (7.2.8) are as follows

$$q_0 = u(0) \quad (\text{prescribed})$$

$$q_1 = q_0(a_1-1) + \frac{1}{\Sigma b_i}$$

$$q_2 = q_0(a_2-a_1) + \frac{a_1}{\Sigma b_i}$$

$$\cdot$$
$$\cdot$$
$$\cdot$$

(7.2.9)

$$q_m = q_0(a_m-a_{m-1}) = \frac{a_{m-1}}{\Sigma b_i}$$

$$q_{m+1} = a_m\left(-q_0 + \frac{1}{\Sigma b_i}\right)$$

$$p_1 = q_0 b_1$$

$$p_2 = q_0(b_2-b_1) = \frac{b_1}{\Sigma b_i}$$

$$\cdot$$
$$\cdot$$
$$\cdot$$

(7.2.10)

$$p_m = q_0(b_m-b_{m-1}) + \frac{b_{m-1}}{\Sigma b_i}$$

$$p_{m+1} = -b_m\left(q_0 - \frac{1}{\Sigma b_i}\right).$$

The controller transfer function is now

$$G_R(z) = \frac{Q(z)}{1-P(z)} = \frac{q_0 + q_1 z^{-1} + \dots + q_{m+1} z^{-(m+1)}}{1 - p_1 z^{-1} - \dots - p_{m+1} z^{-(m+1)}}. \quad (7.2.11)$$

Unlike the controller given by (7.1.14), here the initial manipulated variable $u(0)=q_0$ is given. The second manipulated variable then becomes (see (7.1.7) and 7.2.9))

$$u(1) = q_1 + q_0 = a_1 u(0) + \frac{1}{\Sigma b_i}. \quad (7.2.12)$$

$u(0)$ should not be chosen too small because then $u(1)>u(0)$, which is unsuitable in most cases.

If $u(1) \leq u(0)$ one requires

$$u(0) = q_0 \geq 1/(1-a_1)\Sigma b_i. \quad (7.2.13)$$

Even if $u(1) \leq u(0)$ is specified it is not, of course, certain that for $k \geq 2$ then $|u(k)| < |u(0)|$. Since the calculation of the parameters is relatively simple, one proceeds iteratively, i.e. one varies $u(0)$ so long as there is adequate behaviour. Often the choice $u(1)=u(0)$ gives a good result.

For processes with deadtime $d>0$, one proceeds according to (7.1.22) to (7.1.26). Then, based on the equations corresponding to (7.2.9) and (7.2.10), the transfer function of the deadbeat controller $DB(v+1)$ becomes

$$G_R(z) = \frac{q_0 A(z^{-1})[1 - z^{-1}/\alpha]}{1 - q_0 B(z^{-1}) z^{-d}[1 - z^{-1}/\alpha]} \qquad (7.2.14)$$

with $1/\alpha = 1 - 1/q_0 \Sigma b_i$.

The characteristic equation is thus

$$z^{m+d+1} = 0 \qquad (7.2.16)$$

The following equation can be used in order to choose q_0

$$q_0 = q_{0\ min} + (1-r')(q_{0\ max}-q_{0\ min}) \qquad (0 \le r' \le 1) \qquad (7.2.17)$$

with

$$q_{0\ min} = 1/(1-a_1)\Sigma b_i, q_{0\ max} = 1/\Sigma b_i \qquad (7.2.18)$$

Example 7.2. Deadbeat Controller $DB(v+1)$
For the same process as in example 7.1, Fig. 7.2 shows step responses for given $u(0)=u(1)$ with controller parameters:

$q_0 = 3,810 \quad q_1 = 0 \quad q_2 = -5,884 \quad q_3 = 3,647 \quad q_4 = -0,571$

$p_1 = 0 \quad p_2 = 0,247 \quad p_3 = 0,554 \quad p_4 = 0,244 \quad p_5 = -0,046.$

For a step change of the reference variable the desired deadbeat behaviour is obtained. The manipulated variable $u(0)$ could be decreased to 40% of the value in Fig. 7.1. The control variable needs one more sample time to reach the finite settling time. For a step change of the disturbance v, a relatively well damped behaviour can be seen. The chosen initial manipulated variable $u(0)$ leads to a somewhat worse control performance. However, this deadbeat controller can be applied more generally as it produces smaller amplitudes of the manipulated variable.

The step responses of the deadbeat controllers used in example 7.1 and 7.2 are shown in Fig. 7.3. For the controller $DB(v)$, a negative $u(1)$ follows the initial large positive manipulated variable $u(0)$; this opposite control is required because of the large value of $u(0)$. A positive-valued manipulated variable then follows and, after oscillations have decayed, an integral behaviour with increasing $u(k)$ arises.

Because of the assumption that $u(1)=u(0)$ in closed-loop, and because of the deadtime $d=1$ of the process, and, therefore $p_1=0$, $u(0)$ and $u(1)$ are both equal for the controller $DB(v+1)$. Then a negative value of $u(2)$ occurs, followed by integral behaviour after some oscillations.

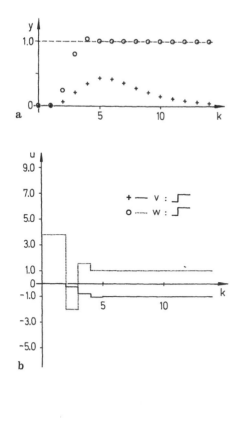

Fig. 7.2. Responses of the control loop with deadbeat controller of $(v+1)^{th}$ order (increased order) and process III to step changes of w and step changes of v. For the design $u(0) = u(1)$, (7.2.13), has been set

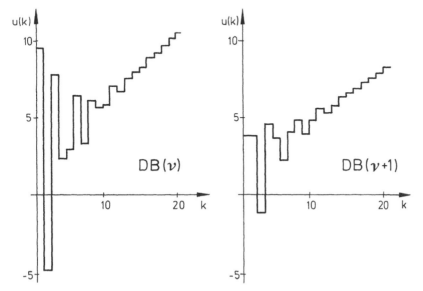

Fig. 7.3. Step responses of control algorithms for finite settling time. Process III

In comparison with the step responses of a PID controller, the deadbeat algorithms generate a larger lead effect which is followed by decaying oscillations to produce the finite settling time of the closed-loop.

If the settling time is increased to $m+2, m+3, \ldots, m+n$, the amplitudes of the manipulated variables can be even reduced. The design of the resulting $DB(v+n)$ controller is described in [7.3].

7.3 Choice of the Sample Time for Deadbeat Controllers

From (7.1.15) the manipulated variable $u(0)$ is inversely proportional to the sum of the numerator parameters of the process model. However, this sum increases (c.f. Table 3.4) with the sample time, so that the magnitude of the manipulated variable can be decreased by increasing the sample time, or vice versa for a given range of a reference variable step a suitable sample time can be determined. Table 7.1 shows the manipulated variable $u(0)$ as a function of the sample time for a process of third-order given by Table 3.4. To avoid $u(0)$ becoming too large, the sample time for the $DB(v)$-controller should be $T_0 \geq 8$ s. This corresponds to

$$T_0/T_\Sigma \geq 0{,}36 \quad \text{or} \quad T_0/T_{95} \geq 0{,}18$$

Here T_Σ is the sum of the time constants and T_{95} the 95% settling time.

If the settling time is increased by one sample interval, i.e. a $DB(v+1)$-controller is used, from (7.2.13) $u(0)$ can be decreased at most by the factor $(1-a_1)$. This means for the example in Table 7.1 a decrease of about 1.8 to 3.3, depending on the sample time. For the $DB(v+1)$-controller the sample time should be $T_0 \geq 5$ s or

$$T_0/T_\Sigma \geq 0{,}22 \quad \text{or} \quad T_0/T_{95} \geq 0{,}11.$$

If that maximum possible $u(0)$ is assumed which can be chosen from the allowable range of the manipulated variable, the sample time of the $DB(v+1)$-controller can be smaller than for the $DB(v)$-controller.

Table 7.2 compares suitable sample times for a parameter-optimized controller and for the deadbeat controllers considered for process III in section 5.4.

The recommended smallest sample time can be about the same for 3-PC-3 and $DB(v+1)$. However, for $DB(v)$ a value twice this can be chosen.

Table 7.1. Influence of the sample time on the manipulated variable $u(0)$ for deadbeat controllers with the third order process of Table 3.7.1

Controller	T_0 in s	2	4	6	8	10
$DB(v)$	$u(0) = q_0$	71.5	13.3	5.75	3.81	2.50
–	$(1-a_1)$	3.25	2.71	2.30	2.00	1.77
$DB(v+1)$	$u(0) = q_0$	22.0	4.91	2.50	1.91	1.41

Table 7.2. Comparison of the sample times for parameter optimized and deadbeat controllers for the example of process III. Assumption $u(0)_{max} \leq 4.5$.

Controller	3PC-3 (PID) ($r = 0$)	DB(v)	DB($v+1$)
T_0/T_{95}	≥ 0.12	≥ 0.2	≥ 0.10

The form of the deadbeat controllers discussed in this section requires particularly few calculations. Therefore, they should be applied when the synthesis must be repeated frequently, e.g. in adaptive control systems. However, as these deadbeat controllers compensate for the poles of the process, as in (7.1.27) and (7.2.14), they should not be applied to processes with poles outside or in the vicinity of the unit circle in the z-plane [7.1]. (This means that the application of deadbeat controllers has to be restricted to asymptotically stable processes (c.f. section 11.1).

7.4 Approximation Through PID-Controllers

The extended deadbeat controller DB($v+1$) can be used to design a PID-controller with relatively small computational effort. By approximating the transient function $u^*(k)$ of the DB($v+1$) controller

$$G_R^*(z) = \frac{u^*(z)}{e_w(z)} = \frac{q_0^* + q_1^* z^{-1} + \ldots + q_{m+1}^* z^{-(m+1)}}{1 - p_{1+d}^* z^{-(d+1)} - \ldots - p_{m+d+1}^* z^{-(m+d+1)}}$$

(7.4.1)

by the transient function $u(k)$ of the PID-controller

$$G_R(z) = \frac{u(z)}{e_w(z)} = \frac{q_0 + q_1 z^{-1} + q_2 z^{-2}}{1 - z^{-1}}$$

(7.4.2)

for $k=0$ and $k\to\infty$, c.f. Figures 5.2 and 7.3. Changing the control deviation $e_w(k)$ after a unit-step $1(k)$ one obtains the transient function of the DB($v+1$) from

$$u^*(k) = \sum_{j=1}^{m+1} p_{j+d}^* u^*(k-d-j) + \sum_{j=0}^{m+1} q_j^* e(k-j)$$

(7.4.3)

for $k=0, 1, 2, \ldots$ assuming $u^*(k)=0$ and $e_w(k)=0$ for $k<0$. Together with (7.2.13) it follows that

$$u(0) = q_0^* = \frac{1}{(1-a_1)\Sigma b_i}.$$

(7.4.4)

The behaviour for large k and consequently the increase or the integral factor is obtained by solving (7.4.3) recursively. The increase δ per sample time T_0 is

$$\delta = \lim_{k \to \infty} [u^*(k) - u^*(k-1)] \qquad (7.4.5)$$

follows by

$$\left.\begin{array}{l}
\delta \approx u^*(M) - u^*(M-1) \\[1.5em]
\delta \approx \dfrac{1}{\xi}[u^*(M) - u^*(M-\xi)]
\end{array}\right\} \qquad (7.4.6)$$

or

with M sufficiently large, for example $M = 4(m+d)$. Alternatively one can calculate for $k > m+d$

$$\delta \approx \Delta u^*(k) = u^*(k) - u^*(k-1) \qquad (7.4.7)$$

and stop if

$$\Delta u^*(k) - \Delta u^*(k-1) < \varepsilon u^*(k) \qquad (7.4.8)$$

with for example $\varepsilon = 0.02$.

The increase δ can also be calculated explicitly as a function of the process parameters and the deadtime as

$$\delta = \lim_{k \to \infty} [u(k) - u(k-1)] = \frac{\Sigma a_i}{[d+1 - q_0^* \Sigma b_i] \Sigma b_i + \Sigma i b_i} \qquad (7.4.9)$$

with $i = 1, 2, \ldots, m$, which follows from (7.4.3). The parameters of the PID-controller 3PC-2 are obtained as follows, c.f. Fig. 5.2

a) $q_0 = q_0^* \qquad (7.4.10)$

b) $q_0 - q_2 = u^*(M) - M\delta \qquad (7.4.11)$
   $q_2 = q_0 - [u^*(M) - M\delta]$

c) $q_0 + q_1 + q_2 = \delta \qquad (7.4.12)$
   $q_1 = \delta - q_0 - q_2.$

The controller characteristics K, c_D and c_I are calculated according (5.2.15). If required, they can be increased or decreased using correction factors, $K_{res} = \varepsilon_K K$; $c_{Dres} = \varepsilon_D c_D$ and $c_{Ires} = \varepsilon_I c_I$ depending on the process.

The computation time for the design is somewhat larger than for $DB(v+1)$ since, besides the calculation of the parameters of $DB(v+1)$ the following is required: the recursive calculation of $u^*(M)$ according (7.4.3), (7.4.6) or (7.4.7) and (7.4.11) and (7.4.12). Compared with numerical parameter optimization the computation time, in general, is considerably smaller with this method.

This direct design method for discrete PID-controllers should be of interest for the following cases:

1. Application of self-tuning control for unique tuning of the controller parameters of PID-controllers.
2. Determination of suitable initial values for numerical parameter optimization.

Provided there are additional assumptions on the course of the step response, this design method can also be applied for controllers with numerator orders $v > 2$.

The described design method can be used for low-pass processes with no or small deadtimes [26.16]. The choice of the sample time, however, depends on the permissible manipulated variables. This generally results in a relatively large sample time, so that applicability is somewhat restricted.

8 State Controller and State Observer

The design of the controllers described in the previous chapters was based on input/output models of the processes in form of difference equations or transfer functions, according Fig. 3.16a. Based on these models, input/output controllers could be designed either by parameter-optimization, chapter 5 or by structure-optimization, chapter 6 and 7. In both cases it is assumed that the closed loop is in a steady state before a disturbance occurs.

The use of internal state variables additionally available in state representation of a process model, Fig. 3.16b), allows to consider the internal relations of the process in the controller. Also the initial conditions of the process can differ from zero which means that they do not have to be in a steady state.

Section 8.1 considers the optimal control of a process from a *given initial state* into the zero state. It is assumed initially that all state variables are measurable. A quadratic performance criterion is used the optimization of which directly leads to the structure and to the parameters of a *state feedback control*. In this case the calculation of the optimal controller requires the solution of a matrix Riccati equation. The design of optimal state controllers for *external disturbances* is described in section 8.2.

The parameters of a linear state controller can also be determined from the *chosen coefficients of the characteristic equation*, section 8.3, which is computationally simple. After transforming the process equation into a diagonal form, the controller parameters can be determined by independently choosing the poles of the closed loop, provided certain requirements are satisfied. This is called *modal state control*, section 8.4. The parameters of the state controller can also be simply determined if the design objective of a *finite settling time* is applied, section 8.5.

If some state variables are unmeasurable they must be reconstructed by an *observer*, section 8.6. The combination of state controllers and observers with the process is considered in section 8.7 for both initial value disturbances and external disturbances. Finally, the design of *observers of reduced order*, section 8.8, and the *selection of free design parameters*, section 8.9, is considered.

As the derivation of state controllers for multivariable control systems and single variable control systems differ only by writing the manipulated and controlled variables in vector form and by using parameter matrices instead of parameter vectors, the general multivariable case is considered in the following sections. As examples, however, only single-input/single-output control systems will be used.

8.1 Optimal State Controllers for Initial Values

It is assumed that the state equation of the process

$$x(k+1) = A x(k) + B u(k) \qquad (8.1.1)$$

with constant parameter matrices A and B is given, together with the initial condition $x(0)$ (see Fig. 8.1.). It is assumed initially that all state variables $x(k)$ are exactly measurable.

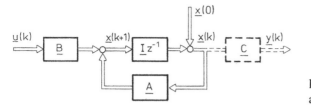

Fig. 8.1. State model of a linear process

Now a controller has to be determined which generates a manipulated variable vector $u(k)$ from the state variable vector $x(k)$ so that the overall system is controlled into the final state $x(N) \approx 0$ and the quadratic performance criterion

$$I = x^T(N) S x(N) + \sum_{k=0}^{N-1} [x^T(k) Q x(k) + u^T(k) R u(k)] \qquad (8.1.2)$$

is minimized. Here

> S is positive semidefinite and symmetric,
> Q is positive semidefinite and symmetric,
> R is positive definite and symmetric,

i.e. $x^T S x \geq 0$, $x^T Q x \geq 0$ and $u^T R u > 0$.

These conditions on the weighting matrices S, Q and R result from the conditions for the existence of the optimum of I and can be discussed as follows. Meaningful solutions in the control engineering sense can only be obtained if all terms have the same sign, e.g. a positive sign. Therefore, all matrices have to be at least positive semidefinite. If $S=0$, i.e. the final state $x(N)$ is not weighted, but $Q \neq 0$, i.e. all states $x(0), \ldots, x(N-1)$ are weighted, a meaningful optimum also exists. That means that if Q is positive definite S can also be positive semidefinite. The converse is also true. One should, however, exclude the case where $S=0$ and $Q=0$, for then the states $x(k)$ would not be weighted and only the manipulated would be weighted by $R \neq 0$, which is nonsense. R has to be positive definite for continuous-time state controllers as R^{-1} is involved in the control law. For discrete-time state controllers, however, this requirement can be relaxed as described later.

As only the case where $x(N) \approx Q$ will be considered in the following $S = Q$ is chosen. In this case, Q should be positive definite. Note that in this problem the influence of the reference variables and external disturbances is ignored, and that the output variables

$$y(k) = C x(k) \tag{8.1.3}$$

are not fed back. Instead, the modification of the process eigen behaviour and stabilization through state feedback is considered. If the optimal manipulated variable is found then

$$\min I = \min_{u(k)} \left\{ x^T(N) Q x(N) + \sum_{k=0}^{N-1} [x^T(k) Q x(k) + u^T(k) R u(k)] \right\}$$

$$k = 0, 1, 2, ..., N-1 \tag{8.1.4}$$

The calculation of the optimal manipulated variable is a problem of dynamic optimization which can be solved by variational calculus, applying the maximum principle of Pontryagin or the Bellman optimization principle [8.1]. The solution outlined below was given by Kalman and Koepcke [8.2] and uses the optimality principle.

Remarks

a) According to the optimality principle of Bellman each final element of an optimal trajectory is also optimal. This means that if the end point is known, one can determine the optimal trajectory in a backward direction.
b) Because of the state equation (8.1.1), $u(k)$ influences the future states $x(k+1)$, $x(k+2), \ldots$. Therefore, one can calculate the optimal $u(k)$ by backward calculation. Hence, (8.1.4) is written as:

$$\min I = \min_{u(k)} \left[\min_{u(N-1)} \{ x^T(N) Q x(N) \right.$$

$$k = 0, 1, ..., N-2 \tag{8.1.5}$$

$$\left. + \sum_{k=0}^{N-1} [x^T(k) Q x(k) + u^T(k) R u(k)] \} \right].$$

Now it yields for the last step in brackets (8.1.5) through transformation of the term

$$\min_{u(N-1)} \{...\} = \sum_{k=0}^{N-1} x^T(k) Q x(k) + \sum_{k=0}^{N-2} u^T(k) R u(k)$$

$$+ \underbrace{\min_{u(N-1)} \{ x^T(N) Q x(N) + u^T(N-1) R u(N-1) \}}_{I_{N-1,N}} \tag{8.1.6}$$

as the first two terms are not influenced by $u(N-1)$ and $I_{N-1,N}$ are the costs of $k=N-1$ to $k=N$ resulting from $u(N-1)$. If the state equation

$$x(N) = A x(N-1) + B u(N-1)$$

or

$$x^T(N) = x^T(N-1)A^T + u^T(N-1)B^T$$

(8.1.7)

is considered as a further condition, it follows from (8.1.6)

$$\begin{aligned} I_{N-1,N} &= \min_{\mathbf{u}(N-1)} \{x^T(N)Qx(N) + u^T(N-1)Ru(N-1)\} \\ &= \min_{\mathbf{u}(N-1)} \{x^T(N-1)A^TQAx(N-1) + 2x^T(N-1)A^TQBu(N-1) \\ &\quad + u^T(N-1)B^TQBu(N-1) + u^T(N-1)Ru(N-1)\} \\ &= x^T(N-1)A^TQAx(N-1) + \min_{\mathbf{u}(N-1)} \{2x^T(N-1)A^TQBu(N-1) \\ &\quad + u^T(N-1)(B^TQB + R)u(N-1)\}. \end{aligned}$$

(8.1.8)

To minimize (8.1.8) the following relations are valid

$$\min_{\mathbf{u}(N-1)} \{...\} = \frac{\partial}{\partial u(N-1)} \{...\} = 0, \quad \frac{\partial^2}{\partial u(N-1)^2} \{...\} > 0.$$

(8.1.9)

Hence, using the rules for taking derivatives of vectors and matrices given in the appendix

$$\frac{\partial}{\partial u(N-1)} \{...\} = 2B^TQ A x(N-1) + 2(B^TQ B + R)u(N-1) = 0$$

and

$$\begin{aligned} u^0(N-1) &= -(B^TQ B + R)^{-1}B^TQ A x(N-1) \\ &= -K_{N-1}x(N-1) \end{aligned}$$

(8.1.10)

Here

$$K_{N-1} = (B^TQ B + R)^{-1}B^TQ A$$

(8.1.11)

and

$$\frac{\partial^2 \{...\}}{\partial u(N-1)^2} = 2(B^TQ B + R) > 0.$$

(8.1.12)

The costs $I_{N-1,N}$ resulting from $u(N-1)$ can therefore be formulated as a function of the initial condition $x(N-1)$ for this stage

$$I_{N-1,N} = x^T(N-1)A^T QA x(N-1)$$

$$- 2x^T(N-1)A^T QB(B^T QB + R)^{-1} B^T QA x(N-1)$$

$$+ x^T(N-1)A^T QB(B^T QB + R)^{-1} B^T QA x(N-1)$$

$$= x^T(N-1)[A^T QA - A^T QB(B^T QB + R)^{-1} B^T QA]x(N-1)$$

$$= x^T(N-1)[A^T QA - K_{N-1}^T(B^T QB + R)K_{N-1}]x(N-1)$$

$$= x^T(N-1)P_{N-1,N} x(N-1). \tag{8.1.13}$$

Here

$$P_{N-1,N} = A^T Q[I - B(B^T Q B + R)^{-1} B^T Q]A$$

$$= A^T Q A - K_{N-1}^T(B^T Q B + R)K_{N-1} \tag{8.1.14}$$

I or min I according to (8.1.5) and (8.1.6) can be given as a function of $x(k)$, $k=0, \ldots, N-1$ and $u(k)$, $k=0, \ldots, N-2$. Thus the unknown $x(N)$ and $u(N-1)$ can be eliminated. In order to perform this elimination, first $I_{N-1,N}$ from (8.1.13) is substituted in (8.1.6), resulting in

$$\min_{u(N-1)} \left\{ x^T(N)Q\, x(N) + \sum_{k=0}^{N-1} [x^T(k)Q\, x(k) + u^T(k)R\, u(k)] \right\}$$

$$= \sum_{k=0}^{N-1} x^T(k)Q\, x(k) + \sum_{k=0}^{N-2} u^T(k)R\, u(k) + x^T(N-1)P_{N-1,N} x(N-1)$$

$$= \sum_{k=0}^{N-2} [x^T(k)Qx(k) + u^T(k)Ru(k)] + \underbrace{x^T(N-1)(P_{N-1,N} + Q)x(N-1)}_{I_{N-1}}. \tag{8.1.15}$$

The abbreviation

$$P_{N-1} = P_{N-1,N} + Q \tag{8.1.16}$$

is introduced so that in (8.1.15)

$$I_{N-1} = I_{N-1,N} + x^T(N-1)Qx(N-1)$$

$$= x^T(N-1)(P_{N-1,N} + Q)x(N-1) = x^T(N-1)P_{N-1}x(N-1) \tag{8.1.17}$$

can be formed. In this abbreviation the costs of the last step and the evaluation of the corresponding initial deviation $x(N-1)$ are included. (This compression allows a simpler formulation of the following equations). If (8.1.16) is introduced into, and if the result is placed into (8.1.5) it follows that

$$\min_{u(k)} I = \min_{u(k)} \left[\min_{u(N-2)} \left\{ \sum_{k=0}^{N-2} [x^T(k) Q x(k) + u^T(k) R u(k)] \right. \right.$$

$$k = 0, ..., N-3 \qquad \left. \left. + x^T(N-1) P_{N-1} x(N-1) \right\} \right] \qquad (8.1.18)$$

Instead of $\min\limits_{u(N-2)}$ now it reads $\min\limits_{u(N-2)}$ as the optimal $u(N-1)$ and the resulting state $x(N)$ have been calculated and substituted. For the term $\min\limits_{u(N-2)} ...$ one obtains by analogy to (8.1.6)

$$\min_{u(N-2)} \{...\} = \sum_{k=0}^{N-2} x^T(k) Q x(k) + \sum_{k=0}^{N-3} u^T(k) R u(k)$$

$$+ \underbrace{\min_{u(N-2)} \{ u^T(N-2) R u(N-2) + x^T(N-1) P_{N-1} x(N-1) \}}_{I_{N-2,N}}.$$

$$\qquad (8.1.19)$$

$I_{N-2,N}$ describes the costs resulting from the last two stages

$$I_{N-2,N} = u^T(N-2) R u(N-2) + x^T(N-1) Q x(N-1) + I_{N-1,N}.$$

If now the state equation is again considered

$$x(N-1) = A x(N-2) + B u(N-2)$$

it follows that

$$I_{N-2,N} = \min_{u(N-2)} \{ u^T(N-2) (R + B^T P_{N-1} B) u(N-2)$$

$$+ 2u^T(N-2) B^T P_{N-1} A x(N-2)$$

$$+ x^T(N-2) A^T P_{N-1} A x(N-2) \}$$

$$= x^T(N-2) A^T P_{N-1} A x(N-2)$$

$$+ \min_{u(N-2)} \{ u^T(N-2) (R + B^T P_{N-1} B) u(N-2)$$

$$+ 2u^T(N-2) B^T P_{N-1} A x(N-2) \}. \qquad (8.1.21)$$

This results by analogy to (8.1.10), in

$$u^0(N-2) = - (R + B^T P_{N-1} B)^{-1} B^T P_{N-1} A x(N-2)$$

$$= - K_{N-2} x(N-2). \qquad (8.1.22)$$

Hence, the controller K_{N-2} becomes

$$K_{N-2} = (R + B^T P_{N-1} B)^{-1} B^T P_{N-1} A \qquad (8.1.23)$$

Therefore the minimal costs $I_{N-2,N}$ for the two last stages become using (8.1.21)

$$I_{N-2,N} = x^T(N-2)A^T P_{N-1} Ax(N-2)$$

$$+ x^T(N-2)A^T P_{N-1} B(R + B^T P_{N-1} B)^{-1} B^T P_{N-1} Ax(N-2)$$

$$- 2x^T(N-2)A^T P_{N-1} B(R + B^T P_{N-1} B)^{-1} B^T P_{N-1} Ax(N-2)$$

$$= x^T(N-2)[A^T P_{N-1} A - A^T P_{N-1} B$$

$$(R + B^T P_{N-1} B)^{-1} B^T P_{N-1} A] x(N-2)$$

$$= x^T(N-2)[A^T P_{N-1} A - K_{N-2}^T (R + B^T P_{N-1} B) K_{N-2}] x(N-2)$$

$$= x^T(N-2) P_{N-2,N} x(N-2) \tag{8.1.24}$$

with

$$P_{N-2,N} = A^T P_{N-1}[I - B(R + B^T P_{N-1} B)^{-1} B^T P_{N-1}] A$$
$$= A^T P_{N-1} A - K_{N-2}^T (R + B^T P_{N-1} B) K_{N-2}. \tag{8.1.25}$$

Now, the minimum of I with respect to $u(N-2)$ can be formulated according to (8.1.19)

$$\min_{u(N-2)} I = \sum_{k=0}^{N-2} x^T(k) Qx(k) + \sum_{k=0}^{N-3} u^T(k) Ru(k)$$

$$+ x^T(N-2) P_{N-2,N} x(N-2)$$

$$= \sum_{k=0}^{N-3} [x^T(k) Qx(k) + u^T(k) Ru(k)]$$

$$+ \underbrace{x^T(N-2)(P_{N-2,N} + Q)x(N-2)}_{I_{N-2}} \tag{8.1.26}$$

If the abbreviation

$$P_{N-2} = P_{N-2,N} + Q \tag{8.1.27}$$

is introduced again, the costs of the two last stages including the weighting of the initial deviation $x(N-2)$ results in

$$I_{N-2} = I_{N-2,N} + x^T(N-2)Q\, x(N-2)$$

$$= x^T(N-2)(P_{N-2,N} + Q)x(N-2)$$

$$= x^T(N-2) P_{N-2} x(N-2). \tag{8.1.28}$$

Considering the state equation, it can now be expressed as a function of $x(k)$ and $u(k)$ with $k=0, \ldots, N-3$. Then $u^0(N-3)$ can be determined, etc.

In general terms, one obtains a linear, *time-variant state controller*

$$u^0(N-j) = -K_{N-j}x(N-j) \quad j=1,2,\ldots,N \qquad (8.1.29)$$

which has a proportional-action negative feedback when applied to the process input via the gain matrix K_{N-j}, Fig. 8.2.

Its parameters are obtained from the recursive equations

$$K_{N-j} = (R + B^T P_{N-j+1} B)^{-1} B^T P_{N-j+1} A \qquad (8.1.30)$$

$$P_{N-j} = Q + A^T P_{N-j+1} A - K^T_{N-j}(R + B^T P_{N-j+1} B) K_{N-j}$$

$$= Q - K^T_{N-j} R \, K_{N-j} + [A - B \, K_{N-j}]^T P_{N-j+1}[A + B \, K_{N-j}]$$

$$= Q + A^T P_{N-j+1}[I - B(R + B^T P_{N-j+1} B)^{-1} B^T P_{N-j+1}]A$$
$$(8.1.31)$$

with $P_N = Q$ as the initial matrix. The last equation is a *matrix Riccati difference equation*. For the value of the performance criterion (8.1.2) one obtains

$$\min_{u(k)} I = I_0 = x^T(0) P_0 x(0) \qquad (8.1.32)$$

with $k=0, 1, \ldots, N-1$. The minimal value of the quadratic criterion can therefore be expressed explicitly as a function of the initial state $x(0)$. As is shown in a following example, K_{N-j} converges for $j=1, 2, \ldots, N$, i.e.

$$K_{N-1}, K_{N-2}, \ldots, K_2, K_1, K_0$$

to a fixed final value when $N \to \infty$

$$K_0 = \bar{K} = \lim_{N \to \infty} K_{N-j}$$

so that in the limit a *timeinvariant state controller*

$$u^0(k) = -\bar{K} x(k) \qquad (8.1.33)$$

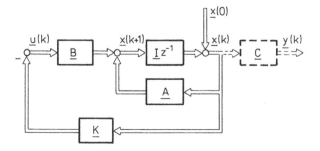

Fig. 8.2. Process state model with optimal state controller K for the control of an initial state deviation $x(0)$. It is assumed that the state vector $x(k)$ is exactly and completely measurable.

is produced. The controller matrix is obtained from

$$\bar{K} = (R + B^T \bar{P} B)^{-1} B^T \bar{P} A \tag{8.1.34}$$

with $\bar{P} = P_0$ as solution of the stationary matrix Riccati equation

$$\bar{P} = \lim_{N \to \infty} P_{N-j} = Q + A^T \bar{P} [I - B(R + B^T \bar{P} B)^{-1} B^T \bar{P}] A. \tag{8.1.35}$$

The solution of this nonlinear equation is obtained from the recursive equation (8.1.31).

The time-invariant state controller of (8.1.33) is most appropriate for practical applications. Because of the matrix inversion,

$$\det[R + B^T \bar{P} B] \neq 0$$

has to be satisfied, but because of the recursive calculation or the existence conditions for the optimum, we must also have:

$$\det[R + B^T Q B] \neq 0 \rightarrow \text{Equation (8.1.12), (8.1.11)}$$

$$\det[R + B^T P_{N-j+1} B] \neq 0 \rightarrow \text{Equation (8.1.30)}$$

This means that the terms in the brackets have to be positive definite. This is satisfied in general by a positive definite matrix R. $R=0$ can, however, also be allowed if the second term $B^T P_{N-j+1} B > 0$ for $j = 1, 2, \ldots, N$ and $Q > 0$. Since P_{N-j+1} is not known a priori, $R > 0$ has to be required in general.

For the closed-loop system from (8.1.1) and (8.1.33) one obtains

$$x(k+1) = [A - B\bar{K}]x(k) \tag{8.1.36}$$

and therefore the characteristic equation becomes

$$\det[zI - A + B\bar{K}] = 0. \tag{8.1.37}$$

This closed system is asymptotically stable if the process (8.1.1) is completely controllable. If it is not completely controllable then the non-controllable part has to have asymptotically stable eigenvalues [8.4].

Example 8.1
This example uses test process III, which is the low-pass third-order process with deadtime described in section 5.4. Table 8.1 gives the coefficients of the matrix P_{N-j}, and Fig. 8.3 shows the controller coefficients k_{N-j}^T as functions of $k = N - j$ (see also example 8.2).

The recursive solution of the matrix Riccati equation was started for $j=0$, with $R=r=1$, $N=29$ and

$$Q = \begin{bmatrix} 0 & 0 & 0 & 0 \\ 0 & 0 & 0 & 0 \\ 0 & 0 & 1 & 0 \\ 0 & 0 & 0 & 0 \end{bmatrix}$$

A value of Q was assumed to give

$$x^T(k)Q\,x(k) = y^2(k).$$

(c.f. section 8.10.1). The coefficients of P_{N-j} and k_{N-j}^T do not change significantly after about ten stages, i.e. the stationary solution is quickly reached.

In this section it was assumed that the state vector $x(k)$ can be measured exactly and completely. This is occasionally true, for example for some mechanical or electrical processes such as aircrafts or electrical networks. However, state variables are often incompletely measurable; they then have to be determined by a reference model or observer (see section 8.6).

8.2 Optimal State Controllers for External Disturbances

The optimal state feedback calculated in the previous section has been designed for the case that a given initial state deviation $x(0)$ of the state variables is controlled into a final state $x(N) \approx 0$. Here, no exterior disturbances were considered as well as the output variable $y(k)$. Only the comparison of $y(k)$ with a reference variable $w(k)$ leads to an actual controller. In order to obtain a useful state controller, externally acting disturbance signals $n(k)$ and reference variables $w(k)$ have to be taken into account, which do not only act at one time point as the initial values, but act with given signals as e.g. step functions, i.e. also constant signals. For this the state feed-

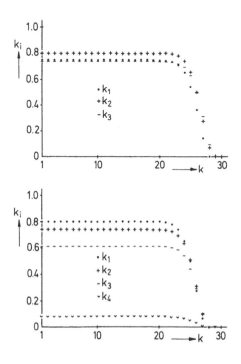

Fig. 8.3. Coefficients of the state controller k_{N-j}^T as functions of $k = N - j$ in the recursive solution of the matrix Riccati equation for process III [8.5]

Table 8.1. Matrix P_{N-j} as a function of k in the recursive solution of the matrix Riccati equation for process III [8.5].

P_{29}	0.0000	0.0000	0.0000	0.0000
	0.0000	0.0000	0.0000	0.0000
	0.0000	0.0000	1.0000	0.0000
	0.0000	0.0000	0.0000	0.0000
P_{28}	0.0000	0.0000	0.0000	0.0000
	0.0000	1.0000	1.5000	0.0650
	0.0000	1.5000	3.3500	0.0975
	0.0000	0.0650	0.0975	0.0042
P_{27}	0.9958	1.4937	1.5385	0.1449
	1.4937	3.2405	3.8078	0.2823
	1.5385	3.8077	5.6270	0.3214
	0.1449	0.2823	0.3214	0.0253
P_{24}	6.6588	6.6533	5.8020	0.7108
	6.6533	7.9940	7.7599	0.8022
	5.8020	7.7599	8.9241	0.7529
	0.7108	0.8022	0.7529	0.0822
P_{21}	7.8748	7.5319	6.4262	0.8132
	7.5319	8.6296	8.2119	0.8763
	6.4262	8.2119	9.2456	0.8056
	0.8132	0.8763	0.8056	0.0908
P_{19}	7.9430	7.5754	6.4540	0.8184
	7.5754	8.6573	8.2296	0.8796
	6.4540	8.2296	9.2570	0.8077
	0.8184	0.8796	0.8077	0.0912
P_1	7.9502	7.5796	6.4564	0.8189
	7.5796	8.6597	8.2310	0.8799
	6.4564	8.2310	9.2578	0.8079
	0.8189	0.8799	0.8079	0.0913

back has to be suitably modified, as in [8.3, 8.4, 8.6, 8.7]. The following representation relies on [8.8].

Now, the case is considered where *constant reference variables* w(k) and *disturbances* n(k) arise. These constant signals can be generated from definite initial values v(0) and y(0) by a *reference variable model* (c.f. Fig. 8.4)

$$
\left.
\begin{aligned}
v(k+1) &= A v(k) + B y(k) \\
y(k+1) &= y(k) \\
w(k) &= C v(k)
\end{aligned}
\right\}
\tag{8.2.1}
$$

Here dim $y(k)$ = dim $w(k)$. For example, if a step change of the reference variable $w(k) = w_0 1(k)$ is required for a process with proportional action, $y(0) = y_0$ is taken, then element (B, A, C) of the reference variable model is excited by a step and generates a response $w_y(k)$ at its output.

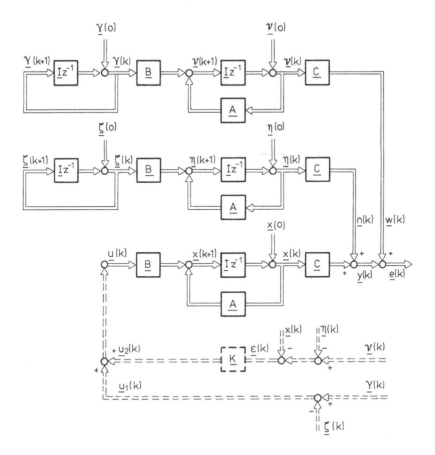

Fig. 8.4. State model of a linear process with a reference and disturbance variable model for the generation of constant reference signals $w(k)$ and disturbances $n(k)$. The resulting state controller is illustrated by dashed lines

The remaining difference to a step function $w(k)$

$$w(k) - w_y(k) = w_v(k)$$

can be generated by suitable initial values $v(0)$. Selecting appropriate values of y_0 and $v(0)$ a step in $w(k)$ can be modelled. The values of y_0 depend on w_0 and on the gain factors of the model (A, B, C). If the gains are unity, $y_0 = w_0$. Giving other values of $v(0)$ can generate other signals $w(k)$ which converge also for $k \to \infty$ to a fixed value w_0.

As with the reference variable model a *disturbance variable model* is assumed to be

$$\left. \begin{aligned} \eta(k+1) &= A\,\eta(k) + B\,\zeta(k) \\ \zeta(k+1) &= \zeta(k) \\ n(k) &= C\eta(k) \end{aligned} \right\} \tag{8.2.2}$$

with dim $\zeta(k) = \dim n(k)$ (see Fig. 8.4). Based on the initial values $\zeta(0)$ and $\eta(0)$ a constant $n(k)$ can be generated.

Using other structures of the preset model with state variables $\gamma(k)$ of $\zeta(k)$ other classes of external signals can be modelled. For example a linearly increasing signal of first-order can be obtained from

$$\left.\begin{array}{l} v(k+1) = A\, v(k) + B\, \gamma_2(k) \\ \gamma_2(k+1) = \gamma_2(k) + \gamma_1(k) \\ \gamma_1(k+1) = \gamma_1(k). \end{array}\right\} \tag{8.2.3}$$

The state variables of the process model, the reference variable and the disturbance model are combined into an error state variable $\varepsilon(k)$, so that for the control deviation $e(k)$ one obtains

$$e_w(k) = w(k) - y(k) - n(k)$$

$$= C\,[v(k) - x(k) - \eta(k)] = C\,\varepsilon(k). \tag{8.2.4}$$

The overall model is then described by

$$\begin{bmatrix} \varepsilon(k+1) \\ \gamma(k+1) - \zeta(k+1) \end{bmatrix} = \begin{bmatrix} A & B \\ 0 & I \end{bmatrix} \begin{bmatrix} \varepsilon(k) \\ \gamma(k) - \zeta(k) \end{bmatrix} - \begin{bmatrix} B \\ 0 \end{bmatrix} u(k). \tag{8.2.5}$$

It is assumed that the state variables $\varepsilon(k)$, $\gamma(k)$ and $\zeta(k)$ are completely measurable. The manipulated variable is now separated into two terms

$$u(k) = u_1(k) + u_2(k). \tag{8.2.6}$$

If

$$u_1(k) = \gamma(k) - \zeta(k) \tag{8.2.7}$$

is taken, the influences of $\gamma(k)$ and of $\zeta(k)$ on $y(k)$ are completely compensated. Therefore, $u_1(k)$ controls the effect of the initial values $\gamma(0)$ and $\zeta(0)$. This corresponds to ideal feedforward control action. Then the partial control $u_2(k)$ has to control the effects of the initial values

$$\varepsilon(0) = v(0) - x(0) - \eta(0) \tag{8.2.8}$$

by using a feedback state controller. Hence one is left with the synthesis of an optimal state feedback controller for the initial values of the system

$$\varepsilon(k+1) = A\,\varepsilon(k) - B\,u_2(k). \tag{8.2.9}$$

Here, the corresponding quadratic performance control

$$I_\varepsilon = \varepsilon^T(N)Q\,\varepsilon(N) + \sum_{k=0}^{N-1} [\varepsilon^T(k)Q\,\varepsilon(k) + u_2^T(k)R\,u_2(k)] \tag{8.2.10}$$

has to be minimized. However, this problem is like that solved in section 8.1, so that the optimal time-invariant state controller given by (8.1.33) is

$$u_2(k) = K \varepsilon(k). \qquad (8.2.11)$$

Unlike the control of initial values $x(0)$ (section 8.1), this state controller controls the system with initial values $\varepsilon(0)$.

The overall controller for constant disturbances is now composed of

— a state controller for initial values $\varepsilon(0) \rightarrow u_2(k)$
— a feedforward control of $y(k)$ and $\zeta(k) \rightarrow u_1(k)$

and therefore with (8.2.6), (8.2.7) and (8.2.11)

$$u(k) = [KI]\begin{bmatrix} \varepsilon(k) \\ y(k) - \zeta(k) \end{bmatrix}. \qquad (8.2.12)$$

This controller is illustrated in Fig. 8.4 by dashed lines.

The overall model of (8.2.5) can be represented by using the abbreviations

$$x^*(k) = \begin{bmatrix} \varepsilon(k) \\ y(k) - \zeta(k) \end{bmatrix}; \qquad A^* = \begin{bmatrix} A & B \\ 0 & I \end{bmatrix}$$

$$B^* = \begin{bmatrix} -B \\ 0 \end{bmatrix}; \qquad C^* = [C \quad 0] \qquad (8.2.13)$$

as follows

$$x^*(k+1) = A^*x^*(k) + B^*u(k) \qquad (8.2.14)$$

$$e_w(k) = C^*x^*(k) \qquad (8.2.15)$$

and the constant state controller becomes

$$u(k) = K^*x^*(k) \qquad (8.2.16)$$

with

$$K^* = [K I]. \qquad (8.2.17)$$

If each output variable $y_i(k)$ has a corresponding command variable $w_i(k)$ and disturbance $n_i(k)$ then

$$\dim x^* = \dim \varepsilon + \dim y = m + r$$

Hence A^* is a $(m+r) \times (m+r)$-matrix.

The characteristic equation of the optimal state control system for stepwise external disturbances is

$$\det [zI - A^* - B^*K^*]$$

$$= \det \begin{bmatrix} zI - A + BK & 0 \\ 0 & (z-1)I \end{bmatrix}$$

$$= \det [z I - A + B K] \; (z-1)^q = 0 \tag{8.2.18}$$

if $\dim \gamma(k) = \dim \zeta(k) = \dim u_1(k) = \dim u(k) = q$.

Assuming the given models for external disturbances, the control system for q manipulated variables aquires q poles at $z=1$, i.e. q "integral actions", which compensate for the offsets. For a single input/single output system the characteristic equation becomes

$$\det [zI - A + bk^T](z-1) = 0 \tag{8.2.19}$$

Hence it has $(m+1)$ poles. Note, that the characteristic equation has poles at $z=1$ as the system is open-loop with respect to the additional state variables $\gamma(k)$ and $\zeta(k)$.

Another way to counteract the offsets due to external disturbances and reference variables is to add a pole at $z=1$ to the process model. For a process with output

$$y(k) = C x(k)$$

this corresponds to the introduction of additional state variables

$$q(k+1) = q(k) + F y(k) \tag{8.2.20}$$

i.e. by adding summing or "integration" action terms. Here F is a diagonal matrix. Using rectangular integration the diagonal elements of F can be interpreted as the ratio of the sample time T_0 to the integration time T_{Ii}, $f_{ii} = T_0/T_{Ii}$. The state controller is then of the form [8.4], [8.5]

$$u(k) = -K x(k) - K_I q(k). \tag{8.2.21}$$

Unlike adding reference variable and disturbance models, the addition of an integration element has the disadvantage that constant disturbances $u_v(k)$ at the process input cannot be controlled without offset. Furthermore, the integration constants f_{ii} can be chosen arbitrarily, because they are not determined by the criterion (8.1.2) used for the controller design. Therefore this integral action does not suit the design requirement for a state controller having a closed-form solution. Better suited methods will be treated in section 8.7.3.

If the state variables $x^*(k)$ cannot be measured exactly as assumed for this section, they have to be reconstructed by a process model. Only then do the advantages of a state control system as in Fig. 8.4 become clear, see section 8.7.2.

8.3 State Controllers with a Given Characteristic Equation

A controllable process with state equation

$$x(k+1) = A x(k) + B u(k) \tag{8.3.1}$$

may be changed by state feedback

$$u(k) = - K x(k) \tag{8.3.2}$$

such that the poles of the total system

$$x(k+1) = [A - B K]x(k) \tag{8.3.3}$$

or the coefficients of the characteristic equation

$$\det [zI - A + B K] = 0 \tag{8.3.4}$$

are given. The procedure of pole assignment of a state feedback control system will be discussed for a single input/single output process. The state equation is transformed into the controller canonical form, as in Table 3.3.

$$x(k+1) = \begin{bmatrix} 0 & 1 & \cdots & 0 \\ \vdots & \vdots & & \vdots \\ 0 & 0 & \cdots & 1 \\ -a_m & -a_{m-1} & \cdots & -a_1 \end{bmatrix} x(k) + \begin{bmatrix} 0 \\ \vdots \\ 0 \\ 1 \end{bmatrix} u(k). \tag{8.3.5}$$

The state feedback is

$$u(k) = -k^T x(k) = - [k_m k_{m-1} \cdots k_1] x(k). \tag{8.3.6}$$

(8.3.6) and (8.3.5) yield

$$x(k+1) = \begin{bmatrix} 0 & 1 & \cdots & 0 \\ \vdots & \vdots & \ddots & \vdots \\ 0 & 0 & & 1 \\ (-a_m - k_m) & (-a_{m-1} - k_{m-1}) & \cdots & (-a_1 - k_1) \end{bmatrix} x(k) \tag{8.3.7}$$

Hence, the characteristic equation is

$$\det (z I - A + b k^T]$$
$$= (a_m + k_m) + (a_{m-1} + k_{m-1})z + \cdots + (a_1 + k_1)z^{m-1} + z^m$$
$$= \alpha_m + \alpha_{m-1}z + \cdots + \alpha_1 z^{m-1} + z^m = 0. \tag{8.3.8}$$

This equation leads to following relationships for the coefficients of the feedback vector k^T

$$k_i = \alpha_i - a_i \quad i = 1, 2, ..., m. \tag{8.3.9}$$

The coefficients k_i are zero if the characteristic equation of the process is not changed by the feedback $\alpha_i = a_i$. The coefficients k_i increase if the coefficients α_i of the closed system are changed in a positive direction compared with the coefficients a_i. Therefore, the manipulated variable $u(k)$ becomes increasingly active when the coefficients α_i are changed by the controller to be further away from the coefficients a_i. The effect of a state feedback on the eigen behaviour can therefore be clearly interpreted.

In *pole assignment* design the poles z_i, $i = 1, \ldots, m$

$$\det [z\,I - A + b\,k^T] = (z-z_1)(z-z_2) \ldots (z-z_m) \tag{8.3.10}$$

are first determined appropriately, and then the α_i are calculated and the k_i are determined from (8.3.9). The multivariable system case is treated for example in [2.19]. See also section 3.5. It should be noted, however, that by placing the poles only single eigen oscillations are determined. As the cooperation of these eigen oscillations and the response to external disturbances is not considered, design methods in which the control and manipulated variables are directly evaluated are generally preferred. The advantage of the above method of pole assignment lies in the especially clear interpretation of the changes of the single coefficients α_i of the characteristic equations caused by the feedback constants k_i. As has been shown in chapter 7, the characteristic equation for *deadbeat control* is $z^m = 0$. (8.3.8) shows that this occurs when $\alpha_i = 0$. This state deadbeat control will be considered in section 8.5.

8.4 Modal State Control

In section 8.3 the state representation in controller canonical form has been used for pole assignment. By changing the coefficients k_i of the feedback matrix, the coefficients α_i of the characteristic equation could be directly influenced. k_i influences only α_i, so that the k_i and α_j of $j \neq i$ are decoupled. In this section the pole assignment of a state control system is considered for state representation in diagonal form. Since then the k_i directly influence the eigenvalues (modes) z_i this is called *modal control*. For multivariable systems modal control was originally described by [8.9]. A more detailed treatment can be found in e.g. [5.17], [8.10].

A linear time invariant process with multi input and output signals

$$x(k+1) = A\,x(k) + B\,u(k) \tag{8.4.1}$$

$$y(k) \quad = C\,x(k) \tag{8.4.2}$$

is considered, and it is assumed that the eigenvalues are distinct. This process is now linearly transformed using (3.6.33)

$$x_t(k) = T\ x(k) \tag{8.4.3}$$

into the form

$$\begin{aligned}
x_t(k+1) &= A_t x_t(k) + B_t u(k)\\
y(k) &= C_t x_t(k)
\end{aligned} \tag{8.4.4}$$

The system

$$A_t = TAT^{-1} = \begin{bmatrix} z_1 & & \cdots & 0 \\ & z_2 & & \\ \vdots & & \ddots & \vdots \\ 0 & & \cdots & z_m \end{bmatrix} = \Lambda \tag{8.4.5}$$

is now in diagonal form and

$$B_t = T\ B \tag{8.4.6}$$

$$C_t = C\ T^{-1} \tag{8.4.7}$$

The characteristic equation of the original process model is

$$\det \{z\ I - A\} = 0 \tag{8.4.8}$$

and the transformed process equation becomes

$$\begin{aligned}
\det\ [zI - \Lambda] &= \det\ [zI - T\ A\ T^{-1}]\\
&= \det\ T\ [zI - A]\ T^{-1} = \det\ [zI - A]\\
&= (z-z_1)\ (z-z_2)\ \cdots\ (z-z_m) = 0.
\end{aligned} \tag{8.4.9}$$

The diagonal elements of Λ are the eigenvalues of (8.4.4) which are identical with the eigenvalues of (8.4.1). The transformation matrix T can be determined as follows [5.17]. (8.4.5) is written in the form

$$A\ T^{-1} = T^{-1}\Lambda. \tag{8.4.10}$$

Then T^{-1} is partitioned into columns

$$T^{-1} = [v_1 v_2 \cdots v_m] \tag{8.4.11}$$

and it follows that

$$A[v_1 v_2 \ldots v_m] = [v_1 v_2 \ldots v_m] \begin{bmatrix} z_1 & & \cdots & 0 \\ & z_2 & & \vdots \\ \vdots & & \ddots & \\ 0 & & \cdots & z_m \end{bmatrix}$$

$$= [z_1 v_1 \quad z_2 v_2 \ldots z_m v_m]. \tag{8.4.12}$$

For each column one obtains

$$A v_i = z_i v_i \qquad i = 1, 2, \ldots, m \qquad (8.4.13)$$

or

$$[z_i I - A] v_i = 0. \qquad (8.4.14)$$

(8.4.14) yields m equations for the m unknown vectors v_i which have m elements $v_{i1}, v_{i2}, \ldots, v_{im}$. If the trivial solution $v_i = 0$ is excluded, there is no unique solution of the equation system (8.4.14). For each i only the direction and not the magnitude of v_i is fixed. The magnitude can be chosen such that in B_t or C_t only elements 0 and 1 appear [2.19]. The vectors v_i are called *eigenvectors*. For a single input/single output process, the state representation in diagonal form corresponds to the partial fraction expansion of the z-transfer function for m different eigenvalues

$$G(z) = c_t^T [zI - A_t]^{-1} b_t$$

$$= \frac{c_{t1} b_{t1}}{z - z_1} + \frac{c_{t2} b_{t2}}{z - z_2} + \ldots + \frac{c_{tm} b_{tm}}{z - z_m}. \qquad (8.4.15)$$

This equation also shows the b_{ti} and c_{ti} cannot be uniquely determined. If e.g. the $b_{ti} = 1$ is chosen then the c_{ti} can be calculated by

$$c_{ti} = [(z - z_i) G(z)]_{z = z_i}$$

(c.f. (3.7.5)). A may have complex conjugate z_i as well as real z_i. If complex conjugate elements must be avoided see e.g. [5.17].

Also in (8.4.4) the control vector $u(k)$ is transformed using

$$u_t(k) = B_t u(k) \qquad (8.4.16)$$

yielding

$$x_t(k+1) = A x_t(k) + u_t(k) \qquad (8.4.17)$$

This process is now extended by the feedback

$$u_t(k) = - K_t x_t(k) \qquad (8.4.18)$$

This results in the homogeneous vector difference equation

$$x_t(k+1) = [A - K_t] x_t(k). \qquad (8.4.19)$$

If K_t is also diagonal

$$K_t = \begin{bmatrix} k_{t1} & & \cdots & 0 \\ & k_{t2} & & \\ \vdots & & \ddots & \vdots \\ 0 & & \cdots & k_{tm} \end{bmatrix} \qquad (8.4.20)$$

the characteristic equation becomes

$$\det [zI - (\Lambda - K_t)]$$
$$= (z - (z_1 - k_{t1}))(z - (z_2 - k_{t2}) \cdots (z - (z_m - k_{tm})) = 0. \qquad (8.4.21)$$

The eigenvalues z_i of the process can now be shifted independently from each other by suitable selection of k_{ti}, since both (8.4.17) and (8.4.18) have diagonal form. This results in m first-order decoupled control loops.

The realizable control vector $u(k)$ is calculated from (8.4.16) and (8.4.6), which yields

$$u(k) = B_t^{-1} u_t(k) = B^{-1} T^{-1} u_t(k). \qquad (8.4.22)$$

Because the inverse of B is involved, this matrix has to be regular, i.e. it has to be quadratic and

$$\det B \neq 0$$

This means that the m eigenvalues of A or Λ can only be influenced independently from each other if m different manipulated variables are at the disposal. The process order and the number of the manipulated variables therefore have to be equal.

The block-diagram structure of the modal state control is shown in Fig. 8.5. The state variables $x_t(k)$ are decoupled using the transformation of (8.4.3); this is called *modal analysis*. The transformed control vector $u_t(k)$ is generated in "separate paths" by the modal controller K_t. The realizable control vector $u(k)$ is then formed by a back transformation (modal synthesis). As regular control matrices B are rare, the modal control described above can rarely be applied.

For multivariable processes of order m with p inputs, i.e. control matrices B of order $(m \times p)$, only p eigenvalues can be influenced independently by a diagonal $(p \times p)$ controller matrix K. The remaining $m - p$ eigenvalues remain unchanged [8.11], [5.17].

A linear process with one input, $p = 1$, is now considered. Its transformed equation is, from (8.4.3)

$$x_t(k+1) = A_t x_t(k) + b_t u(k) \qquad (8.4.23)$$

$$y(k) = c_t^T x_t(k) \qquad (8.4.24)$$

with

$$b_t = T\, b \qquad (8.4.25)$$

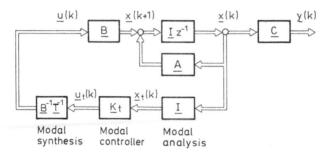

Fig. 8.5. Block diagram of a modal state control system

$$c_t^T = c^T T^{-1}.$$ (8.4.26)

For controlling this transformed process a state feedback

$$u(k) = - k^T x_t(k) = - [k_1 \ k_2 \ ... \ k_m] x_t(k)$$ (8.4.27)

is assumed. Substituting into (8.4.23) leads to

$$x_t(k+1) = [A - b_t k^T] x_t(k) = F x_t(k).$$ (8.4.28)

If the b_{ti} can all be selected to be one, this becomes

$$F = \begin{bmatrix} (z_1 - k_1) & -k_2 & \cdots & -k_m \\ -k_1 & (z_2 - k_2) & & -k_m \\ \vdots & & \ddots & \vdots \\ -k_1 & -k_2 & \cdots & (z_m - k_m) \end{bmatrix}.$$ (8.4.29)

The single state variables are no longer decoupled and the eigenvalues of F change compared with A in a coupled way so that the supposed advantage of modal control cannot be attained by assumption (8.4.27). If, however, a single state variable x_{tj} is fed back by

$$u(k) = - k_j x_{tj}(k)$$

one eigenvalue can be changed independently of the other invariant process eigenvalues

$$F = \begin{bmatrix} z_1 & \cdots & -k_j & \cdots & 0 \\ \vdots & & \vdots & & \vdots \\ 0 & \cdots & (z_j - k_j) & \cdots & 0 \\ \vdots & & \vdots & & \vdots \\ 0 & \cdots & -k_j & \cdots & z_m \end{bmatrix}.$$ (8.4.30)

The characteristic equation is now

$$\det [zI - F] = (z-z_1) \ldots (z-(z_j-k_j)) \ldots (z-z_m). \qquad (8.4.31)$$

It was assumed that the eigenvalues of A are distinct. If there are multiple eigenvalues one must use Jordan matrices instead of diagonal matrices A [8.10].

As modal state control for controller design considers only pole placement, the remarks made at the end of section 8.3 are also valid here. Note, however, that the modal control can advantageously be applied to distributed parameter processes with several manipulated variables [8.11], [3.10], [8.12].

8.5 State Controllers for Finite Settling Time (Deadbeat)

A controllable process of order m with one manipulated variable is considered

$$x(k+1) = A\, x(k) + bu(k). \qquad (8.5.1)$$

It was shown in section 3.6 that this process can be driven from any initial state $x(0)$ to the zero state $x(N)=0$ in $N=m$ steps. The required manipulated variable can be calculated using (3.6.63). It can also be generated by a state feedback

$$u(k) = -k^T x(k) \qquad (8.5.2)$$

Then one obtains

$$x(k+1) = [A - b\, k^T]x(k) = R\, x(k) \qquad (8.5.3)$$

or

$$x(1) = R\, x(0)$$
$$x(2) = R^2 x(0)$$
$$.$$
$$.$$
$$.$$
$$x(N) = R^N x(0).$$

For $x(N)=0$ it follows that:

$$R^N = 0. \qquad (8.5.4)$$

The characteristic equation of the closed system is

$$\det [z\, I - R] = \alpha_m + \alpha_{m-1}z + \ldots + \alpha_1 z^{m-1} + z^m = 0. \qquad (8.5.5)$$

From the Cayley-Hamilton theorem a quadratic matrix satisfies its own characteristic equation, i.e.

$$\alpha_m I + \alpha_{m-1}R + \ldots + \alpha_1 R^{m-1} + R^m = 0. \qquad (8.5.6)$$

(8.5.4) is also satisfied by

$$N = m$$
$$\alpha_1 = \alpha_2 = ... = \alpha_m = 0.$$

The characteristic equation therefore becomes

$$\det[z\,I - R] = z^m = 0. \tag{8.5.7}$$

A multiple pole of order m at $z=0$ characterizes a control loop with deadbeat behaviour (c.f. (7.1.21)).

If the process is given in controllable canonical form as in (8.3.5) then the deadbeat state controller becomes with (8.3.9) and, therefore, with $k_i = -a_i$

$$u(k) = [a_m a_{m-1} ... a_1]x(k). \tag{8.5.8}$$

Using the controllable canonical form all state variables x_i are multiplied by a_i in the state controller and are fed back with opposite sign to the input as in the state model of the process itself, (c.f. Fig. 3.17). Therefore, for m-times, zeros of the first state variable are generated one after another and are shifted forward to the next state variables, so that for $k=m$ all states become zero [2.19]. The deadbeat controller DB(v) described in section 7.1 drives the process from any initial state $x(0) = 0$ in m steps to a constant output of

$$y(m) = y(m-1) = ... = y(\infty)$$

for a constant input of

$$u(m) = u(m+1) = ... = u(\infty).$$

This controller is therefore an "output-deadbeat controller".

The deadbeat controller described in this section drives the process from any initial state $x(0) \neq 0$ to a final state $x(m)=0$. Therefore, it can be called a "deadbeat-controller".

As both systems have the same characteristic equation $z^m=0$, they have the same behaviour for the same initial disturbances $x(0)$, because behaviour after an initial disturbance depends only on the characteristic equation. Hence the deadbeat controller DB(v) also drives the system into the zero state $x(m)=0$ after m steps for any initial state $x(0)$.

8.6 State Observers

As the state variables $x(k)$ are not directly measurable for many processes they have to be determined using measurable quantities. Now we consider the dynamic process

$$x(k+1) = A\,x(k) + B\,u(k)$$
$$y(k) = C\,x(k) \tag{8.6.1}$$

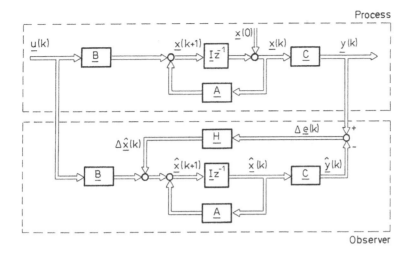

Fig. 8.6. A dynamic process and its state observer

where we assume that only the input vector $u(k)$ and the output vector $y(k)$ can be measured without error and that the state variables $x(k)$ are observable. A model with the same structure is connected in parallel to the process model, as in Fig. 8.6. State corrections $\Delta \hat{x}(k+1)$ are generated by feeding back the difference between the output signals of the model and the process

$$\Delta e(k) = y(k) - \hat{y}(k) \tag{8.6.2}$$

weighted by a matrix H, to the state variables $\hat{x}(k+1)$, so that after convergence the model states follow the process states. This model is called a Luenberger *state observer* [8.13], [8.14]; it is an *identity observer* if a complete model of the process is applied.

The constant observer feedback matrix H must be chosen such that $\hat{x}(k+1)$ approaches $x(k+1)$ asymptotically as $k \to \infty$. Fig. 8.6 leads to the following observer equation

$$\hat{x}(k+1) = A \hat{x}(k) + B u(k) + H \Delta e(k)$$

$$= A \hat{x}(k) + B u(k) + H[y(k) - C \hat{x}(k)]. \tag{8.6.3}$$

or the state error

$$\tilde{x}(k+1) = x(k+1) - \hat{x}(k+1) \tag{8.6.4}$$

and with (8.6.1) and (8.6.3) it follows that

$$\tilde{x}(k+1) = [A - H C]\tilde{x}(k) \tag{8.6.5}$$

Hence, a homogeneous vector difference equation arises. The state variable vector error depends only on the initial error $\tilde{x}(0)$ and is independent of the input $u(k)$. For convergence to a zero state

$$\lim_{k \to \infty} \tilde{x}(k) = 0,$$

(8.6.5) has to be asymptotically stable. Therefore, the characteristic equation

$$\det [z\, I - A + H\, C] = (z - z_1)(z - z_2) \cdots (z - z_m)$$

$$= \gamma_m + \gamma_{m-1} z + \cdots + z^m = 0 \qquad (8.6.6)$$

may have only roots within the unit circle $|z_i| < 1$, $i = 1, 2, \ldots, m$, i.e. only stable observer poles. The poles can be influenced by proper choice of the matrix H. The assignment of this feedback can proceed as in the determination of the state controller matrix. As

$$\det W = \det W^T$$

one obtains

$$\det [z\, I - A + H\, C] = \det [z\, I - A^T + C^T H^T]. \qquad (8.6.7)$$

By comparing the characteristic equation of the state controller with the corresponding process, (3.1.37), it follows that one must modify the design equations of the state controllers

$$A \to A^T; \quad B \to C^T; \quad K \to H^T \qquad (8.6.8)$$

to include the feedback matrix H of the observer. Instead of the process

$$x(k+1) = A\, x(k) + B\, u(k)$$

$$y(k) \quad = C\, x(k)$$

with feedback

$$u(k) = -K\, x(k)$$

to determine the observer poles the "transposed auxiliary process" is introduced

$$\xi(k+1) = A^T \xi(k) + C^T \vartheta(k) \qquad (8.6.9)$$

with feedback

$$\vartheta(k) = -H^T \xi(k) \qquad (8.6.10)$$

so that the equations of the state controller design can be used. The observer matrix H can then be determined, e.g. by:

Determination of the Characteristic Equation According to Section 8.3

For scalar $u(k)$ and $y(k)$ and therefore $H \to h$, the observer equation is

$$\hat{x}(k+1) = [A - h c^T]\hat{x}(k) + b u(k) + h y(k). \tag{8.6.11}$$

Here the canonical form is suitable, due to the controllable canonical form in section 8.3, so that

$$\hat{x}(k+1) = \begin{bmatrix} 0 & 0 & \cdots & (-a_m - h_m) \\ 1 & 0 & & (-a_{m-1} - h_{m-1}) \\ \vdots & & \ddots & \vdots \\ 0 & 0 & 1 & (-a_1 - h_1) \end{bmatrix} \hat{x}(k) + bu(k) + \begin{bmatrix} h_m \\ h_{m-1} \\ \vdots \\ h_1 \end{bmatrix} y(k) \tag{8.6.12}$$

and analogously to (8.3.9) one obtains

$$h_i = \gamma_i - a_i \quad i = 1, 2, ..., m. \tag{8.6.13}$$

where γ_i are the coefficients of (8.6.6) and which must be given.

Deadbeat Behaviour

Choosing

$$h_i = -a_i \tag{8.6.14}$$

the observer attains a minimal settling time and therefore has deadbeat behaviour (c.f. section 8.5).

Minimization of a Quadratic Performance Criterion

In (8.6.8), h can be chosen such that the quadratic criterion

$$I_B = \xi^T(N) Q_b \xi(N) + \sum_{k=0}^{N-1} [\xi^T(k)Q_b\tilde{\xi}(k) + \vartheta^T(k)R_b\vartheta(k)] \tag{8.6.15}$$

is minimized as described in section 8.1. The resulting recursive solution equations are (c.f. (8.1.30) and (8.1.31))

$$\left. \begin{aligned} H_{N-j}^T &= [R_b + C P_{N-j+1}C^T]^{-1}C P_{N-j+1}A^T \\ P_{N-j} &= Q_b + A P_{N-j+1}A^T - H_{N-j}[R_b + C P_{N-j+1}C^T]H_{N-j}^T \end{aligned} \right\} \tag{8.6.16}$$

Hence the behaviour of the observer and therefore its eigenbehaviour can be chosen in several ways. In practical observer realization, the noise that is always present in the output variable limits the attainable settling time.

For the observers described so far, all observer state variables $\hat{x}(k)$ are calculated.

However, some state variables can often be determined directly, e.g. by the output variable $y(k)$, so that reduced-order observers can be derived (see section 8.8).

Fig. 8.6 shows that the state variables of the observer follow the process states with no lag for changes in $u(k)$. However, they lag for initial values $x(0)$. Disturbances affecting the output variable $y(k)$ lead to errors in the observer states. For stochastic disturbances, therefore, states are to be estimated with state estimation methods which leads to Kalman filters, see chapter 22.

8.7 State Controllers with Observers

For the state controllers described in sections 8.1 to 8.5 it was assumed that the state of the process can be measured exactly and completely. However, this is not the case for most processes, so that instead of the actual process state variables $x(k)$ (c.f. (8.1.33)), state variables reconstructed by the observer have to be used by the control law. Hence

$$u(k) = -K \hat{x}(k) \tag{8.7.1}$$

The resulting block diagram is shown in Fig. 8.7.

8.7.1 An Observer for Initial Values

The complete state of the closed control system follows from (8.1.1), (8.6.3) and (8.7.1)

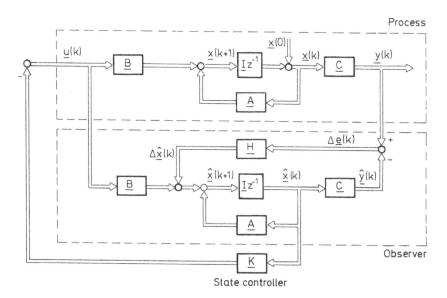

Fig. 8.7. A state controller with an observer for initial values $x(0)$

$$\begin{bmatrix} x(k+1) \\ \hat{x}(k+1) \end{bmatrix} = \begin{bmatrix} A & -BK \\ HC & A-BK-HC \end{bmatrix} \begin{bmatrix} x(k) \\ \hat{x}(k) \end{bmatrix} \qquad (8.7.2)$$

$$y(k) = C x(k). \qquad (8.7.3)$$

$x(k)$ and $\hat{x}(k)$ influence each other. From (8.1.36) the eigenbehaviour of the process and the state feedback without observer is described by

$$x(k+1) = [A - BK]x(k)$$

and from (8.6.5) the eigenbehaviour of the observer by

$$x(k+1) - \hat{x}(k+1) = \tilde{x}(k+1) = [A - HC]\tilde{x}(k)$$

For comparison with these equations, (8.7.2) is transformed by

$$\begin{bmatrix} x(k+1) \\ \tilde{x}(k+1) \end{bmatrix} = \begin{bmatrix} I & 0 \\ I & -I \end{bmatrix} \begin{bmatrix} x(k+1) \\ \hat{x}(k+1) \end{bmatrix} \qquad (8.7.4)$$

giving

$$T^{-1} = \begin{bmatrix} I & 0 \\ I & -I \end{bmatrix} = T$$

$$\begin{bmatrix} x(k+1) \\ \tilde{x}(k+1) \end{bmatrix} = \underbrace{\begin{bmatrix} A-BK & BK \\ 0 & A-HC \end{bmatrix}}_{A^*} \begin{bmatrix} x(k) \\ \tilde{x}(k) \end{bmatrix} \qquad (8.7.5)$$

$$y(k) = [C \quad 0] \begin{bmatrix} x(k) \\ \tilde{x}(k) \end{bmatrix}. \qquad (8.7.6)$$

The eigenbehaviour of this system depends on the characteristic equation

$$\det [z I - A^*] = \det [z I - A + BK] \det [z I - A + HC] = 0. \qquad (8.7.7)$$

Therefore, the poles of the control system with state controller and observer are the poles of the control system without observer together with the poles of the observer. The poles of the controller and the poles of the observer can be determined independently, as they do not influence each other. This is the result of the so-called *separation theorem*. However, it should be noted that, of course, the time behaviour of $x(k)$ is influenced by the observer poles as can be seen from (8.7.5). An observer introduces additional poles and therefore, additional lags into the control system. If an identity observer is used, (8.7.7) shows that the control loop with a process of order m has $2\,m$ poles and consequently is of order $2\,m$.

The lagging influence of an observer can be clearly shown if a state controller for deadbeat behaviour, section 8.5, is combined with an observer with deadbeat behaviour, (8.6.14). Here the characteristic equation becomes

$$\det \left[z\,I - A^*\right] = z^m\,z^m = z^{2m}. \tag{8.7.8}$$

Hence the steady state after a non-zero initial value is reached only after $2\,m$ sampling steps, and not in m steps as with the deadbeat controller according to section 8.5. In this case the deadbeat controller discussed in chapter 6 is faster than the state controller with observer. Section 8.7.2 and 8.8 show how the observer lags can be partially overcome.

8.7.2 Observer for External Disturbances

Section 8.2 described how constant external disturbances can be generated from initial states using extended state models. In order to control constant disturbances the manipulated variable vector $u(k)$ given by (8.2.12) has to be generated by the state vector $\varepsilon(k)$ through a state controller, and by the state vectors $y(k)-\zeta(k)$ using a proportional feedforward control. However, as these state variables are not measurable in general, they have to be determined by an observer. Here it will be assumed, as in the preceding section, that the input variables $u(k)$ and the output variables $y(k)$ can be measured without errors. The overall system described by (8.2.14) and (8.2.15) with the abbreviations (8.2.13) uses an extended state vector $x^*(k)$ containing all state variables of the process and disturbance models. The observer for this state vector is

$$\hat{x}^*(k+1) = A^*\hat{x}^*(k) + B^*u(k) + H^* \left[e_w(k) - C^*\hat{x}^*(k)\right]. \tag{8.7.9}$$

For processes with m state variables and r outputs, the observer feedback matrix H^* has dimension $(m+r)xr$, and can be determined by the methods given in section 8.6.

Using the observer, the controller equation becomes, from (8.2.16)

$$u(k) = K^*\hat{x}^*(k). \tag{8.7.10}$$

Fig. 8.8 shows the resulting block diagram for the case of constant changes of the command variable. Fig. 8.9 shows the corresponding scheme with the abbreviations used in (8.2.13). For changes of disturbances $n(k)$ or reference variables $w(k)$ the unknown state variables $\hat{x}^*(k)$ are first determined by the observer such that the assumed disturbance or reference variable model generates exactly $n(k)$ or $w(k)$ at the output.

Fig. 8.8 indicates that for the manipulated variable $u_1(k)$ results

$$u_1(k) = \hat{y}(k)$$
$$\hat{y}(k+1) = \hat{y}(k) + H^*_{(rxr)}\Delta e(k) \tag{8.7.11}$$

Here $H^*_{(rxr)}$ is the corresponding part of H^*. For a single output variable $(r=1)$ it follows that

$$u_1(z) = \frac{z^{-1}}{1-z^{-1}}h_{m+1}\Delta e(z). \tag{8.7.12}$$

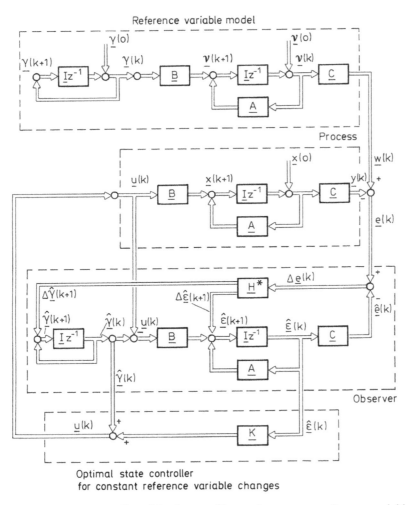

Fig. 8.8. State controller with observer (drawn for constant reference variable changes). Compare with Fig. 8.4.

Therefore the observed state variable $\hat{y}(k)$ leads to a $(m+1)$th state feedback and to the feedforward control part $u_1(k)$ which acts as summation (integration) with respect to $\Delta e(k)$. By introducing the state variable $y(k)$ (or $\zeta(k)$) an "integral action term" is generated in the observer so that offsets disappear. The integration constant is the same as the feedback constant h_{m+1} of the observer. This constant is determined automatically by the observer design and is therefore appropriate for the state control system design method.

Now the eigenbehaviour of the hypothetical, extended process (8.2.14) with extended observer (8.7.9) and state controller (8.7.10) is considered. Analogously to (8.7.5) and (8.7.6) it follows that

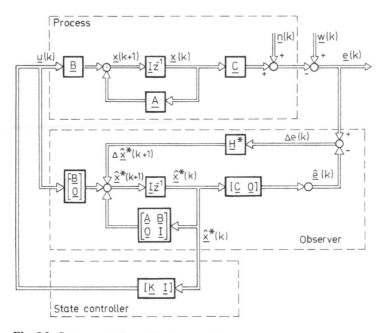

Fig. 8.9. State controller with observer for constant reference variables $w(k)$ and disturbances $n(k)$

$$\begin{bmatrix} x^*(k+1) \\ \tilde{x}^*(k+1) \end{bmatrix} = \begin{bmatrix} A^* + B^* K^* & -B^* K^* \\ 0 & A^* - H^* C^* \end{bmatrix} \begin{bmatrix} x^*(k) \\ \tilde{x}(k) \end{bmatrix} \tag{8.7.13}$$

$$y(k) = [C^* \quad 0] \begin{bmatrix} x^*(k) \\ \tilde{x}^*(k) \end{bmatrix}.$$

The characteristic equation is therefore

$$\det [z\,I - A^* - B^* K^*]\det [z\,I - A^* + H^* C^*] = 0. \tag{8.7.14}$$

The controller and observer poles appear separately in this hypothetical system. The real behaviour is obtained by coupling the process of (8.6.1) with the extended observer (8.7.9), and the controller of (8.7.10) and (8.2.4)

$$\begin{bmatrix} x(k+1) \\ \hat{x}^*(k+1) \end{bmatrix} = \begin{bmatrix} A & BK^* \\ -H^* C & (A^* + B^* K^* - H^* C^*) \end{bmatrix} \begin{bmatrix} x(k) \\ \hat{x}^*(k) \end{bmatrix}$$

$$+ \begin{bmatrix} 0 \\ H^* \end{bmatrix} (w(k) - n(k)). \tag{8.7.15}$$

Hence, after z-transformation

$$[z\,I - A]x(z) = B\,K^* \hat{x}^*(z) \tag{8.7.16}$$

$$[z I - (A^* + B^* K^* - H^* C^*)]\hat{x}^*(z) = - H^* C^* x(z) + H^*(w(z) - n(z))$$
(8.7.17)

and after elimination of $\hat{x}^*(z)$ one obtains

$$[z I - A + B K^* [z I - (A^* + B^* K^* - H^* C^*)]^{-1} H^* C^*] x(z)$$

$$= B K^* [z I - (A^* + B^* K^* - H^* C^*)^{-1} H^* (w(z) - n(z)).$$
(8.7.18)

With $y(z) = C x(z)$ the reference of the disturbance behaviour can be calculated. In both cases the poles are given by

$$\det [z I - A + B K^* [z I - (A^* + B^* K^* - H^* C^*)]^{-1} H^* C^*] = 0.$$
(8.7.19)

Controller and observer poles no longer appear separately. The dynamics of the observer are part of the reference or the disturbance behaviour of the overall system.

It should be mentioned that a state controller with observer can be designed so that the dynamics of the observer do not influence the reference behaviour. Here the reference variable $w(k)$ is introduced only after the observer and the state controller using feedforward control with $u(k)$ [2.19]. However, a direct comparison of the control variables and the reference variables can no longer be made, the parameters of the feedforward control element depend on the process parameters, and offsets can arise if the process parameters are not exactly known or changing. The design described above does not have these disadvantages as the control deviations $e(k)$ are formed before the observer and the state controller using a direct comparison of the reference and the control variables. Therefore, no offset can occur for step changes in the external disturbances because the poles at $z = 1$. The resulting delays of the observer can be partially overcome as shown in the following.

In Fig. 8.10 the behaviour of the controlled and manipulated variable for the process III and the state controller designed for external disturbances is shown for a step change in the disturbance $n(k)$. No offset in the controlled variable arises. The manipulated variable, however, is changed only after one sampling interval. This delay occurs as all changes $\Delta e(k)$ of the control deviation have to pass one element z^{-1} in the observer before a change in the manipulated variable can happen (see Fig. 8.8).

Therefore the observer, unlike the optimized parameter and compensating controller, causes the manipulated variable to have undesirable delays. The initial delay, however, can be avoided. Then for the observer of Fig. 8.8 all state variables are reconstructed, although one state variable can be measured directly if it is identical to the output variable $\hat{y}(k)$. This is the case for a state representation in observer canonical form (see Fig. 3.19). Instead of the delayed part of the control

$$u_m(k) = k_m \hat{x}_m(k) = k_m \hat{y}(k)$$

one uses the undelayed signal

$$u'_m(k) = k_m y(k)$$
(8.7.20)

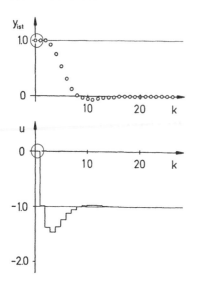

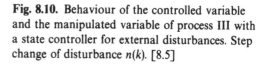

Fig. 8.10. Behaviour of the controlled variable and the manipulated variable of process III with a state controller for external disturbances. Step change of disturbance $n(k)$. [8.5]

(see Figures 8.11 and 8.12). An undelayed output variable $\hat{y}(k)$ can also be included by using a reduced-order observer (see section 8.8).

Example 8.2

As an example, the design of a state controller described in section 8.2 with an observer for external disturbances described in section 8.7.2 is now considered for process III (c.f. section 5.4.1). The state representation of test process III is chosen in observable canonical form. As the process has a deadtime $d=1$, if follows from (3.6.44) and (3.6.45)

$$
\begin{bmatrix} x_1(k+1) \\ x_2(k+1) \\ x_3(k+1) \\ x_4(k+1) \end{bmatrix} = \begin{bmatrix} 0 & 0 & -a_3 & b_3 \\ 1 & 0 & -a_2 & b_2 \\ 0 & 1 & -a_1 & b_1 \\ 0 & 0 & 0 & 0 \end{bmatrix} \begin{bmatrix} x_1(k) \\ x_2(k) \\ x_3(k) \\ x_4(k) \end{bmatrix} + \begin{bmatrix} 0 \\ 0 \\ 0 \\ 1 \end{bmatrix} u(k)
$$

or

$$ x_d(k+1) = A_d x_d(k) + b_d u(k) $$

$$ y(k) = [0\,0\,1\,0] \begin{bmatrix} x_1(k) \\ x_2(k) \\ x_3(k) \\ x_4(k) \end{bmatrix} $$

or

$$ y(k) = c_d^T x_d(k). $$

A block diagram is shown in Fig. 8.11. The observer for step changes of external variables $w(k)$ or $n(k)$ has its parameters given by (8.7.9) and (8.2.13)

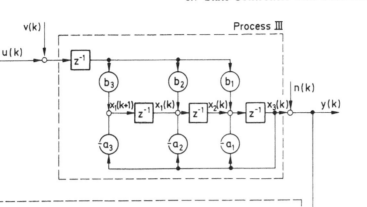

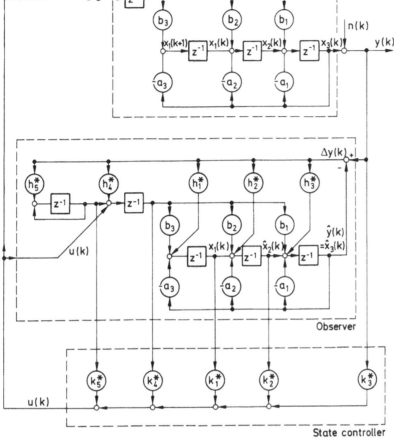

Fig. 8.11. Block diagram for process III with state controller and observer for external disturbances, with bypass of the initial delay

$$
A^* = \begin{bmatrix} 0 & 0 & -a_3 & b_3 & \vline & 0 \\ 1 & 0 & -a_2 & b_2 & \vline & 0 \\ 0 & 1 & -a_1 & b_1 & \vline & 0 \\ 0 & 0 & 0 & 0 & \vline & 1 \\ \hline 0 & 0 & 0 & 0 & \vline & 1 \end{bmatrix} \qquad b^* = \begin{bmatrix} 0 \\ 0 \\ 0 \\ 1 \\ \hline 0 \end{bmatrix}
$$

$$c^{*T} = [0 \quad 0 \quad 1 \quad 0 \ \vline \ 0]$$

$$h^{*T} = [h_1^* \quad h_2^* \quad h_3^* \quad h_4^* \ \vline \ h_5^*].$$

The calculation of the feedback constants h^* of the observer is performed by minimizing the quadratic performance criterion of (8.6.15) for the transposed observer, (8.6.9) and (8.6.10), by

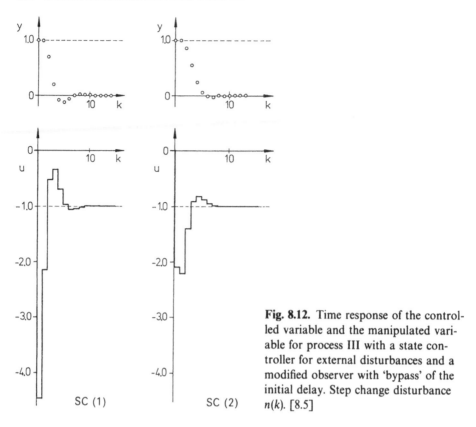

Fig. 8.12. Time response of the controlled variable and the manipulated variable for process III with a state controller for external disturbances and a modified observer with 'bypass' of the initial delay. Step change disturbance $n(k)$. [8.5]

using the recursive solution of the matrix Riccati equation (8.6.16). If the weighting coefficients are chosen to be:

$$Q_B = \begin{bmatrix} 1 & 0 & 0 & 0 & 0 \\ 0 & 1 & 0 & 0 & 0 \\ 0 & 0 & 1 & 0 & 0 \\ 0 & 0 & 0 & 1 & 0 \\ 0 & 0 & 0 & 0 & 25 \end{bmatrix} ; \; r_B = 5$$

then

$$h^{*T} = [0{,}061 \; -0{,}418 \; 0{,}984 \; 1{,}217 \; 1{,}217].$$

results. When designing the state controller of (8.2.17), (8.2.11) or (8.1.33) via a recursive solution of the matrix Riccati equation, (8.1.30) and (8.1.31), the state variable weighting Q of (8.10.4) in the quadratic performance criterion (8.1.2) is chosen such that only the controlled variable $y(k)$ is given unit weight, i.e.

$$Q = c_d c_d^T = \begin{bmatrix} 0 & 0 & 0 & 0 \\ 0 & 0 & 0 & 0 \\ 0 & 0 & 1 & 0 \\ 0 & 0 & 0 & 0 \end{bmatrix}$$

For a weighting of $r=0.043$ on the manipulated variable the state controller becomes:

$$k^{*T} = [4,828 \quad 5,029 \quad 4,475 \quad 0,532 \quad 1,000]$$

and for $r=0.18$

$$k^{*T} = [2,526 \quad 2,445 \quad 2,097 \quad 0,263 \quad 1,000]$$

Fig. 8.12 shows the time responses of the resulting control and the manipulated variables.

The algorithms required for one sample of this process of total order $m+d=4$ are
— observer output error

$$\Delta e(k-1) = y(k-1) - \hat{x}_3(k-1)$$

— state estimates

$$\begin{aligned}
\hat{x}_1(k) &= -a_3\hat{x}_3(k-1) + b_3\hat{x}_4(k-1) + h_1^*\Delta e(k-1) \\
\hat{x}_2(k) &= \hat{x}_1(k-1) - a_2\hat{x}_3(k-1) + b_2\hat{x}_4(k-1) + h_2^*\Delta e(k-1) \\
\hat{x}_3(k) &= \hat{x}_2(k-1) - a_1\hat{x}_3(k-1) + b_1\hat{x}_4(k-1) + h_3^*\Delta e(k-1) \\
\hat{x}_4(k) &= \hat{x}_5(k-1) + u(k-1) + h_4^*\Delta e(k-1) \\
\hat{x}_5(k) &= \hat{x}_5(k-1) + h_5^*\Delta e(k-1)
\end{aligned}$$

— manipulated variable
— without 'bypass' of the observer delay z^{-1} for $y(k)$
$$u(k) = k_1^*\hat{x}_1(k) + k_2^*\hat{x}_2(k) + k_3^*\hat{x}_3(k) + k_4^*\hat{x}_4(k) + k_5^*\hat{x}_5(k)$$
— with 'bypass' of the observer delay z^{-1} for $y(k)$
$$u(k) = k_1^*\hat{x}_1(k) + k_2^*\hat{x}_2(k) + k_3^*y(k) + k_4^*\hat{x}_4(k) + k_5^*\hat{x}_5(k).$$

The following calculations are required for each sample

15 multiplications
16 summations.

Here $k_5^* = 1$ is taken into account. For realization in a digital computer 8 shift operations of the variables must be added.

8.7.3 Introduction of Integral Action Elements into the State Controller

Besides introducing an *integral acting part into the observer*, section 8.7.2 and adding a *parallel integral acting part* to the state controller, section 8.2, it is possible to *introduce integral acting parts into the process model* and the closed-loop design for the such extended model. Here the integral acting part can be located in front or after the process model [5.26].

Process Model with Subsequent Integral Action

The process model is extended by an integral acting part (see Fig. 8.13)

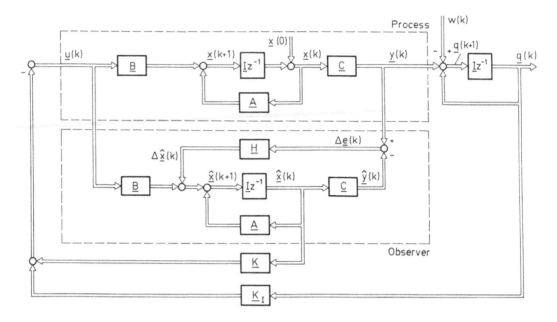

Fig. 8.13. State controller with observer and integral acting part located after the process

$$x(k+1) = A\,x(k) + B\,u(k)$$

$$q(k+1) = I\,q(k) + w(k) - C\,x(k)$$

(8.7.21)

Hence the extended model becomes

$$\begin{bmatrix} x(k+1) \\ q(k+1) \end{bmatrix} = \begin{bmatrix} A & 0 \\ -C & I \end{bmatrix}\begin{bmatrix} x(k) \\ q(k) \end{bmatrix} + \begin{bmatrix} B \\ 0 \end{bmatrix}u(k) + \begin{bmatrix} 0 \\ I \end{bmatrix}w(k)$$

or

$$\varphi(k+1) = A_\varphi\,\varphi(k) + B_\varphi u(k) + E\,w(k)$$

$$y(k) \quad = C_\varphi\varphi(k)$$

(8.7.22)

The influence of the reference variables $w(k)$ is considered a state variable disturbance which will be neglected for the controller design. As treated in section 8.1 and 8.3, a state controller

$$K_\varphi = [\,K\,K_I\,]$$

(8.7.23)

can be designed, using A_φ and B_φ.
  The control law then is

$$u(k) = -\,K\,\hat{x}(k) - K_I\,q(k)$$

(8.7.24)

In order to avoid a one sample lag by the I-element when changing the reference variables $w(k)$, the integral acting part can be replaced by the following PI-algorithm

$$u_I(k) = L_{iq}(k+1) + (I - L_I)q(k) \qquad (8.7.25)$$

compare section 8.7.4.

Process Model with Preset PI-Element

The process model is extended by a preset proportionally integral acting part, see Fig. 8.14

$$\begin{aligned}
x(k+1) &= A\,x(k) + B\,u(k) \\
q(k+1) &= I\,q(k) + u_\varepsilon(k) \\
u(k) &= q(k+1) = q(k) + u_\varepsilon(k)
\end{aligned} \Biggr\} \qquad (8.7.26)$$

(The PI-structure is used to avoid the shift delay of an I-element).
   Hence the extended process model becomes

$$\begin{bmatrix} x(k+1) \\ q(k+1) \end{bmatrix} = \begin{bmatrix} A & B \\ 0 & I \end{bmatrix} \begin{bmatrix} x(k) \\ q(k) \end{bmatrix} + \begin{bmatrix} B \\ I \end{bmatrix} u_\varepsilon(k) \qquad (8.7.27)$$

or

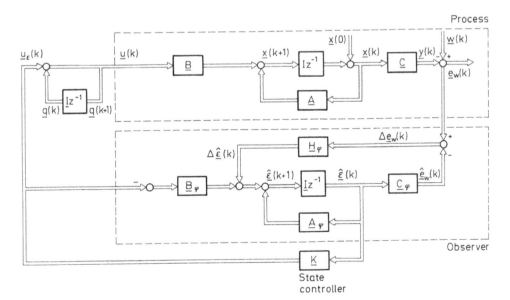

Fig. 8.14. State controller with observer and preset proportional-plus-integral element

$$\left.\begin{array}{l} \varphi(k+1) = A_\varphi\,\varphi(k) + B_\varphi u_\varepsilon(k) \\[2mm] y(k) = [C\ \ 0]\left[\begin{array}{c} x(k) \\ q(k) \end{array}\right] = C_\varphi\varphi(k) \end{array}\right\} \qquad (8.7.28)$$

The block diagram shows that because of the reference value $w(k)$ the control difference

$$e_w(k) = w(k) - y(k) \qquad (8.7.29)$$

is used as output variable and therefore reference variable for the observer. By using the reference variable model according to (8.2.1) the reference variable $w(k)$ can be written as

$$\left.\begin{array}{l} v(k+1) = A_\varphi v(k) + B_\varphi \gamma(k) \\[2mm] w(k) = C_\varphi v(k) \end{array}\right\} \qquad (8.7.30)$$

leading to

$$e_w(k) = C_\varphi[v(k) - \varphi(k)] = C_\varphi\varepsilon(k). \qquad (8.7.31)$$

For the state design as overall model according to (8.7.30) and (8.7.28) the process model is

$$\left.\begin{array}{l} \varepsilon(k+1) = A_\varphi\varepsilon(k) + B_\varphi[\gamma(k) - u_\varepsilon(k)] \\[2mm] e_w(k) = C_\varphi\varepsilon(k) \end{array}\right\} \qquad (8.7.32)$$

For a step change in $w(k) = 1(k)$, $y(k)$ is set only as initial value $y(0) = y_0$. Hence $y(k) = 0$ for $k \geq 1$, because of the I-element already included in (8.7.30). $y(0)$ is an initial value disturbance and therefore the models for the observer and controller design become

$$\left.\begin{array}{l} \varepsilon(k+1) = A_\varphi\varepsilon(k) - B_\varphi u_\varepsilon(k) \\[2mm] e_w(k) = C_\varphi\varepsilon(k). \end{array}\right\} \qquad (8.7.33)$$

Hence the observer equation

$$\hat{\varepsilon}(k+1) = A_\varphi\hat{\varepsilon}(k) - B_\varphi u_\varepsilon(k) + H_\varphi[e_w(k) - C_\varphi\hat{\varepsilon}(k)] \qquad (8.7.34)$$

and the control law

$$u_\varepsilon(k) = K\,\hat{\varepsilon}(k) \qquad (8.7.35)$$

where K is calculated for the model (8.7.33) according to the methods in section 8.1 and 8.3.

Unlike Fig. 8.13, the I-element also exists in the observer, if this design is used.

8.7.4 Measures to Minimize Observer Delays

In section 8.7.2 it was already shown how for the controller the shift delay which is caused by the observer can be avoided. This was accomplished by using the measured output variable $y(k)$ instead of the observed $\hat{y}(k)$. If a minimization of this delay is also desired for the other states, one can proceed according to Fig. 8.15 and include the state variable $\hat{x}(k+1)$ which has been predicted because of the measured difference $y(k) - \hat{y}(k)$ [5.26].

$$u(k) = -K[L\hat{x}(k+1) + [I - L]\hat{x}(k)]$$

$$L = diag\ (l_i);\ l_i \leq 1$$

With identical weighting L of all states

$$L = lI \quad l \leq 1$$

this is valid for the most simple case.

The less precise the prediction of $\hat{x}(k+1)$, e.g. because of disturbance signals or numerical influence, the smaller the weighting factors l_i should be chosen.

8.8 State Observer of Reduced-Order

In the *identity observer* of section 8.6 all state variables $x(k)$ are reconstructed. However, if some state variables are directly measurable they need not be

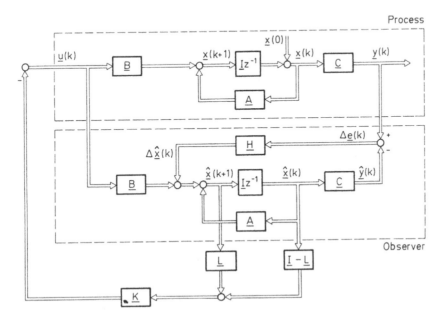

Fig. 8.15. Minimization of the observer delay through additional feedback of $x(k+1)$

calculated. For example, in an mth-order process with one input and one output one state variable can directly be calculated from the measurable output $y(k)$, so that only $(m-1)$ state variables have to be determined by the observer. An observer whose order is lower than the order of the process model is called an *observer of reduced-order* (see [8.13, 8.15]). The following provides the derivation of a reduced observer using [8.15] and [2.19]. The process is assumed to be

$$x(k+1) = A\,x(k) + B\,u(k) \qquad (8.8.1)$$

$$y(k) = C\,x(k) \qquad (8.8.2)$$

The dimensions of the vectors are

$$x(k) : (m \times 1)$$
$$u(k) : (p \times 1)$$
$$y(k) : (r \times 1).$$

For r independent measurable output variables $y(k)$, r state variables can be calculated directly. Therefore, the state vector $x(k)$ is partitioned into a directly calculable part $x_b(k)$ and an observable part $x_a(k)$

$$\begin{bmatrix} x_a(k+1) \\ x_b(k+1) \end{bmatrix} = \begin{bmatrix} A_{11} & A_{12} \\ A_{21} & A_{22} \end{bmatrix} \begin{bmatrix} x_a(k) \\ x_b(k) \end{bmatrix} + \begin{bmatrix} B_1 \\ B_2 \end{bmatrix} u(k) \qquad (8.8.3)$$

$$y(k) = [C_1 \quad C_2] \begin{bmatrix} x_a(k) \\ x_b(k) \end{bmatrix}. \qquad (8.8.4)$$

The directly calculable state vector $x_b(k)$ is replaced by $y(k)$. Then, a state vector v is obtained by the linear transformation

$$v = \begin{bmatrix} x_a \\ y \end{bmatrix} = Tx = \begin{bmatrix} T_{11} & T_{12} \\ T_{21} & T_{22} \end{bmatrix} \begin{bmatrix} x_a \\ x_b \end{bmatrix}. \qquad (8.8.5)$$

It follows from (8.8.4) that $T_{21} = C_1$ and $T_{22} = C_2$. As $x_a(k)$ remains unchanged, $T_{11} = I$, and is independent of $x_b(k)$, $T_{12} = 0$. Therefore, the transformation matrix is given by

$$T = \begin{bmatrix} I & 0 \\ C_1 & C_2 \end{bmatrix} \qquad (8.8.6)$$

and the transformed process is

$$v(k+1) = A_t v(k) + B_t u(k) \qquad (8.8.7)$$

$$y(k) = C_t v(k) \qquad (8.8.8)$$

Hence it follows from (3.2.32) that

$$A_t = T A T^{-1}$$
$$B_t = T B$$
$$C_t = C T^{-1} = [0 \; I]. \qquad \Bigg\} \qquad (8.8.9)$$

If (8.8.7) is partitioned as in (8.8.3) we obtain

$$x_a(k+1) = A_{t11}x_a(k) + A_{t12}y(k) + B_{t1}u(k) \qquad (8.8.10)$$

$$y(k+1) = A_{t21}x_a(k) + A_{t22}y(k) + B_{t2}u(k). \qquad (8.8.11)$$

In (8.8.10) an identity observer of order $(m-r)$ is now used

$$\hat{x}_a(k+1) = A_{t11}\hat{x}_a(k) + A_{t12}y(k) + B_{t1}u(k) + H\,e_t(k) \qquad (8.8.12)$$

(c.f. (8.6.3)). With the identity observer of order m the output error given by (8.6.2) is used for the error between the observer and the process. However, as the reduced-order observer does not calculate explicitly $\hat{y}(k)$, and furthermore as $y(k)$ contains no information concerning $\hat{x}_a(k)$, the observer error $e_t(k)$ must be redefined. Here, (8.8.11) can be used because it yields an equation error $e_t(k)$ if $\hat{x}_a(k)$ is not yet adapted to the measurable variables $y(k)$, $y(k+1)$ and $u(k)$

$$e_t(k) = y(k+1) - \underbrace{A_{t21}\hat{x}_a(k) - A_{t22}y(k) - B_{t2}u(k)}_{\hat{y}(k+1)}; \qquad (8.8.13)$$

From (8.8.12) and (8.8.13) the observer becomes

$$\hat{x}_a(k+1) = A_{t11}\hat{x}_a(k) + A_{t12}y(k) + B_{t1}u(k)$$
$$+ H[y(k+1) - A_{t22}y(k) - B_{t2}u(k) - A_{t21}\hat{x}_a(k)] \qquad (8.8.14)$$

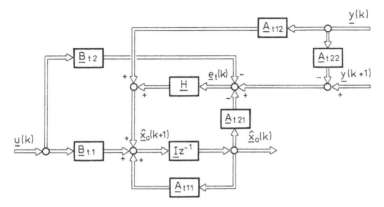

Fig. 8.16. Block diagram of a reduced-order observer given by (8.8.14)

Its block diagram is shown in Fig. 8.16. In (8.8.14), $y(k+1)$ is unknown at time k. Fig. 8.16 shows that, with respect to the output $\hat{x}_a(k)$, nothing changes if the signal path

$$\hat{x}_a(z) = H z^{-1} y(z) z$$

is replaced by

$$\hat{x}_a(z) = H y(z).$$

However, one must introduce new observer state variables

$$\hat{\mu}(k) = \hat{x}_a(k) - H y(k) \qquad (8.8.15)$$

From Fig. 8.17 the observer of reduced-order is

$$\hat{\mu}(k+1) = A_{t11}\hat{\mu}(k) + [A_{t12} + A_{t11}H] y(k) + B_{t1}u(k)$$
$$+ H[-A_{t21}H y(k) - A_{t22}y(k) - B_{t2}u(k) - A_{t21}\hat{\mu}(k)]$$

or

$$\hat{\mu}(k+1) = [A_{t11} - H A_{t21}]\hat{\mu}(k)$$
$$+ [A_{t12} - H A_{t22} + A_{t11}H - H A_{t21}H]y(k)$$
$$+ [B_{t1} - H B_{t2}]u(k). \qquad (8.8.16)$$

The state variables to be observed are obtained from

$$\hat{x}_a(k) = \hat{\mu}(k) + H y(k) \qquad (8.8.17)$$

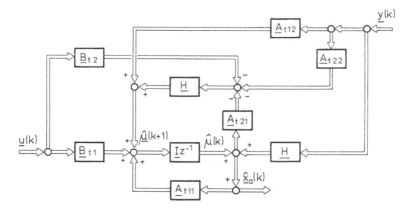

Fig. 8.17. Modified block diagram of a reduced-order observer given by (8.8.15)

and finally the overall state vector is

$$\hat{v}(k) = \begin{bmatrix} \hat{x}_a(k) \\ y(k) \end{bmatrix} = \begin{bmatrix} I & H \\ 0 & I \end{bmatrix} \begin{bmatrix} \hat{\mu}(k) \\ y(k) \end{bmatrix}. \qquad (8.8.18)$$

Considering the estimation error of the reduced observer

$$\tilde{x}_a(k+1) = x_a(k+1) - \hat{x}_a(k+1) \qquad (8.8.19)$$

it follows from (8.8.10) to (8.8.13) that

$$\tilde{x}_a(k+1) = [A_{t11} - H A_{t21}]\tilde{x}_a(k). \qquad (8.8.20)$$

Compared with the identity observer, in this homogeneous error difference equation one takes the transformed part system matrix A_{t11} which belongs to the state vector x_a, rather than the system matrix A. Instead of the output matrix C one takes the transformed part system matrix A_{t21}, yielding the relationship between $x_a(k)$ and $y(k+1)$ given by (8.8.11). The characteristic equation of the reduced-order observer is written as

$$\det[z I - A_{t11} + H A_{t21}] = (z-z_1)(z-z_2)\dots(z-z_{m-r}) = 0. \qquad (8.8.21)$$

The observer poles can be determined using section 8.6.

The advantages of a reduced-order observer compared with the identity observer of section 8.6 are in its lower order (reduced by the number r of the directly measurable output variables) and in the use of current output variables $y(k)$ with no delay, hence avoiding the delays described in section 8.7. These advantages, however, are offset by the increased computations required. Moreover, an additional equation, (8.8.17), arises in the calculation of the state variables to be estimated. In digital computer realizations reduced-order observers are usually preferred if relatively many state variables are directly measurable. In all other cases, e.g. for processes with only one measurable input and output, the identity observer modified according to section 8.7 is better as the design is simpler and more transparent and the potential saving of operational calculations is comparatively small.

8.9 State Variable Reconstruction

In order to determine the state variables, the previously considered state observers feedback the output error $\Delta e(k)$ thus forming a dynamic system which can be suitably designed by the choice of the feedback constant H.

Another possibility to calculate the state variables directly from the measured signals follows from the observability equations in section 3.6.8, [2.19, 8.17, 8.18].

For a single input/single output process according to (3.6.68)

$$y_m = Q_B x(k) + S u_m \qquad (8.9.1)$$

For an observable system, see (3.6.69)

$$x(k) = Q_B^{-1} [y_m - S u_m] \tag{8.9.2}$$

results. If the state representation is now chosen according to the observable canonical form, it follows that, due to the setting of A and C, c.f. Table 3.3 and (3.6.68)

$$Q_B = I \tag{8.9.3}$$

and from (8.9.2)

$$x(k) = y_m - S u_m \tag{8.9.4}$$

In order to have in y_m and u_m rather past values instead of future ones, this equation is written for m sample steps earlier

$$x(k-m) = y'_m - S u'_m \tag{8.9.5}$$

with

$$\left.\begin{array}{l} y_m'^T = [y(k-m) \ y(k-m+1) \ ... \ y(k-1)] \\ u_m'^T = [u(k-1) \ u(k-2) \ ... \ u(k-m)] \end{array}\right\} \tag{8.9.6}$$

From this, with (3.6.49) or (3.6.59), the required state at time instant k can be determined due to the known input signal and the known state $x(k-m)$ as

$$x(k) = A^m x(k-m) + \sum_{i=1}^{m} A^{i-1} b u(k-i) \tag{8.9.7}$$

Introducing (8.9.5) leads to

$$\hat{x}(k) = A^m [y'_m - S u'_m]$$

$$+ A^{m-1} b u(k-m) + A^{m-2} b u(k-m+1) + ...$$

$$+ A b u(k-2) + b u(k-1) \tag{8.9.8}$$

Thus the state variable vector $x(k)$ can be calculated using the measured signals $u(k-1), \ldots, u(k-m)$ and $y(k-1), \ldots, y(k-m)$. Using the terms introduced in section 3.6.8 the states thus have become "reconstructed". This process is called *state reconstruction*.

Applying the controllability matrix Q_s, (3.6.62) can be written as follows

$$\hat{x}(k) = A^m [y'_m - S u'_m] + Q_s u'_m \tag{8.9.9}$$

Compare [2.19]. For deadtime processes (8.9.8) becomes

$$\hat{x}(k+d) = A^{m+d}[y'_m - Su'_m]$$
$$+ A^{m+d-1}b\, u(k-d-m) + A^{m+d-2}b\, u(k-d-m+1)$$

with   (8.9.10)

$$u'_m = [u(k-d-1)\ u(k-d-2)\ ...\ u(k-d-m)]$$ (8.9.11)

The principle of state reconstruction is one of calculating the m states $x(k-m)$ for the former time instant $(k-m)$ from the m input/output signal values $(k-m)$ to $(k-1)$ with (8.9.5). Then the state $x(k)$ is determined in front from the (initial) value $x(k-m)$ and the known input signal according to (8.9.7). Thus, after each sample step all states are calculated anew from the input- and output signals.

This state determination method hence is not recursive and does not require feedback design as the state variable observer does. Therefore, design parameter need not to be determined.

For multivariable processes state reconstruction becomes

$$\hat{x}(k) = A^m x(k-m) + \sum_{i=1}^{m} A^{i-1}B\, u(k-i)$$ (8.9.12)

where m is the maximum order of the under systems 1 to r which are located in y, c.f. [8.18].

The state variable construction can also be given for the observable canonical form. Then, however, Q_B^{-1} is written in (8.9.8) using (8.9.2) as (8.9.3) is not valid [5.26]. In addition, it was shown in [5.26] that state reconstruction corresponds to reduced-order observers with minimal settling time (deadbeat behaviour). This therefore results in a relatively high sensitivity for higher frequency disturbance signals in the measured output signal so that prefiltering of the output signal should always be provided.

8.10 Choice of Weighting Matrices and Sample Time

If the controller is not designed for finite settling time (deadbeat behaviour), then comparatively many free parameters have to be carefully chosen in designing state controllers, compared with other structure optimal controllers. For a design with no performance criterion, either the coefficients of the characteristic equation (section 8.3) or the eigenvalues (section 8.4) have to be chosen. The quadratic optimal state controller assumes the choice of the weighting matrices Q for the state variables and R for the manipulated variables. The free parameters in the observer design have also to be chosen; these parameters again are either coefficients of the characteristic equation or weighting matrices Q_b and R_b of a quadratic performance criterion (section 8.6). In addition, the parameters of the assumed external disturbance (section 8.2) and the sampling time also influence the design, as for other controllers. These relatively many free parameters for the design of state controllers mean on the one hand there is an especially great ability for the adaptation to the process to be controlled; whilst on the other hand there always exists a certain freedom of choice in selection if there are too many parameters. Therefore, the design of state controllers is rarely performed in one step, but rather is performed iteratively, using evaluations of the resulting control behaviour as described in chapter 4. In special cases, however, the number of free parameters can be reduced.

8.10.1 Weighting Matrices for State Controllers and Observers

In the state controller design using the performance criterion of (8.1.2), the manipulated variables can be weighted separately, so that R can be taken to be the diagonal matrix

$$R = \begin{bmatrix} r_1 & 0 & \cdots & 0 \\ 0 & r_2 & & \\ \vdots & & \ddots & \vdots \\ 0 & & \cdots & r_p \end{bmatrix} \qquad (8.10.1)$$

To give a positive R, the elements r_i must be positive for all $i = 1, 2, \ldots, p$. In special cases where R can be positive semi-definite certain r_i can be zero (see section 8.1).

Individual state variables can also be weighted in general by a diagonal matrix Q.

$$Q = \begin{bmatrix} q_1 & 0 & \cdots & 0 \\ 0 & q_2 & & \\ \vdots & & \ddots & \vdots \\ 0 & & \cdots & q_r \end{bmatrix} \qquad (8.10.2)$$

Q has to be positive definite, so that $q_i > 0$ (see section 8.1).

If only the output variables $y(k)$ have to be weighted by a diagonal matrix L as for parameter-optimized controllers using the quadratic performance criterion (5.2.6) with (8.1.2) it then follows that

$$x^T(k) \; Q \, x(k) \; = \; y^T(k) \; L \; y(k)$$

and as with (8.1.3)

$$y^T(k) \; L \; y(k) \; = \; x^T(k) \; C^T L \; C \, x(k)$$

it also follows that

$$Q = C^T L \; C. \qquad (8.10.3)$$

Hence, for a single input/output process with $L = 1$

$$R = r$$
$$Q = c \, c^T. \qquad (8.10.4)$$

Note, that in optimal state controller design, the squared manipulated variable $u^2(k)$ is weighted by r, unlike (5.2.6) where $\Delta u^2(k) = [u(k) - u(\infty)]^2$ is weighted. With proportional acting processes, however, a state controller has $u(\infty) = 0$ giving $u(k) = \Delta u(k)$ (because of the assumed initial value disturbance $x(0)$), so that there is no difference, in principle.

In observer design using the quadratic performance criterion (8.6.15) for the transposed system (8.6.9) and (8.6.10), the weighting matrices Q_b and R_b can be assumed for state controllers. Generally, however, one would design a faster observer in comparison with the process, i.e. the elements of R_b are chosen smaller than those of Q_b.

8.10.2 Choice of the Sample Time

In choosing an appropriate sample time T_0 there seems to exist a further possibility compared with other controllers by taking into account the analytical relations between the optimal control performance and the process parameters. From (8.1.32) one then obtains

$$I_{opt}(T_0) = x^T(0) \; P_0(T_0) \; x(0)$$

where $P_0 = P$ is a stationary solution of the matrix Riccati equation (8.1.31). An analytical solution giving the cost function in terms of the sample time, however, is complicated [8.16]. It can be shown (for small sampling times T_0) that the costs $I_{opt}(T_0)$ increase monotonically with increasing T_0 if the process is controllable. This is the case for processes with reals poles, but not for processes with complex-conjugate poles if the sampling is close to the half-period of the natural frequencies [8.16]. Generally, the smallest costs are attained for $T_0 = 0$, i.e. for the continuous state controller and for very small sample times the control performance differs only little from the continuous case. Only for larger sample times does the control performance deteriorate significantly.

Current experience is that the sample time for state controllers can be chosen using the rules given in section 5.7 and 7.3.

With state controllers, as with deadbeat controllers, there is a relationship between the required manipulated variable changes and the sample time, if a disturbance has to be controlled completely in a definite time period. If the restricted range of the manipulated variable is given, the required sample time can be pre-determined [2.19].

9 Controllers for Processes with Large Deadtime

The controller designs of the preceding chapters automatically took the process deadtime into account. This was straightforward because deadtime can be simply included in process models using discrete-time signals—one of their advantages compared with models with continuous-time signals. Therefore, controllers for processes with deadtimes can be designed directly using the methods previously considered.

Processes with *small deadtime* compared with other process dynamics have already been discussed in some examples. A small deadtime can replace several small time constants or can describe a real transport delay. If, however, the *deadtime is large* compared with other process dynamics some particular cases arise which are considered in this chapter. Large deadtimes are exclusively pure transport delays. Therefore one has to distinguish between pure deadtime processes and those which have additional dynamics.

9.1 Models for Processes with Deadtime

A *pure deadtime* of duration $T_t = dT_0$ can be described by the transfer function

$$G_P(z) = \frac{y(z)}{u(z)} = bz^{-d} \quad d = 1,2, \dots \tag{9.1.1}$$

or the difference equation

$$y(k) = bu(k-d) \tag{9.1.2}$$

Here T_t is an integer multiple of the sample time T_0.
Processes with additional process dynamics have the transfer function

$$G_P(z) = \frac{y(z)}{u(z)} = \frac{B(z^{-1})}{A(z^{-1})} z^{-d} = \frac{b_1 z^{-1} + \dots + b_m z^{-m}}{1 + a_1 z^{-1} + \dots + a_m z^{-m}} z^{-d} \tag{9.1.3}$$

(or the corresponding difference equation —see (3.4.13)). (9.1.1) follows either by the replacement of d by $d' = d - 1$, $b_1 = b$ and $b_2, \dots, b_m = 0$ and $a_1, \dots, a_m = 0$

$$G_P(z) = b_1 z^{-1} z^{-d'} = b\, z^{-d} \tag{9.1.4}$$

or simply by taking $d=0$ in z^{-d}, $m=d$ in $B(z^{-1})$; $b_m=b_d=b$ and $b_1, \ldots, b_{m-1}=0$ and $a_1, \ldots, a_m=0$ (or $B(z^{-1})=z^{-d}$ and $A(z^{-1})=1$)

$$G_P(z) = b_m z^{-m} = b\, z^{-d}. \tag{9.1.5}$$

When considering the state representation of single input/single output processes there are several ways to add a deadtime (see section 3.6.4).

— Deadtime at the input

$$x(k+1) = A\, x(k) + b\, u(k-d)$$
$$y(k) = c^T x(k) \tag{9.1.6}$$

— Deadtime included in the system matrix A (c.f. (3.6.48))

$$x(k+1) = A\, x(k) + b\, u(k)$$
$$y(k) = c^T x(k) \tag{9.1.7}$$

— Deadtime at the output

$$x(k+1) = A\, x(k) + b\, u(k)$$
$$y(k) = c^T x(k-d) \text{ oder } y(k+d) = c^T x(k). \tag{9.1.8}$$

In all cases A can have different canonical forms (see section 3.6.3). For (9.1.6) and (9.1.8) A has dimension $m \times m$, but for (9.1.7) the dimension is $(m+d) \times (m+d)$. If the deadtime is included in the system matrix A, d more state variables must be taken into account. Though the input/output behaviour of all processes is the same, for state controller design the various cases must be distinguished as they lead to different controllers. Inclusion of the deadtime at the input or at the output depends on the technological structure of the process and in general can be easily determined. For a pure deadtime, including the deadtime within the controllable canonical form system matrix one obtains (3.6.43) with a state vector $x(k)$ of dimension d. In contrast, for (9.1.6) and (9.1.8), $A=a=0$ results and d has to be replaced by $d'=d-1$. In this case one can no longer use a state representation.

Note that as well as deadtime at the input or the output deadtime can also arise between the state variables. In the continuous case, vector difference differential equations of the form

$$x(t) = A_1 x(t) + A_2 x(t-T_{tA}) + B\, u(t)$$
$$y(t) = C\, x(t).$$

result. For discrete-time signals, however, these deadtime systems can be reduced to (9.1.7) by extending the state vector and the system matrix A.

9.2 Deterministic Controllers for Deadtime Processes

There is a substantial literature discussing the design of controllers for deadtime processes with *continuous signals*; see e.g. [9.1] to [9.7] and [5.14]. As well as parameter-optimized controllers with proportional and integral behaviour, the predictor controller proposed by Reswick [9.1] has been studied. Here, a very small settling time can be achieved by using a model of the deadtime process in the feedback of the controller. The disadvantages of this predictor controller and its modification (see [5.14]) have been its relatively high technological cost and high sensitivity of the difference between the assumed and the real deadtime. The general conclusion is that the use of proportional plus integral controllers which approximate the behaviour of a predictor controller are recommended. Digital computers have overcome the disadvantage of high cost. Therefore the control of processes with (large) deadtime but *discrete-time signals* is again discussed below.

9.2.1 Processes with Large Deadtime and Additional Dynamics

The parameter-optimized controllers of chapter 5 in the form of the structure adapted deadbeat controller of chapter 7, and the state controller of chapter 8, can be used for controlling processes with large deadtime. The structure of the parameter-optimized controllers iPC can be the same, except that the controller parameters may change considerably. The deadbeat controllers DB(v) and DB($v+1$) have already been derived for processes with deadtime. In the case of state controllers the addition of the deadtime in the state model plays a role. Some additions are made in this section to the controller design considered earlier.

Predictor Controller (PREC)

Firstly the predictor controller [9.1] which was specially designed for deadtime processes is considered for discrete-time signals. In the original derivation a transfer element $G_{ER}(z)$ was placed in parallel with the process $G_p(z)$ such that the overall transfer function is equal to the process gain K_p. The parallel transfer element $G_{ER}(z)$ is changed to an internal feedback of the controller $G_R(z)$ [5.14]. When $G'_R \to \infty$ (6.2.4) produces a cancellation controller with closed-loop transfer function

$$G_w(z) = \frac{1}{K_P} G_P(z) = \frac{1}{K_P} \frac{B(z^{-1})}{A(z^{-1})} z^{-d} \qquad (9.2.1)$$

The closed-loop transfer function is then equal to the process transfer function with unity gain — a reasonable requirement for deadtime processes. The *predictor controller* then becomes, using (6.2.4) and (9.2.1)

$$G_R(z) = \frac{1}{K_P - G_P(z)} = \frac{A(z^{-1})}{K_P A(z^{-1}) - B(z^{-1}) z^{-d}}$$

$$= \frac{1 + a_1 z^{-1} + ... + a_m z^{-m}}{K_P + K_P a_1 z^{-1} + ... + (K_P a_{1+d} - b_1) z^{-(1+d)} + ... - b_m z^{-(m+d)}}$$

$$(9.2.2)$$

The characteristic equation follows from (9.2.1) as

$$z^d z^m A(z^{-1}) = z^d[z^m + a_1 z^{m-1} + \ldots + a_{m-1} z + a_m] = 0. \qquad (9.2.3)$$

The characteristic equations of the process and the closed-loop are identical. Therefore, the predictor controller can only be applied to asymptotically stable processes. In order to decrease the large sensitivity of the predicator controller of Reswick (see section (9.2.2)) to changes in the deadtime, Smith introduced a modification [5.14, 9.2–9.4] so that the closed-loop behaviour

$$G_w(z) = \frac{1}{K_P} G_P(z) \cdot G'(z) \qquad (9.2.4)$$

has an additional delay $G'(z)$. The *modified predictor controller* is the

$$G_R(z) = \frac{G'(z)}{K_P - G_P(z) G'(z)} \qquad (9.2.5)$$

A first-order lag can be chosen for $G'(z)$.

State Controller (SC)

If the deadtime d cannot be included in the system matrix A (see (9.1.7)), but can only represent delayed inputs $u(k-d)$ or delayed state variables $x(k-d)$, as in (9.1.6) and (9.1.8), the advantage of state controllers in the feedback of all state variables cannot be achieved. In designing state controllers for processes with deadtime the deadtime should be included in the system matrix A. However, for large deadtimes the order $(m+d) \times (m+d)$ of the matrix A also becomes large. An advantage is that the design of the state controller does not change. As can be seen from (3.6.43) and (3.6.48) only A, b and c^T change compared with processes with time lags.

For structure-optimized input/output controllers for processes with deadtime the order of the numerator of the transfer function depends only on the process order m, and is equal to m for the controllers DB(v) and PREC and $(m-1)$ for the minimum variance controller MV3-d (see chapter 14). The deadtime influences only the numerator order and is equal to $(m+d)$ or $(m+d-1)$ (see Table 9.1)

Table 9.1. Non-zero parameters of deadbeat controllers, predictive controllers and minimum variance controllers (chapter 14) for processes of order $m \geq 1$ and deadtime d

	q_0	q_1	$\cdots$	q_{m-1}	q_m	p_0	p_1	$\cdots$	p_{1+d}	$\cdots$	p_{m+d-1}	p_{m+d}
DB (v)	×	×	×	×	×	×	—		×		×	×
PREC	×	×	×	×	×	×	×		×		×	×
MV3-d	×	×	×	—	×	×	×		×		×	—

9.2.2 Pure Deadtime Processes

Input/output controller (deadtime-, predictor- and PI-controller).
   The structure optimal *input/output controller* for pure deadtime processes

$$G_P(z) = \frac{y(z)}{u(z)} = b\,z^{-d} \tag{9.2.6}$$

is given by the corresponding controller equations for time lag processes of order m and deadtime d using (9.1.4) or (9.1.5). Both cancellation controllers — the *deadbeat controller* DB(v) and the *predictor controller* PREC — give the same transfer function

$$G_R(z) = \frac{1}{b}\frac{1}{1-z^{-d}} \tag{9.2.7}$$

or the difference equation

$$u(k) = u(k-d) + q_0 e(k) \tag{9.2.8}$$

with $q_0 = 1/b$. The current manipulated variable $u(k)$ is calculated from the manipulated variable $u(k-d)$, delayed by the deadtime, and the preset control error $e(k)$. The transient response of the cancellation controller given by (9.2.2) is shown in Fig. 9.1. As already remarked at the beginning of this chapter, the deadtime controller can be approximated by a PI-controller as shown by Fig. 9.1.
   Then one obtains the control algorithm

$$u(k) = u(k-1) + q'_0\, e(k) + q'_1\, e(k-1)$$

with parameters

$$\left.\begin{aligned} q'_0 &= \frac{q_0}{2} = \frac{1}{2b} \\[2mm] q'_1 &= q_0\left[\frac{1}{d} - \frac{1}{2}\right] = -\frac{1}{2b}\frac{(d-2)}{d} \end{aligned}\right\} \tag{9.2.9}$$

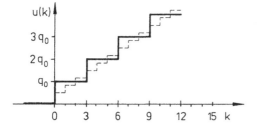

Fig. 9.1. Transient response of the deadtime cancellation controller $u(k)=u(k-d)+q_0e(k)$ for $d=3$. Dashed line: approximation by a PI-controller from (9.2.9)

or the characteristic values, as in section 5.2.1

$$K = q'_0 = \frac{1}{2b} \qquad\qquad \text{gain factor}$$

$$c_I = \frac{q'_0 + q'_1}{q'_0} = \frac{2}{d} \qquad\qquad \text{integration factor.}$$

For $d \geq 5$ these values correspond well with the optimal factors because of parameter optimization [5.8].

Now the characteristic equation of the resulting feedback loop is considered, for both exact and inexact deadtimes in order to discuss the sensitivity to the deadtime. For the cancellation controller and exactly chosen deadime one obtains

$$1 + G_R(z)\, G_P(z) = 0 \rightarrow z^d = 0. \qquad\qquad (9.2.10)$$

The characteristic equation of the closed-loop is the same as that of the process. The sensitivity of the deadtime controller for inexactly assumed deadtime can be seen from the characteristic equation. If the controller is designed for a deadtime d and is applied to a process with deadtime $d+1$ (i.e. the assumed deadtime is too small) the characteristic equation becomes

$$z^{d+1} - z + 1 = 0. \qquad\qquad (9.2.11)$$

For $d \geq 1$ the roots are on or outside the unit circle, giving rise to instability (c.f. Table 9.2). If the process has a deadtime $d-1$ one gets:

$$z^d + z - 1 = 0. \qquad\qquad (9.2.12)$$

In this case instability occurs for $d \geq 2$ (Table 9.2). Table 9.3 shows the largest magnitudes of the unstable roots for $d = 1, 2, 5, 10$ and 20; for very large deadtime the feedback loop with the cancellation controller is so sensitive to changes of the process deadtime by one sampling unit that instability is induced. Therefore in these cases this controller can only be applied if the deadtime is known exactly. If a PI-controller (2PC-2) is used the characteristic equation becomes

$$z^{d+1} - z^d + bq'_0 z + q'_1 b = 0 \qquad\qquad (9.2.13)$$

Table 9.2. Magnitudes $|z_i|$ of the roots of the characteristic equations (9.2.1), (9.2.11), (9.2.12) for deadtime processes with the controller $G_R(z) = 1/(1-z^{-d})$. (Underlined values are outside unit circle)

Process	$d=1$		$d=2$			$d=5$				
z^{-d}	0		0	0		0	0	0	0	0
$z^{-(d+1)}$	1.0	1.0	1.325	0.869	0.869	1.126	1.126	1.050	1.050	0.846
$z^{-(d-1)}$	0.5		1.62	0.618		1.151	1.151	1.000	0.755	

Table 9.3. Largest magnitudes $|z_i|_{max}$ of the roots of the characteristic equation for deadtime processes with controller $G_R(z) = 1/(1 - z^{-d})$

Process	$d = 1$	2	5	10	20
z^{-d}	0	0	0	0	0
$z^{-(d+1)}$	1.0	1.320	1.126	1.068	1.034
$z^{-(d-1)}$	0.5	1.618	1.151	1.076	1.036

with parameters given by (9.2.9)

$$2\,z^{d+1} - 2\,z^d + z - \frac{d-2}{d} = 0. \tag{9.2.14}$$

If the process deadtime changes from d to $d+1$, then

$$2\,z^{d+2} - 2\,z^{d+1} + z - \frac{d-2}{d} = 0. \tag{9.2.15}$$

For a change from d to $d-1$ one obtains

$$2\,z^d - 2\,z^{d-1} + z - \frac{d-2}{d} = 0. \tag{9.2.16}$$

Tables 9.4 and 9.5 show the magnitudes of the resulting roots. If the case $d=1$ is excluded, no instability occurs. Therefore the feedback loop with the PI-controller

Table 9.4. Magnitudes $|z_i|$ of the roots of the characteristic equations (9.2.14), (9.2.15), (9.2.16) for deadtime processes with the PI-controller of (9.2.9). The underlined values are located outside of the unit circle

Process	$d = 1$			$d = 2$			$d = 5$				
z^{-d}	0.707	0.707		0.707			0.886	0.886	0.829	0.829	0.701
$z^{-(d+1)}$	1.065	1.065	0.441	0.941	0.941	0.565	0.923	0.923	0.858	0.858	0.856	0.856
$z^{-(d-1)}$	0.333			0.500	0		0.796	0.796	0.789	0.789	0.760

Table 9.5. Largest magnitudes $|z_i|_{max}$ of the roots of the characteristic equation for deadtime processes with the PI-controller of (9.2.9)

Process	$d = 1$	2	5	10	20
z^{-d}	0.707	0.707	0.866	0.938	0.970
$z^{-(d+1)}$	1.065	0.941	0.923	0.951	0.974
$z^{-(d-1)}$	0.333	0.500	0.796	0.923	0.967

is less sensitive to changes in deadtime. Only for PI-controller designed for $d=1$ does instability arise when a process with $d=2$ is connected. Furthermore, it can be observed that the largest magnitudes of the roots of the characteristic equation increase if the deadtime of the process is assumed too small. As then the stability margin is smaller, it is better to choose the *deadtime too large than too small* if a PI-controller has to be designed. Section 14.3 considers controllers for pure deadtime and stochastic disturbances designed according to the *minimum variance principle*.

State Controller

If the deadtime as in (9.1.7) and (3.6.43) is included in the system matrix, and assuming that all state variables are directly measurable, one obtains the following characteristic equation corresponding to the pure deadtime process (see 8.3.8)

$$\det[z\,I\,-\,A\,+\,b\,k^T] \,=\, k_d\,+\,k_{d-1}z\,+\,...\,+\,k_1 z^{d-1}\,+\,z^d$$

$$= (z-z_1)\,(z-z_2)\,...\,(z-z_d)\,=\,0. \qquad (9.2.17)$$

If the characteristic equation has to be the same as for the input/output cancellation controllers, i.e. $z^d=0$, then all proportional feedback terms have to be zero, i.e. $k_i=0$, $i=1\,...\,d$. The reason is that an open loop deadtime process with reducing state feedback has the smallest settling time for initial values $x(0)$, corresponding to deadbeat behaviour $(\rightarrow z^d=0)$. If the state feedback is not to vanish, the poles z_i in (9.2.17) must be non-zero. As the state variables introduced in the process model (3.6.43) cannot, in general, be measured directly the state variables have to be estimated using an observer. Then the question arises as to whether the state controllers with the state observers of sections 8.6 and 8.7 or state estimators of sections 23.3, 15.2 and 15.3 have advantages over the input/output controllers discussed above. This is considered in the following sections using the results of digital computer simulations.

9.3 Comparison of the Control Performance and the Sensitivity of Different Controllers for Deadtime Processes

To compare the control performance and the sensitivity to inexact deadtime of various control algorithms and processes with large deadtime, the control behaviour was simulated with a process computer (program package CADCA, described in chapter 30) [5.8]. Two processes have been investigated, a pure deadtime process

$$G_P(z^{-1}) = \frac{y(z)}{u(z)} = z^{-d} \quad \text{with} \quad d = 10 \qquad (9.3.1)$$

and the low-pass process III (see (5.4.4) and Appendix) with deadtime $d=10$

$$G_P(z^{-1}) = \frac{y(z)}{u(z)} = \frac{b_1 z^{-1} + b_2 z^{-2} + b_3 z^{-3}}{1 + a_1 z^{-1} + a_2 z^{-2} + a_3 z^{-3}} z^{-d}. \qquad (9.3.2)$$

The resulting closed-loop behaviour for step changes of the set point is shown in Fig. 9.2 and Fig. 9.3. The root-mean-squared (or rms) control error

$$S_e = \sqrt{\frac{1}{M+1} \sum_{k=0}^{M} e_w^2(k)} \qquad (9.3.3)$$

and the rms changes of the manipulated variable

$$S_u = \sqrt{\frac{1}{M+1} \sum_{k=0}^{M} [u(k) - u(\infty)]^2} \qquad (9.3.4)$$

are shown in Fig. 9.3 for $M = 100$ and for the deadtime d_E chosen for the design (which is exact for $d_E = d = 10$, too small for $s_E = 8$ and 9, or too large for $d_E = 11$ and 12). Table 9.6 shows the resulting controller parameters for $d_E = d = 10$. The results can be summarized as follows:

Pure Deadtime Process

For the pure deadtime process the controller 2 PC-2 with PI-behaviour shows — within the class of parameter-optimized *control algorithms* – a somewhat better control performance than the controller 3-PC-3 with PID-behaviour, as better damped control variable and manipulated variable can be observed. The gain K in both cases is about 0.5. A weighting of the manipulated variable with $r > 0$ has only an insignificant influence on the resulting control behaviour. The sensitivity of these parameter-optimized controllers to errors in the assumed deadtime is smaller than that of all the other controllers. The best possible behaviour of the control variable is produced by the *deadbeat controller* DB(v) or the identical *predictor controller* PREC. The modified deadbeat controller DB($v+1$) reaches the new steady state one sampling unit later. Neither the deadbeat controllers nor the predictor controller, however, can be recommended as instability occurs if the deadtime is used for the design differs from the real deadtime. Well-damped control behaviour is produced by a *state controller* with observer. Here $u(0) = 0$, as in the optimization of the quadratic performance criterion (8.1.2) the state feedback k_d and also all k_i for $i = 1$ to $d - 1$ become zero. Only k_{d+1}, the feedback of the state variable $y(k) = x_{d+1}(k)$ of the augmented observer becomes $k_{d+1} = 1$ (compare Fig. 8.8, Fig. 8.11 and Example 8.2). This state controller is independent on the choice of the weighting r of the manipulated variable. The sensitivity arising from a wrong deadtime is, however, for $|\Delta d| = |d_E - d| = 1$ greater than for parameter-optimized controllers. For $|\Delta d| > 1$ instability has been observed. Therefore for a pure deadtime process with a relatively well-known deadtime ($|\Delta d| \leq 1$), a state controller with modified observer can be recommended, but for an inexactly known or changing deadtime $|\Delta d| > 1$, a parameter-optimized controller with PI-behaviour is to be preferred.

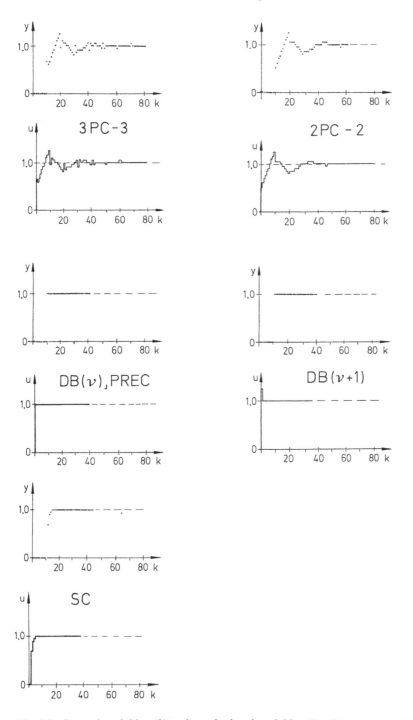

Fig. 9.2. Control variable $y(k)$ and manipulated variable $u(k)$ of the *pure deadtime process* $G_P(z) = z^{-d}$ with $d = 10$ and a step change in the set point

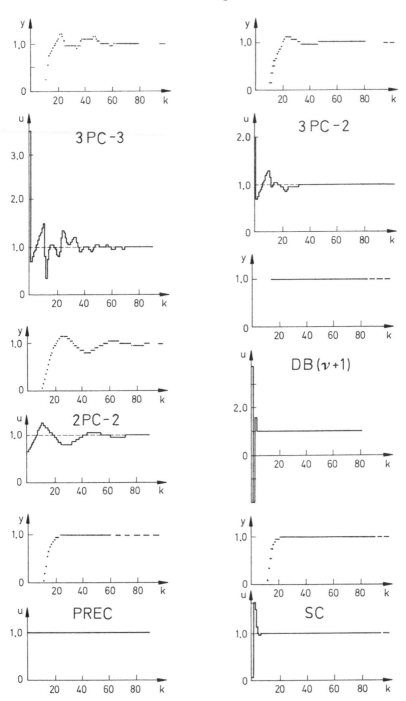

Fig. 9.3. Control variable $y(k)$ and manipulated variable $u(k)$ for the *process* III *with deadtime* $d = 10$, $G_p(z) = B(z^{-1})z^{-d}/A(z^{-1})$, for a step change in the set point

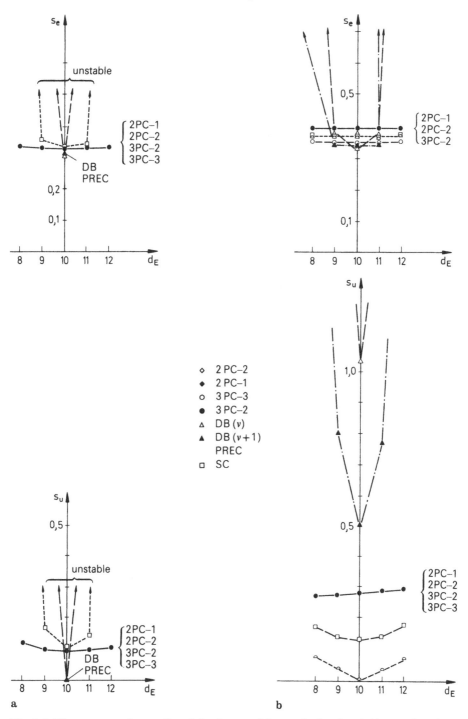

Fig. 9.4. The rms control error S_e and the change of the manipulated variable S_u as functions of the deadtime d_E used for the design. **a** pure deadtime process z^{-d} with $d = 10$. **b** process III with deadtime $d = 10$

Table 9.6. Controller parameters for the investigated processes with large deadtime

Controller parameters	$G_P = z^{-d}$ $d = 10$		$G_P = \dfrac{B(z^{-1})}{A(z^{-1})} z^{-d}$ (Proc. III with $d = 10$)		
	3PC-3 $(r=0)$	2PC-2 $(r=0)$	3PC-2 $(r=0)$	3PC-2	2PC-2 $(r=0)$
q_0	0.6423	0.5198	3.4742	2.0000	0.6279
q_1	−0.6961	−0.4394	−6.2365	−3.3057	−0.5714
q_2	0.1372	0	2.8483	1.3749	0
K	0.5052	0.5198	0.6258	0.6251	0.6280
c_D	0.2715	0	4.5515	2.1993	0
c_I	0.1651	0.1547	0.1373	0.1106	0.092
	DB$(v+1)$	DB(v)	DB$(v+1)$	DB(v)	PREC
q_0	1.25	1	3.8810	9.5238	1
q_1	−0.25	0	−0.1747	−14.2762	−1.4990
q_2	0	0	−5.7265	6.7048	0.7040
q_3	0	0	3.5845	−0.9524	−0.1000
q_4	0	0	−0.5643	0	0
p_0	1	1	1	1	1.0000
p_1	0	0	0	0	−1.4990
p_2	0	0	0	0	0.7040
p_3	0	0	0	0	−0.1000
p_4	0	0	0	0	0
p_d	0	−1.25	0	0	0
p_{d+1}	0	0	−0.2523	−0.6190	0.0650
p_{d+2}	0	0	−0.5531	−0.4571	0.0480
p_{d+3}	0	0	−0.2398	−0.0762	−0.0451
p_{d+4}	0	0	0.0451	0	−
	SC $(r=1)$*		SC $(r=1)$*		
k_1	0		0.0680		
k_2	0		0.0473		
k_3	0		0.0327		
k_4	0		0.0807		
k_5	0		0.0691		
k_6	0		0.0551		
k_7	0		0.0420		
k_8	0		0.0311		
k_9	0		0.0226		* $r_b = 5$;
k_{10}	0		0.0161		Q_b see
k_{11}	1.0		0.0114		example 8.2.
k_{12}	−		0.0080		
k_{13}	−		0.0056		
k_{14}	−		1.0		

Low Pass Process III with Large Deadtime

The *parameter-optimized control algorithms* with PI-behaviour (2 PC-2, 2 PC-1) designed with $r=0$ lead to relatively undamped control variables and manipulated variables. The control variable settles faster using the controller 3 PC-3 with PID-behaviour. However, large changes of the manipulated variable are required. A controller 3 PC-2 with $q_0=2$ leads to the best behaviour among the group of parameter-optimized controllers, with good damping of the controlled and manipulated variables. There is little sensitivity of all parameter-optimized controllers to inexactly chosen deadtime. The controlled variable of the *deadbeat controller* reaches the new steady state faster than all other controllers. However, the changes of the manipulated variable are excessive using $DB(v)$ and also large with $DB(v+1)$. The sensitivity to inexactly chosen deadtime is greatest for the deadbeat controllers. In the example, deadtime errors of $\Delta d_E = \pm 1$ can be permitted, especially for $DB(v+1)$. However, larger deadtime errors result in poor control behaviour. The *predictor controller* models the behaviour of the process itself. The manipulated variable immediately reaches the new steady state. The *state controller* with an observer designed for $r=1$ results in a very well-damped control variable compared to the predictor controller. The manipulated variable $u(0)$ is still small because of the small value of the state feedback k_3, as shown in Table 9.6 and Fig. 8.11. The manipulated variable has its largest value for $k=1$ and then approaches the new steady state value with good damping. Both the predictor controller and the state controller with observer have about the same small sensitivity.

The best control performance for this low pass process with large time delay can therefore be achieved with the state controller, the predictor controller or the parameter-optimized controller 3 PC-2 (or 3 PC-3 with $r\approx 1$). The predictor controller leads to the smallest, the 3 PC-2 to the largest and the state controller to average changes in the manipulated variable. A comparison of the control performance shows that the open loop transient response to set point changes can hardly be changed compared with the transient response of the process itself if very large variations of the input have to be avoided. For larger changes of the manipulated variable, as for deadbeat controllers, one can reach a smaller settling time. However, this leads to a higher sensitivity to deadtime. Therefore deadbeat controllers cannot be recommended in general for processes with large deadtime. As the predictor controller can be applied only to asymptotically stable processes, state controllers with observer and parameter-optimized controllers with PID- or PI-behaviour are preferred for low pass processes with large deadtime. If not known precisely, the deadtime should preferably be assumed too large rather than too small [5.8].

Since for processes with large deadtime the possibilities to counteract disturbances are limited, one should always try to use additionally feedforward control (see chapter 17).

10 Sensitivity and Robustness with Constant Controllers

The preceding controller design methods assumed that the process model is exactly known. However, this is never the case in practice. In theoretical modelling as well as in experimental identification one must always take into account both the small and often the large differences between the derived process model and the real process behaviour. If, for simplicity, it is assumed that the structure and the order of the process model are chosen exactly then these differences are manifested as parameter errors. Moreover, during most cases of normal operation, changes of process behaviour arise for example through changes of the operating point (the load) or changes of the energy- mass- or momentum storages or flows. When designing controllers one must therefore assume

— the assumed process model is inexact;
— the process behaviour changes with time during operation.

This chapter will regard how changes of the process influence the closed-loop behaviour and how they can be considered for the design of the controllers. Here parameter changes with reference to a nominal parameter vector Θ_n are assumed. Then the closed-loop behaviour for parameter vectors

$$\Theta = \Theta_n + \Delta\Theta$$

is of interest if a constant controller is assumed.

It is further assumed that the order of the process model does not change and that the parameters change slowly compared with the closed-loop dynamics. This last assumption means that the process can be regarded as quasi-time-invariant. If the parameter changes are small, then *sensitivity methods* can be used (see section 10.1). If the parameter sensitivity is known, then for controller design both good control performance and small parameter sensitivity can be required. The resulting controllers then are referred to as *insensitive controllers*, (see section 10.2). However, for larger parameter changes this sensitivity design is unsuitable. Instead, one has to assume several process models with different parameter vectors

$$\Theta_1, \Theta_2, ..., \Theta_M$$

and should try to design a *common constant controller* which for those processes will maintain stability and a certain control performance range. The resulting controllers are called *"robust controllers"*. Section 10.3 and 10.4 will discuss this problem.

The approach for the design of robust controllers is more general than the one for insensitive controllers. However, it can be expected that for medium-sized parameter changes similar properties of the resulting insensitive and robust control systems arise.

10.1 On the Sensitivity of Closed-loop Systems

Compared with feedforward control systems, feedback control systems have the ability not only to decrease the influence of disturbances on the output variable but also to decrease the influence of process parameter changes on the output. In order to demonstrate this well-known property [10.1], a feedback control and a feedforward control as in Figures 6.1 and 6.2 are examined. It is assumed that the process and controller transfer functions are $G_p(z)$ and $G_R(z)$ and that the feedforward element has a transfer function $G_S(z)$. Both systems are designed for the nominal process parameter vector Θ_n, so that the same input signal $w(k)$ generates the same output signal $y(k)$. The process $G_p(z)$ is assumed to be asymptotically stable so that after the decay of the free response both processes are in the same steady state before $w(k)$ changes. The closed-loop input/output behaviour corresponding to the nominal working point is described by

$$G_w(\Theta_n,z) = \frac{y(z)}{w(z)} = \frac{G_R(z)G_P(\Theta_n,z)}{1+G_R(z)G_P(\Theta_n,z)}. \tag{10.1.1}$$

The feedforward control with the same input/output behaviour has the transfer function

$$G_S(\Theta_n,z) = \frac{u(z)}{w(z)} = \frac{G_R(z)}{1+G_R(z)G_P(\Theta_n,z)} \tag{10.1.2}$$

The process parameter vector now changes by an infinitesimal value $d\Theta$. For the control loop it follows by differentiation of the process parameter vector that

$$\left.\frac{\partial G_w(\Theta_n,z)}{\partial \Theta}\right|_R = \frac{G_R(z)}{[1+G_R(z)G_P(\Theta_n,z)]^2}\frac{\partial G_p(\Theta_n,z)}{\partial \Theta}. \tag{10.1.3}$$

Accordingly, for the feedforward control one obtains

$$\left.\frac{\partial G_w(\Theta_n,z)}{\partial \Theta}\right|_S = G_S(\Theta_n,z)\frac{\partial G_P(\Theta_n,z)}{\partial \Theta}$$

$$= \frac{G_R(z)}{1+G_R(z)G_P(\Theta_n,z)}\frac{\partial G_P(\Theta_n,z)}{\partial \Theta}. \tag{10.1.4}$$

As for both systems one has

$$\frac{\partial y(z)}{\partial \Theta} = \frac{\partial G_w(\Theta_n,z)}{\partial \Theta}w(z) \tag{10.1.5}$$

it follows that

$$\left.\frac{\partial y(z)}{\partial \Theta}\right|_R = R(\Theta_n, z) \left.\frac{\partial y(z)}{\partial \Theta}\right|_S \tag{10.1.6}$$

with

$$R(\Theta_n, z) = \frac{1}{1 + G_R(z) G_P(\Theta_n, z)} \tag{10.1.7}$$

as the dynamic control factor. $\partial y / \partial \Theta$ is called the *parameter sensitivity* of the output variable y. As can be seen from (10.1.6), the relative parameter sensitivities of the feedback control and the feedforward control depend on the frequency ω of the signal $w(k)$. If $|R(z)| < 1$ the feedback control has a smaller parameter sensitivity than the feedforward control, but if $|R(z)| > 1$ the opposite is the case. In general, however, feedback control systems are designed so that in the significant frequency range $(0 \leq \omega \leq \omega_{max})$ the magnitude of the dynamic control factor $|R(z)| < 1$ is less than unity to achieve good control performance. Therefore in most cases the parameter sensitivity of feedback control systems is smaller than that of feedforward controllers. The parameter sensitivity increases with the exciting frequency and is therefore smallest at $\omega = 0$, i.e. in the steady state.

The same equation gives both the ratio of the parameter sensitivity of the feedback and the feedforward control, and the ratio of the influences of the disturbance $n(k)$ on the output variable $y(k)$

$$\left.\frac{y(z)}{n(z)}\right|_R = R(z) \left.\frac{y(z)}{n(z)}\right|_S . \tag{10.1.8}$$

From (10.1.3) and (10.1.1) it follows further that for the feedback control

$$\frac{dG_w(\Theta_n, z)}{G_w(\Theta_n, z)} = S(\Theta_n, z) \frac{dG_P(\Theta_n, z)}{G_P(\Theta_n, z)} \tag{10.1.9}$$

with the *sensitivity function* $S(\Theta_n, z)$ of the feedback control

$$S(\Theta_n, z) = R(\Theta_n, z) = \frac{1}{1 + G_R(z) G_P(\Theta_n, z)} . \tag{10.1.10}$$

This sensitivity function shows how relative changes of the input/output behaviour of a closed-loop depend on changes of the process transfer function. Since this ratio is the same as the ratio of the parameter sensitivity of feedback and feedforward control the remarks given above can also be transferred to this case. The sensitivity function can also be used for non-parametric models. Small sensitivity of the closed-loop behaviour after set point changes can be obtained by choosing a small dynamic control factor $|R(\Theta_n, z)|$ in the significant frequency range $0 \leq \omega \leq \omega_{max}$ for the exciting external signals $n(z)$ or $y_w(z) = G_R(z) G_P(z) w(z)$. Note, that the parameter sensitivity of the output variable and the sensitivity function represent the time functions $\partial y(k)/\partial \Theta$ or $s(k)$ after back transformation in the original region. For a process in state representation

$$x(k+1) = A\,x(k) + b\,u_p(k) + b\,u_w(k) \qquad (10.1.11)$$

with corresponding state controller

$$u_R(k) = -k^T x(k) \qquad (10.1.12)$$

or the corresponding closed-loop transfer function

$$G_R(z)\,G_P(z) = \frac{u_R(z)}{u_P(z)} = k^T[zI - A]^{-1}b \qquad (10.1.13)$$

the dynamic control factor can now be defined as $R'(z) = u_P(z)/u_w(z)$ with $u_p = u_R$,

$$R'(z) = \frac{u_P(z)}{u_w(z)} = \frac{1}{1 + k^T[zI - A]^{-1}b}. \qquad (10.1.14)$$

In [8.10] it is shown that (10.1.6) gives the parameter sensitivity of the open-loop variable $y(k) = c^T x(k)$ of the state controller, but with $R'(z)$ instead of $R(z)$. Optimal state controllers for continuous signals always have smaller parameter sensitivities than feedforward controllers, [8.4] (page 314), [10.2], [10.8] (page 126). State controllers with observers and state controllers for discrete time signals do not obey this rule [8.4] (page 419 and 520).

The sensitivity function $S(\Theta_n, z)$ of (10.1.10) expresses the influence of relative changes of the process transfer function. Absolute changes of the closed-loop behaviour for set point changes follow from (10.1.9) and $G_w = RG_RG_P$

$$|\Delta G_w(\Theta_n, z)| = |R(\Theta_n, z)|^2 |G_R(z)| \,|\Delta G_P(\Theta_n, z)|. \qquad (10.1.15)$$

The influence of process changes $|\Delta G_P|$ on $|\Delta G_w|$ are enforced by $|R|^2|G_R|$. Compare the corresponding relation for the signals

$$|y(z)| = |R(z)| \,|n(z)| \qquad (10.1.16)$$

This can be used to evaluate the control performance. The changes $|\Delta G_w|$ influence $|y(z)|$ linearly as follows

$$|\Delta y(\Theta_n, z)| = |\Delta G_w(\Theta_n, z)| \,|w(z)|. \qquad (10.1.17)$$

The amplification of the term $|\Delta G_P|$ by $|R|^2|G_R|$ implies that changes of $|\Delta G_P|$ have a great influence in frequency ranges II and III of the dynamic control factor, as shown in Fig. 11.5. For very low frequencies and controllers with integral action $|R|^2|G_R| \sim |R|$. This means an influence such as that for the control performance given by (10.1.16). An insensitive control can be obtained in general by making $|R(z)|$ as small as possible, particularly for the higher frequencies of range I and for the ranges II and III should disturbances arise in these ranges.

From Fig. 11.6 and Table 11.5 it follows that for feedback controls which are insensitive to low-frequency disturbances the weight of the manipulated variable r must be small, leading to a strong control action. Disturbance signal components $n(k)$ in the vicinity of the resonance frequency, however, require a decrease of the resonance peak and therefore a smaller value of r, or a weaker control action. This again shows that steps towards an insensitive control depend on the disturbance signal spectrum. If $|R(z)|^2$ is considered, Figures 11.6 and 11.7 show that, for different controllers high sensitivity to process changes arises with the following controllers: Range I: 2PC-2. Range II: 2PC-2, DB(v) and SC. Small sensitivities are obtained for range I: SC, and for range II: DB($v+1$). Note, however, that parameter-optimized and deadbeat controllers have been designed for step changes of the set point, i.e. for a small excitation in ranges II and III. For step changes of $w(k)$ these results agree essentially with the sensitivity investigations of section 11.3.

Until now, only some sensitivity measures have been considered. Other common parameter sensitivity measures given for a nominal parameter vector Θ_n are

sensitivity of a state variable
(trajectories)
$$\sigma_x = \frac{\partial x}{\partial \Theta} \qquad (10.1.18)$$

sensitivity of the performance
$$\sigma_I = \frac{\partial I}{\partial \Theta} \qquad (10.1.19)$$

Sensitivity of an eigenvalue
$$\sigma_\lambda = \frac{\partial \lambda i}{\partial \Theta}. \qquad (10.1.20)$$

The sensitivity of the output variables follows from the state variable sensitivity function

$$\sigma_y = \frac{\partial y}{\partial \Theta} = \frac{\partial}{\partial \Theta} [c^T x] = \frac{\partial x}{\partial \Theta} c. \qquad (10.1.21)$$

10.2 Insensitive Control Systems

The parameter sensitivity can be taken into account in the controller design by adding to the performance criterion I_n positive semi-definite functions $f(\sigma) \geq 0$ for the nominal point so that

$$I_{n\sigma} = I_n + f(\sigma) = I_n + I_\sigma \qquad (10.2.1)$$

is minimized. If the parameter sensitivity of the control performance criterion

$$I_\sigma = x^T \sigma_I = x^T \frac{\partial I_n}{\partial \Theta} \qquad (10.2.2)$$

is used, in which κ are weighting factors, the criterion

$$I_{n\sigma} = I_n + x^T \frac{\partial I_n}{\partial \Theta} \qquad (10.2.3)$$

has to be minimized instead of I_n. Here, either the structure of the controller designed for I_n can be simply adopted and possibly free design parameters are varied in such a way that the extended $I_{n\sigma}$ is minimized (insensitivity through varying the design of usual controllers) or the extended criterion I_n is used to determine a structurally changed controller (insensitivity through additional dynamic feedback).

Most of the publications on the design of insensitive control systems deal with continuous-time systems, see e.g. [10.7, 10.9–10.11]. The general procedure of some methods which can be transferred to discrete-time systems will briefly described in the following section.

10.2.1 Insensitivity Through Additional Dynamic Feedback

A process with state equation

$$x(k+1) = A(\boldsymbol{\Theta})x(k) + B(\boldsymbol{\Theta})u(k)$$
$$= A_n x(k) + B_n u(k) \tag{10.2.4}$$

is considered. The trajectory sensitivity for a parameter Θ_i is written with

$$\sigma_{xi}(k) = \frac{\partial x(k)}{\partial \Theta_i}\bigg|_n = \sigma_i(k) \quad i = 1, 2, ..., p \tag{10.2.5}$$

as follows

$$\sigma_i(k+1) = A_i x(k) + A_n \sigma_i(k) + B_i u(k) + B_n \frac{\partial u(k)}{\partial \Theta_i}\bigg|_n \tag{10.2.6}$$

using

$$A_i = \frac{\partial A}{\partial \Theta_i}\bigg|_n ; \quad B_i = \frac{\partial B}{\partial \Theta_i}\bigg|_n ; \quad \sigma_i(0) = 0 \tag{10.2.7}$$

Hence for each process parameter a state sensitivity model emerges according to (10.2.6).

In order to not only keep small the state variables but also the sensitivities the state variables $x(k)$ as well as the sensitivities $\sigma_i(k)$ are solved by negative feedback [10.12] forming

$$u(k) = -K x(k) - \sum_{i=1}^p K_i \sigma_i(k) \tag{10.2.8}$$

Introducing into (10.2.4) yields for the feedback process

$$x(k+1) = [A_n - B_n K]x(k) - B_n \sum_{i=1}^p K_i \sigma_i(k). \tag{10.2.9}$$

The sensitivity of this closed-loop system results from applying (10.2.5) to (10.2.9)

$$\sigma_i(k+1) = [A_i - B_iK]x(k) + [A_n - B_nK]\sigma_i(k)$$

$$- B_i \sum_{l=1}^{p} K_l\sigma_l(k) - B_n \sum_{l=1}^{p} K_l\frac{\partial\sigma_l(k)}{\partial\Theta_i} \qquad (10.2.10)$$

The previous term contains second-order derivations and can therefore be neglected.

Note, that $\sigma_i(k) = \sigma_l(k)$, $i = l = 1, 2, \ldots, p$.

Considering (10.2.8)

$$\tilde{\sigma}_i(k+1) = [A_n - B_nK]\tilde{\sigma}_i(k) + A_ix(k) + B_iu(k) \qquad (10.2.11)$$

$$u(k) = -Kx(k) - \sum_{l=1}^{p} K_l\tilde{\sigma}_l(k)$$

$$= -Kx(k) - K_\sigma\tilde{\sigma}(k). \qquad (10.2.12)$$

Here (10.2.11) represents the sensitivity model of the closed-loop system. Using (10.2.4) and the extended state vector

$$z^T(k) = [x^T(k)\tilde{\sigma}_1^T(k)\ldots\tilde{\sigma}_p^T(k)]$$

$$= [x^T(k)\tilde{\sigma}^T(k)] \qquad (10.2.13)$$

a system of overall order $n(1+p)$

$$\left.\begin{array}{c} z(k+1) = \tilde{A}z(k) + \tilde{B}u(k) \\[2mm] u(k) = -K_zz(k) \end{array}\right\} \qquad (10.2.14)$$

can be formed, see Fig. 10.1.

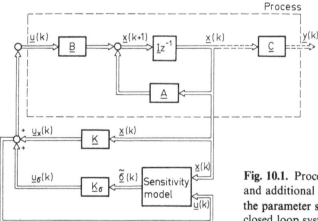

Fig. 10.1. Process with state feedback K and additional dynamic feedback K_σ of the parameter sensitivity $\tilde{\sigma}(k)$ of the closed loop system

The following performance criterion

$$I_{n\sigma} = \sum_{k=0}^{\infty} [z^T(k)Q_z z(k) + u^T(k)R\,u(k)] \qquad (10.2.15)$$

can be related to this system evaluating the state variables $x(k)$ as well as the trajectory sensitivity $\sigma_i(k)$. Optimization of I_σ therefore leads to a compromise between control performance and parameter sensitivity. K_z are determined iteratively through repeated solving and reintroducing the corresponding matrix Riccati equation (K can be found in (10.2.11) as well as in (10.2.12)). Note that the computational effort is large because:

a) for each variable parameter Θ_i a sensitivity model has to be realized
b) the order of the overall system augments by n with each variable parameter.

The sensitivity feedback according to (10.2.12) introduces an *additional dynamic feedback* with delayed character as can be seen from (10.2.11). At the same time the state feedback gain K is diminished compared with the usual design. This results in a weaker controller action. (For a first-order process one obtains a proportional and a parallel proportionally delayed first-order feedback control instead of a purely proportional feedback control. For higher-order processes the proportionally delayed feedback control will also be of higher order. This causes (after the proportionally changed manipulated variable) a delayed adjustment to a final manipulated value, thus corresponding roughly with the behaviour of a PI controller).

Modifications of this design method, especially with respect to avoid iterative computing procedures, are described in [10.13, 10.14]. However, these insensitive control systems with additional dynamic feedback have not, in general, been successful because of the significantly large effort in design and realization required as well as the requirement for relatively precise knowledge of the process.

10.2.2 Insensitivity Through Variation of the Design of General Controllers

A lower realization effort can be obtained by neglecting the sensitivity feedback and using a general controller structure.

For the common state feedback control

$$u(k) = -K\,x(k) \qquad (10.2.16)$$

is valid, leading to

$$\left.\frac{\partial u(k)}{\partial \Theta_i}\right|_n = -K\sigma_i(k). \qquad (10.2.17)$$

These equations are introduced into (10.2.4) and (10.2.6)

$$x(k+1) = [A_n - B_nK]x(k) \qquad (10.2.18)$$

$$\sigma_i(k+1) = [A_n - B_nK]\sigma_i(k) + [A_i - B_iK]x(k). \qquad (10.2.19)$$

K is now determined in such a way that the criteria

$$I_{n\sigma} = I_n + I_\sigma \qquad (10.2.20)$$

$$I_n = \sum_{k=0}^{\infty} [x^T(k)Q\,x(k) + u^T(k)R\,u(k)] \qquad (10.2.21)$$

$$I_\sigma = \sum_{k=0}^{\infty} \sum_{i=1}^{p} \sigma_i^T(k)Q_\sigma\sigma_i(k) \qquad (10.2.22)$$

is minimized. In [10.5] this is done by numerical parameter optimization.
    Besides I_n the sensitivity of this criterion with $F \neq Q$ is used in [10.16]

$$\sigma_{Ii} = \frac{\partial I_1}{\partial \Theta_i}; \; I_1 = \sum_{k=0}^{\infty} x^T(k)F\,x(k) \qquad (10.2.22)$$

and

$$I_{n\sigma} = \beta_1 I_n + \beta_2 tr\,[\sigma_{Ii}d\Theta_i] \qquad (10.2.23)$$

$$\beta_1 + \beta_2 = 1, \quad 0 \leq \beta_1 \leq 1$$

is minimized by iterative determination of K. However, the computational effort of these methods is relatively high.
    A less complicated approach is to use the *variation of the design parameters* in the usual controller design, e.g. the weighting matrices Q and R in the performance criterion (10.2.21). For example,

$$Q_1 = \mu Q \quad \mu > 1 \qquad (10.2.24)$$

can be chosen and μ be varied as a function of the sensitivity criterion (10.2.22), [10.17, 10.18]. Here the sensitivity criterion (10.2.22) can be determined by a Lyapunov equation so that:

$$Q_1 = Q + \mu P \quad (\mu > 0). \qquad (10.2.25)$$

Q_1 is varied through an iterative procedure until the sensitivity reduction has reached a lower limit [10.19, 10.20]. A comparison of various design methods of insensitive control systems is given in [10.11, 10.18] and [10.20] using two simulation examples with continuous-time signals (magnetic-field monorail, aircraft with jet engines and canards). There it was shown that the improved parameter sensitivity could only be obtained by taking into account a longer settling time, that the variation methods of the Q-matrix lead to similar results than the methods with additional dynamic feedback and that the methods of reducing the parameter sensitivity can also be applied for large parameter changes [10.24]. The influence of different excitation signals (see section 10.1), however, was not considered. It must

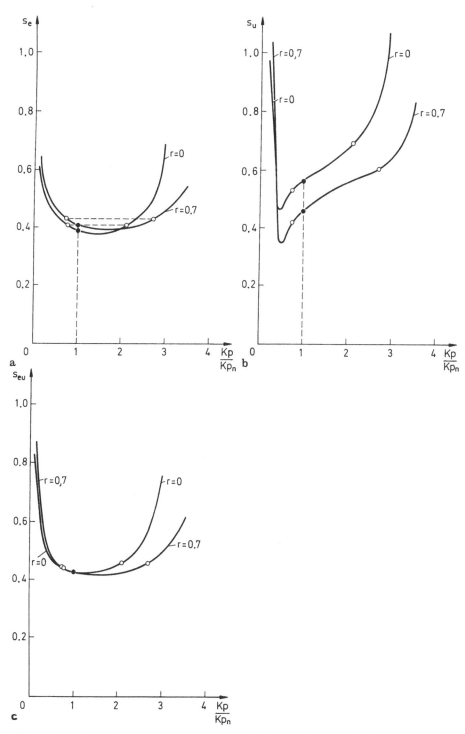

Fig. 10.2a–c. Control performance for process gain changes. Process VI with PID-controller.
a Quadratic mean value of the control deviation S_e; **b** Quadratic mean value of the
manipulated variable deviation S_u; **c** Mixed quadratic criterion $S'_{eu} = \sqrt{S_e^2 + 0.1 S_u^2}$

be stressed that the possibilities of varying the time behaviour of control- and manipulated variables are in general limited.

This is to be shown by the following example:

Example 10.1 Control performance of process gain changes for variation of the design parameters.

For test process VI (low pass, $m = 3$, $T_0 = 4$ s), see Appendix, a PID-control performance criterion

$$S_{eu}^2 = \sum_{k=0}^{M} [e^2(k) + r \Delta u^2(k)], \; M = 60$$

was designed through numerical parameter optimization for $r = 0$ and $r = 0.7$. The following controller parameters result (program package CADCA)

$$r = 0 : q_0 = 4,0781; \; q_1 = -6,0313; \; q_2 = 2,2813$$
$$K = 1,80; \quad c_D = 1,27; \quad c_I = 0,18$$

$$r = 0,7: q_0 = 3,0021; \; q_1 = -4,2005; \; q_2 = 1,4835$$
$$K = 1,52; \quad c_D = 0,98; \quad c_I = 0,19.$$

Fig. 10.2 represents for a variation of the process gain K_p/K_{pn} according to (5.4.10) and (5.4.13)

$$S_e = \sqrt{\overline{e^2(k)}} \quad \text{und} \quad S_u = \sqrt{\overline{\Delta u^2(k)}}$$

If a nominal point K_p/K_{pn} with control performance S_{en} is assumed and if the control performance may deteriorate for 5% that is $S_e/S_{en} \leq 1.05$, then the following variations areas of the permissible process gain result

$$r = 0: \qquad 0,75 \leq K_p/K_{pn} \leq 2,1$$
$$r = 0,7: \qquad 0,71 \leq K_p/K_{pn} \leq 2,7$$

The resulting changes of the manipulating effort S_u are larger (up to 30% approximately). If only S_e is considered, it follows that:

— The design with $r = 0$ furnishes a better performance at the nominal point. The permissible process changes, however, are smaller (the control loop is more sensitive or less robust).
— The more sensitive design $(r = 0)$, however, yields a better performance for range $K_p/K_{pn} \leq 1.8$. The least sensitive design $(r = 0.7)$ gives a better performance only for larger process changes $K_p/K_{pn} > 1.9$.

For comparison, Fig. 10.2c represents the control performance behaviour for the same control performance measure S'_{eu} $(r = 0.1$ for all cases). For about the same performance at the nominal point, for $r = 0.7$ larger process gains could be permitted.

This example also shows, that the results may depend essentially on the mode of the control performance criterion.

10.3 On the Robustness of Control Systems

The design methods for insensitive control systems assume a control behaviour for a nominal point and try to minimize the first derivative of a measure for the control behaviour after the variable parameter (sensitivity). With this, however, only the control behaviour for small parameter changes (the local control behaviour) is specifically influenced. Now, also the second and perhaps higher derivations could be considered, yet an increasing effort and a nevertheless restricted parameter range make this procedure appear rather a poor choice, whereas the design of a constant controller for several different parameter vectors $\Theta_1, \Theta_2, \ldots, \Theta_M$ can be used more generally. Dependent on the operating point the parameter of a process can change continually in finite order. This, however, can also be done in a discontinuous way, as for example with processes for which breafing the sound barrier plays an important role (valves, airplanes) or for differently connected plants and equal controllers (e.g. heating and cooling for air conditioning systems) or through switching a part of several interconnected control systems from automatic to manual mode or through adding other users (e.g. more inductive users in electrical networks) etc.

In these cases, for simplicity and reliability reasons a constant controller is desired which furnishes for all parameter vectors stability and a control behaviour within a certain performance range. These controllers are referred to as *robust controllers*.

In order to evaluate the robustness of control systems the parameter range is discretized in such a way that M process models e.g. of the form

$$\left.\begin{array}{ll} x(k+1) = A(\Theta_j)x(k) + B(\Theta_j)u(k) \\[2mm] y(k) \quad = C(\Theta_j)x(k) \quad j = 1, 2, ..., M \end{array}\right\} \tag{10.3.1}$$

are obtained. Hence, for each operating point j a local simplified process model is assumed (e.g. linearized) and presumed that the process parameters Θ_j change only slowly compared with the signals. With this procedure nonlinear or slowly time-varying processes can be described approximately.

Now the robustness of the standard controller designed for one parameter vector is of interest. For the state feedback with *continuous-time signals* designed with a Matrix–Riccati–Equation

$$u(t) = -k^T x(t) \tag{10.3.2}$$

it has been shown in [10.21, 10.8] that the frequency response of the open-loop system

$$G_0(i\omega) = k^T [i\omega I - A]^{-1} b \tag{10.3.3}$$

fulfills the inequality for the so-called return difference

$$|1 - G_0(i\omega)| \geq 1 \tag{10.3.4}$$

This means that the distance of any point on the locus of the open loop function $G_0(i\omega)$ from the critical point $-1 + i0$ of the Nyquist–Stability-Criterion is at least 1. $G_0(i\omega)$ thus always avoids a circular locus around $-1 + i0$ and radius 1 independent of how large the feedback gains are. This means that the critical point $-1 + i0$ is always located inside this circle. Therefore the optimal state feedback control has an infinitely large amplitude margin and a phase margin of at least $60°$ (with $|G_0(i\omega)| = 1$). It has been shown in [10.22] that the same properties are valid for each channel in multivariable systems. [10.23] can also be used.

The infinitely large gain margin which holds for optimal state feedback (10.3.2) with the feedback control gains k^T, signifies that the closed-loop system remains asymptotically stable for infinitely large control gains. This means that with

$$k_\mu^T = \mu k^T \tag{10.3.5}$$

the system

$$\dot{x}(t) = [A - \mu bk^T]x(t) \tag{10.3.6}$$

is asymptotically stable for all

$$1 \leqq \mu < \infty. \tag{10.3.7}$$

If, in (10.3.2) k^T is replaced by k_μ^T according to (10.3.6), then this yields together with (10.3.4)

$$\left.\begin{array}{c} |1 - \mu G_0(i\omega)| \geqq 1 \\[2mm] \left|\dfrac{1}{\mu} - G_0(i\omega)\right| \geqq \dfrac{1}{\mu} \end{array}\right\} \tag{10.3.8}$$

or

The previous equation signifies, that $\mu G_0(i\omega)$ represents a transformed locus with the axis multiplied by $\frac{1}{\mu}$. Then the formerly critical point $-1 + i0$ is transformed to $-\frac{1}{\mu} + i0$. $G_0(i\omega)$, however, does not circle the $-1 + i0$ point with radius 1 including the point $-2 + i0$. Hence asymptotic stability is valid for $\mu > 1/2$ [10.8].

In addition to (10.3.7) another stable range exists as follows

$$1/2 < \mu \leqq 1. \tag{10.3.9}$$

The optimal state feedback control thus is asymptotically stable for feedback gains in the range

$$1/2 < \mu < \infty. \tag{10.3.10}$$

This also holds for time-variant and nonlinear feedback gains k_μ^T [10.8].

The previous considerations assumed a time invariant process A, b with changed feedback k_μ^T. It follows from (10.3.6) that even for a changed process A, b_μ or $A_\mu b$ an asymptotically stable behaviour and constant feedback is obtained [8.8], if

$$b_\mu = \mu b \quad 1/2 < \mu < \infty \tag{10.3.11}$$

or

$$A_\mu = \frac{1}{\mu} A \quad 0 \leqq \frac{1}{\mu} < 2 \tag{10.3.12}$$

(The previous equation results after introducing a new timescale $t' = \mu t$).

An optimal Matrix-Riccati-State controller designed for a process modelled by A, b is asymptotically stable, if the parameters b_i change for the factors $0.5 < \mu \leqq \infty$ or (with diagonal form of A) the eigenvalues λ_i by the factors $0 \leqq 1/\mu < 2$. Thus, the optimal state feedback control has significant robustness properties (assuming the exclusion of a state observer). Also compare the discussion at the end of section 11.1.

For state feedback with *discrete-time control signals* of the form

$$u(k) = -K x(k) \tag{10.3.13}$$

it has been demonstrated graphically in [10.23], that the optimal state controller designed in section 8.2 as

$$K = [R + B^T P B]^{-1} B^T P A \tag{10.3.14}$$

with $R = \text{diag}[r_1, \ldots, r_p]$ is stable. If the perturbation included in Fig. 10.3

$$L(z)u(z) = \begin{bmatrix} L_1(z)u_1(z) \\ \cdot \\ \cdot \\ \cdot \\ L_p(z)u_p(z) \end{bmatrix} \tag{10.3.15}$$

has a locus $L_i(z)$ for each channel which is located inside a circle in the complex plane with

$$\text{central point:} \frac{1}{1-\gamma_i^2}; \quad \text{radius:} \frac{\gamma_i}{1-\gamma_i^2}$$

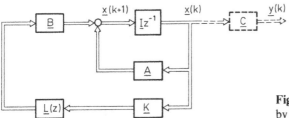

Fig. 10.3. Modified state feedback by the perturbation $L(z)$

with

$$\gamma_i^2 = \frac{r_i}{r_i + \lambda_{max}[B^T K \, B]} ; \; i = 1, ..., p \qquad (10.3.16)$$

Here $\lambda_{max}[\dots]$ stands for the maximum eigenvalue of $[\dots]$. (This has been shown in [10.23] for a nonlinear function L).

If, as a special case, gain factors μ_i are considered for the perturbation, then the optimal state feedback is stable for

$$\frac{1}{1+\gamma_i} \leq \mu_i \leq \frac{1}{1-\gamma_i} \qquad (i = 1, ..., p). \qquad (10.3.17)$$

Correspondingly this yields, for pure time delays with unity gain

$$|\varphi_i| \leq 2 \arcsin\left(\frac{\gamma_i}{2}\right). \qquad (10.3.18)$$

Now consider a state feedback with a single manipulated variable in which the gains are changed by the scale factor μ

$$k_\mu^T = \mu k^T. \qquad (10.3.19)$$

Then the system operated with the quadratic optimal state feedback

$$k^T = [r + b^T P \, b]^{-1} b^T P \, A \qquad (10.3.20)$$

$$x(k+1) = [A - \mu b \, k^T] x(k) \qquad (10.3.21)$$

is *stable*, if μ with

$$\gamma^2 = \frac{r}{r + b^T P \, b} \qquad (10.3.22)$$

fulfils the inequality (10.3.17). This and the following results were derived in [10.29].

Unlike the continuous-time controller case, μ depends on the design parameter r, the matrix P and the process parameters b. For prescribed process parameters the matrix P of the Matrix-Riccati-Equation (8.1.35) is influenced by the weighting factors Q and r of the performance criterion. The upper limit of μ is then the larger and the lower limit the smaller, the larger r and the smaller trace Q, i.e. the weaker will be the control action. For $r = 0$, $\mu = 1$. For $r \gg b^T P b$ $\gamma \approx 1$ and from (10.3.17) it follows that

$$1/2 < \mu < \infty$$

which is the appropriate range of the continuous-time system. These results for changed feedback gains and for a constant process can be, as with continuous signals, directly transformed to constant feedback and changed process with

$$b_\mu = \mu b \qquad (10.3.23)$$

Next the influence of the factor μ on the performance criterion is considered. It remains to determine the new weighting factors Q_μ and r_μ of the quadratic criterion

$$I_\mu = \sum_{k=0}^{\infty} [x^T(k)Q_\mu x^T(k) + r_\mu u^2(k)] \qquad (10.3.24)$$

The weighting parameters have to be selected to correspond to an *optimally changed* system.

If $k_\mu^T = \mu k^T$ is changed, according to (10.3.21) then it follows that [10.29]

$$r_\mu = r + (1-\mu)b^T P b \geq 0 \qquad (10.3.25)$$

$$Q_\mu = \mu Q + \mu(\mu-1)A^T P b k \qquad (10.3.26)$$

in the range

$$\mu \leq \frac{r + b^T P b}{b^T P b}. \qquad (10.3.27)$$

Now the process is to be changed and to be operated with the unchanged feedback k^T. For

$$b_\mu = \mu b \qquad (10.3.28)$$

then the performance criterion (10.3.24) is valid with

$$r_\mu = r + (1-\mu)b^T P b \qquad (10.3.29)$$

$$Q_\mu = Q + \left(1-\frac{1}{\mu}\right)A^T P b k \qquad (10.3.30)$$

in the range of μ as in (10.3.27). Hence r_μ is equal to the value given by (10.3.25). Q_μ, however, must not be changed. Correspondingly, one obtains for

$$A_\mu = \frac{1}{\mu}A \qquad (10.3.31)$$

the equations

$$r_\mu = r + (1-\mu)b^T P b \qquad (10.3.32)$$

$$Q_\mu = Q + (\mu-1)P + \left(1-\frac{1}{\mu}\right)A^T P A \qquad (10.3.33)$$

satisfying the range of μ as given by (10.3.27).

The comparison of (10.3.27) with (10.3.22) and (10.3.17) shows that the range tolerated to fulfil the performance criterion μ is smaller, see Fig. 10.4.

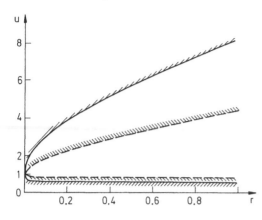

Fig. 10.4. Permissible changes in factor μ of the optimal state feedback k^T for test process VI dependent on the manipulated variable weighting r for $Q=\mathrm{diag}[1, 1, 5]$
——— stability limits,
------ optimality limits

This figure evidently shows that the permissible changes increase as the manipulated variable weighting r increases.

In summary, it proves that the optimal state feedback (for stable processes) is the more robust

— the larger the manipulated variable weighting r
— the smaller the state variable weighting q_i

which corresponds to the weaker the feedback control action. For $r=0$, $\mu=1$ so that changes are not allowed (unrobust behaviour). The discrete-time state feedback is thus less robust than the system with continuous-time full state feedback. For $r\gg b^T Pb$ the stability limit approaches that of the continuous-time state feedback system.

These results, however, are only valid for full state feedback control without state observers. If observers or Kalman filters are included, the robust ranges become substantially smaller because of feedback delay, see [10.28]. Nevertheless, the results presented give valuable advice for the design of robust control systems.

10.4 Robust Control Systems

It is expected that, with a robust control system design technique a constant controller for given M process models can be determined systematically according to (10.3.1). Following requirements have to be satisfied:

a) *Stabilization*: the constant controller must stabilize the process for all M parameter vectors Θ_j (or M processes)
b) *Performance requirement*: A determined control performance must be attained for all parameter vectors Θ_j
c) For discretized parameter vectors Θ_j the requirements for a) and b) have also to be met for the intermediate values.

Up to now most methods have referred to continuous-time signals. The older methods transform tolerances in the time domain into frequency response and operate with "tubes" in frequency responses [10.25, 10.28]. Later design methods proceed almost exclusively from state representation. Former surveys can be found e.g. in [2.10, 10.23, 10.26–10.28]. The following contains a brief discussion of methods for suitable parametric models based on a combination of continuous-time and sampled signals.

Optimal Robust State Controller for Multivariable Process Models

One of the first systematic design methods for constant state controllers for processes with large parameter changes is described in [10.29] and [10.30]. To begin with it is demonstrated that for the case of continuous-time signals the process

$$\dot{x}(t) = A\,x(t) + B\,u(t) \tag{10.4.1}$$

can be stabilized by all state feedback methods according to the control

$$u(t) = -\,K\,x(t) \tag{10.4.2}$$

which belong to the quantity

$$K = K_0 + L\,B^T P \tag{10.4.3}$$

Here, K_0 is an arbitrarily real constant, $L=L^T$ is positive definite and P is the solution of the Lyapunov-equation given by

$$P[A - B\,K_0] + [A - B\,K_0]^T P - 2\,P\,B\,L\,B^T P + S = 0 \tag{10.4.4}$$

which is positive definite in S. If (10.4.4) is assumed to be the Matrix–Riccati-Equation of an equivalent optimal state feedback law

$$u(t) = -\,[K_0 + 2\,L\,B^T P]x(t) \tag{10.4.5}$$

then the performance measure

$$I = \frac{1}{2}\,x^T(0)\,P\,x(0). \tag{10.4.6}$$

becomes valid.

Thus the solution P of (10.4.4) is the cost matrix of an optimal state feedback system. This optimal state feedback is asymptotically stable for positive definite P [10.8], showing that state feedback leads to asymptotically state behaviour according to (10.4.3).

Using (10.4.3), a process is equipped with constant feedback (10.4.1) in M operating points M, hence for m different terms (A, B)

$$\bar{K} = K_0 + \sum_{j=1}^{M} L_j B_j^T P_j \tag{10.4.7}$$

Here, P_j result from M coupled Matrix–Riccati-Equations. A common stabilizing constant feedback $\bar{K}$ exists, if all P_j are positive definite.

Assuming the existence of a constant state feedback matrix controller which stabilizes all operating points an "in average" optimal state feedback can be determined minimizing the overall criterion

$$
\left.
\begin{array}{l}
\bar{I} = \sum\limits_{j=1}^{M} \varepsilon_j I_j \quad 0 \leq \varepsilon_j \leq 1 \\[4mm]
\sum\limits_{j=1}^{M} \varepsilon_j = 1
\end{array}
\right\}
\tag{10.4.8}
$$

with I_j as the respective local quadratic control performance criteria

$$
I_j = \frac{1}{2} \int_0^\infty [x_j^T(t) Q_j x_j(t) + u_j^T(t) R_j u_j(t)] dt
\tag{10.4.9}
$$

which are weighted with ε_j for each operating point. Using (10.4.7) a constant feedback gain is

$$
\bar{K} = \bar{R}^{-1} \bar{B}^T \bar{P} = \left[\sum\limits_{j=1}^{M} \varepsilon_j R_j \right]^{-1} \sum\limits_{j=1}^{M} \varepsilon_j B_j^T P_j
\tag{10.4.10}
$$

This is formed with $K_0 = 0$. The P_j results from the solution of the M coupled Matrix–Riccati-Equations. A permissible solution then exists, if all matrices P_j are positive definite. $\bar{K}$ can be taken as the optimal feedback of a transformed system

$$
\bar{x}(t) = \bar{A} x(t) + \bar{B} u(t)
\tag{10.4.11}
$$

with

$$
\left.
\begin{array}{l}
\bar{A} = \left[\sum\limits_{j=1}^{M} \varepsilon_j P_j \right]^{-1} \sum\limits_{j=1}^{M} \varepsilon_j P_j A_j \\[6mm]
\bar{B} = \left[\sum\limits_{j=1}^{M} \varepsilon_j P_j \right]^{-1} \sum\limits_{j=1}^{M} \varepsilon_j P_j B_j
\end{array}
\right\}
\tag{10.4.12}
$$

It is not necessary for the transformed system to have a particular physical meaning. More details can be found in [10.30] and also in [10.28].

Examples for application (360° pendulum, heat exchanger) are given in [10.30] demonstrating the attainable control performances for large parameter variations. These methods can be applied directly for small sampling times. The result is then that of a discretized continuous robust feedback system.

Robust Controllers Through Variation of the Design Parameters

Another possibility results directly from analysing the robustness properties of constant controllers according to section 10.2. An "average" process behaviour is assumed and the "process models" are stated for extreme operating points. Then the *weighting factors* Q and R in the performance criterion are varied through an *iterative procedure* until the control performances e.g. according to (10.4.8) reach a minimum.

This, by the way is a method for controller tuning which, in principle, has been applied for a long time. A well-known procedure for controlling power plants is to set the controllers well-damped for approximately 40% load. Generally, this leads in the region of 20 to 100% load, to a satisfactory control performance. It was shown in [10.32] through simulation that a constant PI-controller results for the steam temperature at 35% load, if the common intersection of the local control performance ranges $0.9 \leq I/I_{\mathrm{opt}} \leq 1$ is used.

With reference to methods for the design of insensitive controllers through *variation of the design parameters* according to section 10.2.2 a performance criterion

$$I_{nv} = I_n + I_v \tag{10.4.13}$$

can be used (c.f. (10.2.20)), where one of the discrete operating points is formed as the *nominal case* with the following performance criterion

$$I_n = \sum_{k=0}^{\infty} [x_0^T(k) Q \, x_0(k) + u_0^T(k) R \, u_0(k)] \tag{10.4.14}$$

A performance criterion for the trajectory deviations,

$$I_v = \sum_{k=0}^{\infty} \sum_{i=1}^{p} \Delta x_i^T(k) Q_{vi} \Delta x_i(k) \tag{10.4.15}$$

is used in addition (c.f. [10.24]). Here, for the trajectory errors it follows that

$$\Delta x_i(k) = x_i(k) - x_0(k) \tag{10.4.16}$$

on referring to (10.2.6)

$$\Delta x_i(k+1) = A_0 \Delta x_i(k) + \Delta A_i[x_0(k) + \Delta x_i(k)]$$

$$+ B_0 \Delta u_i(k) + \Delta B_i[u_0(k) + \Delta u_i(k)]. \tag{10.4.17}$$

becomes valid.

The individual performance criteria are calculated and I_{nv} is minimized through systematic changes in the matrix Q in (10.3.14). An example in [10.24] demonstrates for continuous-time signals that compared with the design with sensitivity methods significant differences do not exist. Hence, this method requires the assignment of a nominal process model and it can be considered as an enlargement of the sensitivity methods.

Robust Controllers Through Pole Assignment

For some processes as e.g. aeroplanes, servo-systems or electrical networks it is known where the poles of the closed-loop system may be located. Therefore a pole area can be assigned in the stable range and it can be determined for M different process models in which areas the respective controller parameter are located. The parameter of a robust controller then are obtained from a common intersection [10.26], [10.27] with perhaps considering further criteria. The pole assignment method does not require a prescribed nominal process. A common stable controller, however, must exist. Relevant conditions can be found in [10.30].

11 Comparison of Different Controllers for Deterministic Disturbances

At the end of part B the various design methods and the resulting controllers or control algorithms for linear processes with and without deadtime are compared. Section 11.1 considers the controller structures and, in particular, the resulting poles and zeros of the closed-loop system. The control performance for two test processes are then compared quantitatively for different controllers in sections 11.2 and 11.3. The dynamic control factors for different controllers are compared in section 11.4. Finally section 11.5 draws conclusions as to the application of the various control algorithms.

11.1 Comparison of Controller Structures, Poles and Zeros

The transfer functions for the general linear input/output controller for the process and reference value w and the disturbance signals n and u as the closed-loop system inputs are given in (6.1.1) to (6.1.6). The orders v and μ of the individual controllers and the poles and zeros of the transfer functions of the closed-loop are considered in the following. The analysis also applies for mismatches between the dynamics of the actual process and process model as this is relevant to the practical application. The polynomials of $G_*(z)$ thus must be written with positive exponents $z^0, z^1, z^2, \ .$

Based on the controller

$$G_R(z) = \frac{Q(z)}{P(z)} \tag{11.1.1}$$

and the process

$$G_P(z) = \frac{B_0(z)}{A_0(z)z^d} \tag{11.1.2}$$

(see 6.2.9) results the general transfer function

$$G_*(z) = \frac{\mathscr{B}_*(z)}{\mathscr{A}(z)} \tag{11.1.3}$$

with the closed-loop characteristic equation

$$\mathcal{A}(z) = P(z)A_0(z)z^d + Q(z)B_0(z)$$

$$= (z - z_{\alpha 1})(z - z_{\alpha 2}) \dots (z - z_{\alpha \ell}) = 0 \qquad (11.1.4)$$

In this $z_{\alpha i}$ are the poles. The zeros of $G_*(z)$ i.e.

$$\mathcal{B}_*(z) = (z - z_{\beta 1})(z - z_{\beta 2}) \dots (z - z_{\beta s}) = 0 \qquad (11.1.5)$$

follow from

$$\left. \begin{array}{l} \mathcal{B}_w(z) = Q(z) \, B_0(z) = 0 \\ \mathcal{B}_n(z) = P(z) \, A_0(z)z^d = 0 \\ \mathcal{B}_u(z) = P(z) \, B_0(z) = 0. \end{array} \right\} \qquad (11.1.6)$$

They depend on the input location of the external signal. The subsequent discussion considers the existence and placement of poles and zeros for the closed-loop system. This procedure is also done for state controllers.

11.1.1 General Linear Controller for Specified Poles

Section 6.1 has already shown that the smallest possible order numbers are $v = m$ and $\mu = m + d$ and that, with pole assignment the zeros of $\mathcal{B}_*(z)$ are also assigned. Resulting from (6.1.10) the order of the characteristic equation becomes $l = 2m + d$. The general linear controller can be applied for any locations of poles and zeros of the process.

11.1.2 Low Order Parameter-optimized Controllers

For low-order parameter-optimized controllers, for example the 3PC-3 controller with PID-behaviour

$$G_R(z) = \frac{Q(z^{-1})}{P(z^{-1})} = \frac{q_0 + q_1 z^{-1} + q_2 z^{-2}}{1 - z^{-1}}$$

one must remember that, in contrast with the general parameter-optimized controller of (6.1.1), there are $l = m + d + 2$ poles and that because there are only 3 free controller parameters the coefficients of the characteristic equation for process orders $m > 1 - d$ are not independent. Furthermore, the zeros of $G_n(z)$ and $G_u(z)$ are dictated by the process and the controller pole at $z = 1$, as in (11.1.6). Only some zeros of $G_w(z)$ can be influenced by the controller parameters.

11.1.3 General Cancellation Controller

It was already shown in section 6.2 that pole-zero cancellation controllers with a given closed-loop transfer function for set point changes $G_w(z)$ (6.2.4) lead to the characteristic equation given by

$$A(z) = A_0(z)z^d B(z) + [A(z)z^d B_0(z) - A_0(z)z^d B(z)]G_w(z) = 0 \tag{11.1.7}$$

For an approximate agreement of the process and its model one has with $G_w(z) = \mathcal{B}_{wo}(z)/\mathcal{A}_{wo}(z)$

$$A_0(z)z^d B(z)A_{wo}(z) = 0. \tag{11.1.8}$$

Therefore general cancellation controllers can only be applied to processes with poles and zeros inside the unit circle. For certain closed-loop responses these restrictions can be relaxed as with the deadbeat and the predictor controllers.

11.1.4 Deadbeat Controller

For the deadbeat controller DB(v) one has

$$G_R(z) = \frac{Q(z^{-1})}{P(z^{-1})} = \frac{q_0 A(z^{-1})}{1 - q_0 B(z^{-1})z^{-d}}$$

(see (7.1.27)); after expansion with $z^{(m+d)}$ in the nominator and the denominator

$$G_R(z) = \frac{Q(z)}{P(z)} = \frac{q_0 A(z)z^d}{z^{(m+d)} - q_0 B(z)} \tag{11.1.9}$$

Here $A(z)z^d$ and $B(z)$ are polynomials of the process model. The characteristic equation becomes with (11.1.4)

$$\mathcal{A}(z) = z^{(m+d)}A_0(z)z^d - q_0 A_0(z)z^d B(z) + q_0 A(z)z^d B_0(z) = 0. \tag{11.1.10}$$

If the process and the model approximately agree so that $A_0(z)z^d$ and $B(z) \approx B_0(z)$ one then obtains:

$$\mathcal{A}(z) \approx z^{(m+d)}A_0(z)z^d = 0. \tag{11.1.11}$$

For the zeros it yields

$$\left. \begin{aligned} \mathcal{B}_w(z) &= q_0 A(z)z^d B_0(z) = 0 \\ \mathcal{B}_n(z) &= [z^{(m+d)} - q_0 B(z)]A_0(z)z^d = 0 \\ \mathcal{B}_u(z) &= [z^{(m+d)} - q_0 B(z)]B_0(z) = 0 \end{aligned} \right\} \tag{11.1.12}$$

and, using (11.1.11), the transfer functions are

$$G_w(z) = \frac{q_0 B_0(z) A(z) z^d}{z^{(m+d)} A_0(z) z^d} = \frac{q_0 B_0(z) A(z)}{z^{(m+d)} A_0(z)}$$

$$G_n(z) = \frac{[z^{(m+d)} - q_0 B(z)] A_0(z) z^d}{z^{(m+d)} A_0(z) z^d}$$

$$= \frac{[z^{(m+d)} - q_0 B(z)]}{z^{(m+d)}} = \frac{P(z)}{z^{(m+d)}}$$

$$(11.1.13)$$

$$G_u(z) = \frac{[z^{(m+d)} - q_0 B(z)] B_0(z)}{z^{(m+d)} A_0(z) z^d} = \frac{P(z)}{z^{(m+d)}} G_P(z).$$

If there is an exact match between the process model and actual process, i.e. $A(z) = A_0(z)$ and $B(z) = B_0(z)$, the polynomial $A_0(z)$ in $G_w(z)$ cancels so that

$$G_w(z) = \frac{q_0 B_0(z)}{z^{(m+d)}}.$$

$$(11.1.14)$$

and

$$\mathscr{A}(z) = 1 + G_R(z) G_P(z) = z^{(m+d)} = 0.$$

$$(11.1.15)$$

Deadbeat behaviour exists only for an exact match of process model and process. If there is no agreement the free oscillations decay besides to $z^{(m+d)}$ delayed by $A_0(z) z^d$, as shown by (11.1.11). Therefore deadbeat processes may only be applied to processes with poles sufficiently inside the unit circle in the z-plane, i.e. for asymptotically stable processes. The zeros of the transfer functions of the closed-loop are mainly determined by the zeros of the process. As can be seen from (11.1.10) the differences $\Delta B(z) = B(z) - B_0(z)$ between the numerator polynomial of the process and the process model influence the characteristic equation as follows

$$\mathscr{A}(z) = A_0(z) z^d [z^{(m+d)} - q_0 \Delta B(z)] = 0.$$

$$(11.1.16)$$

with $A_0(z) = A(z)$. Small changes $\Delta B(z)$ do not have a serious effect on stability. The zeros of the process can therefore be placed outside the unit circle, since they are not compensated by the deadbeat controller

11.1.5 Predictor Controller

The predictor controller is given by (9.2.2)

$$G_R(z) = \frac{Q(z^{-1})}{P(z^{-1})} = \frac{A(z^{-1})}{K_P A(z^{-1}) - B(z^{-1}) z^{-d}}$$

or alternatively by

$$G_R(z) = \frac{Q(z)}{P(z)} = \frac{A(z) z^d}{K_P A(z) z^d - B(z)}$$

$$(11.1.17)$$

which leads to the characteristic equation given by (11.1.4)

$$\mathcal{A}(z) = K_P A(z) z^d A_0(z) z^d - A_0(z) B(z) z^d + A(z) B_0(z) z^d = 0.$$
$$(11.1.18)$$

If the process and the process model approximately agree then

$$\mathcal{A}(z) \approx K_P A(z) z^d A_0(z) z^d = 0.$$
$$(11.1.19)$$

The transfer functions then become

$$G_w(z) = \frac{B_0(z)}{K_P A_0(z) z^d}$$

$$\left. G_n(z) = \frac{[K_P A(z) z^d - B(z)]}{K_P A(z) z^d} = \frac{P(z)}{K_P A(z) z^d} \right\} \qquad (11.1.20)$$

$$G_u(z) = G_n(z) G_P(z).$$

For $G_w(z)$ or $G_n(z)$ the poles $A(z)z^d$ or $A_0(z)z^d$ are always cancelled by the corresponding zeros. Closed loops using predictor control are only stable for asymptotically stable processes, as (11.1.19) shows. The process zeros can thus lie outside the unit circle. The zeros of the closed-loop system correspond only to the closed-loop response to set point changes as dictated by the process zeros. If the poles of the process are sufficiently within the unit circle small differences $\Delta Bz = B(z) - B_0(z)$ do not influence the stability (11.1.18).

11.1.6 State Controller

For a simple state control system with one controlled, one manipulated variable and the state controller

$$u(k) = - k^T x(k)$$

with no external disturbances it yields

$$x(k+1) = [A - b k^T] x(k)$$
$$y(k) = c^T x(k). \qquad (11.1.21)$$

The control is now influenced by an external disturbance signal $v(k)$ as

$$x(k+1) = [A - b k^T] x(k) + f \, v(k). \qquad (11.1.22)$$

If $f = b$ the disturbance is at the process input. By appropriate choice of f each variable can be disturbed. The transfer function of the process alone is, from (3.6.56)

$$G_P(z) = \frac{y(z)}{u(z)} = c^T [zI - A]^{-1} b = \frac{b_m + \dots + b_1 z^m}{a_m + \dots + a_1 z^m} = \frac{B(z)}{A(z)}.$$
$$(11.1.23)$$

Hence, for the state control system it follows that

$$G_v(z) = \frac{y(z)}{v(z)} = c^T[zI - A + b\,k^T]^{-1}f$$

$$= c^T \frac{\text{adj}[zI - A + b\,k^T]}{\det[zI - A + b\,k^T]}f = \frac{\mathscr{B}(z)}{\mathscr{A}(z)}. \tag{11.1.24}$$

Now (8.3.8) gives the characteristic equation

$$\mathscr{A}(z) = (a_m + k_m) + (a_{m-1} + k_{m-1})z + ... + z^m$$

$$= \alpha_m + \alpha_{m-1}z + ... + z^m = 0. \tag{11.1.25}$$

By suitable choice of k_i, arbitrary α_i for arbitrary a_i can be generated. Unstable processes can be stabilized. If the state control system is disturbed at the input, $f = b$, then the zeros of $G_v(z)$ given by

$$\mathscr{B}(z) = c^T\text{adj}[zI - A + b\,k^T]b \tag{11.1.26}$$

do not change compared with the process (see (11.1.23) and (11.1.24)). This holds because the parameters of the denominator polynomials $\mathscr{B}(z) = B(z)$ are contained either in c^T or in b, depending on the assumed canonical state representation. If, on the other hand, the disturbance influences one state variable the zeros of the transfer function $G_v(z)$ are also influenced by the state controller.

Example 11.1:
The process order m is 2, and controllable canonical form is chosen. Then it follows that

$$\mathscr{B}(z) = [b_2\ b_1]\begin{bmatrix} z + (a_1 + k_1) & 1 \\ -(a_2 + k_2) & z \end{bmatrix}f.$$

With $f = b = \begin{bmatrix} 0 \\ 1 \end{bmatrix}$ it holds

$$\mathscr{B}(z) = b_1 z + b_2 = B(z)$$

which is the process numerator polynomial. Choosing $f = \begin{bmatrix} 1 \\ 0 \end{bmatrix}$ it becomes

$$\mathscr{B}(z) = b_2 z + [b_2(a_1 + k_1) - b_1(a_2 + k_2)].$$

The choice of the poles of the control law also determines the zeros in the latter case.

The poles of state control systems with observers were considered in section 8.7. The observer adds further poles and zeros to the control system, (see (8.7.7), (8.7.18) and (8.7.19)). If the external disturbances are exactly measurable and can be input directly to the observer, and if the observer and process are in direct agreement, then the observer does not furnish additional poles for then $\Delta e(k) = 0$, (Fig. 8.7). If in this case the disturbance arises at the process input the zeros do not change either, and so $\mathscr{B}(z) = B(z)$.

In Table 11.1 the most important structural properties of various controllers are summarized for the process

Table 11.1. Structural properties of different deterministic controllers $G_R(z)=Q(z^{-1})/P(z^{-1})$
$A^-(z)$: Process poles near or outside the unit circle
$B^-(z)$: Process zeros near or outside the unit circle
n: no y: yes
* for exact agreement of process and process model

	controller	abbrev.	orders of controllers $Q(z^{-1})$	orders of controllers $P(z^{-1})$	char. eq. $\mathscr{A}(z)=0$*	zeros $\mathscr{B}_*(z)$ $G_w(z)$	zeros $\mathscr{B}_*(z)$ $G_n(z)$	zeros $\mathscr{B}_*(z)$ $G_u(z)$	risk of instability at $A^-(z)$	risk of instability at $B^-(z)$
input/output controlers	general linear controller	LC	m	$m+d$	$2m+d$	QB	PAz^d	PB	n	n
	param. opt. low order controller	3PC-3 (PID)	2	1	$m+d+2$	QB	PAz^d	PB	n	n
	general cancell. controller	CC	$\geqq m+1$	$\geqq m+d+1$	$\geqq 2m+d$	QB	PAz^d	PB	y	y
	deadbeat controller	DB(v)	m	$m+d$	$m+d$	q_0B	P	PB	y	n
	predictor controller	PREC	m	$m+d$	$m+d$ or $2(m+d)$	B	P	PB	y	n
state controllers	state contr. with no observer	SC w.n.o.	–	—	$m+d$	–	–	B	n	n
	state contr. with observ.	SC w.o	–	–	$2(m+d)$	fixed by process, controller and observer			n	n
									n	n

$B(z^{-1})z^{-d}/A(z^{-1})$. The *input/output controllers* have orders $v \geqq m+d$, and $\mu \geqq m+d$ if they are matched to the process in a structurally optimal way. The order of the characteristic equation and therefore the number of the poles is different; the smallest number is $(m+d)$ for the exactly matched deadbeat controller. The zeros of the processes appear in all cases as zeros of $G_w(z)$ and $G_u(z)$ as well. Furthermore the controller poles $P(z) = 0$ become the zeros of $G_n(z)$ and $G_u(z)$.

For linear processes the general linear controller and the parameter-optimized controller of low-order can be applied in general. The deadbeat controller and the predictor controller may only be applied to processes with poles within the unit circle, and the general cancellation controller only for processes with both poles and zeros within the unit circle. For *state controllers* with no observer the controller vector k^T has at least an order of $(m+d)$. The order of the characteristic equation is also $(m+d)$ and is therefore smaller than that of input/output controllers with the exception of the deadbeat controller. This advantage, however, disappears if an observer has to be used. The state controllers are applicable to a very wide class of processes.

If, however, $P(z^{-1}) = 1$ is set for the general linear controller and

$$G_R(z) = Q(z^{-1}) = q_0 + q_1 z^{-1} + \ldots + q_m z^{-m} \qquad (11.1.27)$$

that means a proportionally acting controller with m-times differencing (PD_m-controller), then one obtains with $d = 0$

$$G_u(z) = \frac{B(z^{-1})}{A(z^{-1}) + Q(z^{-1})B(z^{-1})}. \qquad (11.1.28)$$

Like with the state controller only the zeros of the process appear. Since differencing can be performed only after measuring the output variable and not, as with the state controller already in the process, $2m$ poles emerge which is double the amount compared with the state controller.

Example 11.2:
When being disturbed at the input with a PD_2-controller according (11.1.28), the closed-loop transfer function for $m = 2$ is

$$G_u(z) = \frac{b_1 z^{-1} + b_2 z^{-2}}{1 + (a_1 + q_0 b_1)z^{-1} + (a_2 + q_0 b_2 + q_1 b_1)z^{-2} + (q_1 b_2 + q_2 b_1)z^{-3} + q_2 b_2 z^{-4}}$$

and with a state controller (as in example 11.1) one obtains

$$G_u(z) = \frac{b_1 z^{-1} + b_2 z^{-2}}{1 + (a_1 + k_1)z^{-1} + (a_2 + k_2)z^{-2}}.$$

For the same amount of zeros, the PD_2-controller generates four poles, the state controller only two in the closed loop.

Thus, the state controller is able to generate a m-times differentiating behaviour without using additional poles. Apart from an excellent damping behaviour, this leads to a very large stability margin (amplitude margin, phase margin) and to especially robust properties for large parameter changes of the process, compare section 10.3. For nonmeasurable states and when using observers some of these positive properties, however, get lost.

11.2 Characteristic Values for Performance Comparison

The last section summarizes the structural differences of the various controllers; this section compares the most important controllers with respect to the control performance. The word 'performance' is taken to mean the following properties: The actual control performance and the required effort of the manipulated variable, the sensitivity to an inexactly known process model, the computational effort between sampling times and in the numerical part of the synthesis. Since a quantitative comparison is impossible without specifying particular systems, the two test processes described in section 5.4.2 and in the Appendix are used:

Process II: Second-order, non-minimum phase behaviour $T_0 = 2$ sec.
Process III: Third-order with deadtime, low pass behaviour, $T_0 = 4$ sec.

The properties of following control algorithms are compared:

1. *Parameter-optimized control algorithms of low order*
   2PC–2, PI-behaviour with no prescribed manipulated variable $\left.\right\}$ (5.2.20)
   2PC–1, PI-behaviour with prescribed manipulated variable $\left.\right\}$
   3PC–3, PID-behaviour with no prescribed manipulated variable $\left.\right\}$ (5.2.10)
   3PC–2, PID-behaviour with prescribed manipulated variable $\left.\right\}$
2. *Control algorithms for finite settling time* (deadbeat)
   DB(v), v-th order, with no prescribed manipulated variable   (7.1.26)
   DB($v+1$), ($v+1$)-th order, with prescribed manipulated variable   (7.2.11)
3. *State control algorithms with observer for external disturbances*
   SC-1, small weight r on the manipulated variable $\left.\right\}$ (8.7.9, 8.7.10)
   SC-2, larger weight r on the manipulated variable $\left.\right\}$ Fig. 8.7.5

The control algorithms are investigated for the single input/single output control systems of Fig. 5.1. The comparison is performed particularly with regard to the computer-aided design of algorithms by the process computer itself [8.5]. As process computers often have to perform other tasks the computational time for the synthesis should be small. Furthermore, the required storage should not be too large considering the capacity of smaller process computers and micro computers. A further criterion is the computation time of the algorithms between two samples. Not only the computational burden of the synthesis but also that required during operation have to be considered in connection with characteristic values of the control problem such as, for example, the control performance, required manipulation range, sensitivity to inexact process models and to parameter changes of the process.

In order to compare the control performance, the following characteristic values are used:

a) The root-mean squared control error

$$S_e = \sqrt{\overline{e^2(k)}} = \sqrt{\frac{1}{M+1} \sum_{k=0}^{M} e^2(k)}; \quad M = 63 \qquad (11.2.1)$$

b) The root-mean squared change in the manipulated variable (manipulating effort)

$$S_u = \sqrt{\overline{u^2(k)}} = \sqrt{\frac{1}{M+1} \sum_{k=0}^{M} \Delta u^2(k)} \qquad (11.2.2)$$

where $\Delta u(k) = u(k) - u(\infty)$.

c) Value of the quadratic performance criterion of (5.2.6) for $r = 0.1$ and 0.25.
d) Overshoot

$$y_m = y_{max}(k) - w(k) \tag{11.2.3}$$

e) Control settling time k_3 for $|e(k)| \leq 0.03|w(\infty)|$
   or $|e(k)| \leq 0.03|v(\infty)|$
f) Manipulated variable $u(0)$ for a step change in setpoint $w(k)$
g) Sensitivity with respect to an inexact process model

$$\varepsilon_1 = \sigma_{\delta_y}/\sigma_{\delta_g} \tag{11.2.4}$$

This value is described at the end of section 11.3.
For judging the computational effort between two samples the following measures will be used:
h) Number of additions and subtractions: ℓ_{add}
   Number of multiplications and divisions: ℓ_{mult}
   number of operations $\ell_\Sigma + \ell_{add} + \ell_{mult}$.

11.3 Comparison of the Performance of the Control Algorithms

The control algorithms considered in this chapter have been designed for a step change of the set point $w(k)$. This also corresponds to a step change of the disturbance $n(k)$ at the output of the process. The resulting frequency spectrum of this input contains high frequency components compared with the dynamics of the process. The control behaviour will also be described for a step change of the disturbance $v(k)$ at the process input. The weighting r of the manipulated variable has to be discussed separately. For the parameter-optimized control algorithms 2PC-2 and 3PC-3, $r = 0$ was set in the quadratic criterion (5.2.6) to provide for relatively large manipulated variables. For control algorithms 2PC-1 and 3PC-2 the weighting $r = 0$ was assumed. Here the first manipulated variable $u(0)$ was chosen such that $u(1) \approx u(0)$ (c.f. (5.2.31)). This gives relatively small values of the manipulated variable. For the design of state controllers the weighting matrix Q was chosen to give criterion (5.2.6). Furthermore, $R = r$ was chosen to give the same manipulated variable $u(0)$ as for control algorithms 3PC-3 and 3PC-2, so that a direct comparison is possible. The manipulated variable $u(0)$ for the deadbeat control algorithm DB($v+1$) was set to give $u(0) = u(1)$ in order to minimize the manipulated variable changes. The characteristic values of all algorithms are summarized in Table 11.2.

The time responses of the controlled and manipulated variables are shown in Figures 11.1 and 11.2 for the three main important control algorithms and with processes II and III for step changes the setpoint. Fig. 11.3 shows a graphical representation of the characteristic values given in section 11.2 for the processes II (□) and III (o) and stepwise set point changes $w(k)$ (lefthand side) or stepwise disturbance changes $v(k)$ (righthand side). These figures show several properties of the single control algorithms (for these processes under consideration).

Table 11.2. Parameters of the investigated control algorithms

Para-meters	Controlalgorithms							
	Process II				Process III			
	2PC-1	2PC-2	3PC-2	3PC-3	2PC-1	2PC-2	3PC-2	3PC-3
q_0	2.00	1.364	2.00	3.485	2.00	1.615	2.00	4.562
q_1	-1.886	-1.229	-2.596	-5.433	-1.802	-1.405	-2.400	-7.200
q_2	0	0	0.753	2.150	0	0	0.649	3.033
K	2.00	1.364	1.247	1.335	2.00	1.615	1.351	1.534
c_D	0.0	0.0	0.604	1.610	0.0	0.0	0.480	1.977
c_I	0.057	0.099	0.126	0.151	0.099	0.129	0.184	0.257
	DB($v+1$)		DB(v)		DB($v+1$)		DB(v)	
q_0	5.840		14.084		3.810		9.523	
q_1	-0.078		-20.070		-0.001		-14.285	
q_2	-8.851		6.985		-5.884		6.714	
q_3	4.089				3.647		-0.952	
q_4					0.571			
p_1	-0.595		-1.436		0		0	
p_2	0.169		2.436		0.247		0.619	
p_3	1.426		$-$		0.554		0.457	
p_4	$-$		$-$		0.244		-0.076	
p_5	$-$		$-$		-0.046		$-$	
	SC (1)		SC (2)		SC (1)		SC (2)	
k_1	4.157		2.398		4.828		2.526	
k_2	3.441		1.983		5.029		2.445	
k_3	1.0		1.0		4.475		2.097	
k_4	$-$		$-$		0.532		0.263	
k_5	$-$		$-$		1.532		1.263	

Behaviour for Setpoint Changes w(k)

For step changes of the setpoint (the case for which the design has been made) the most important results are summarized below:

Process III (lowpass-behaviour)

3PC–3 (PID-behaviour)

The choice of $r = 0$ results in a large $u(0)$ and a relatively weakly damped behaviour. The rms control error S_e is relatively large. The overshoot y_m and the settling time k_3 have average values.

3PC-2 (PID-behaviour, with a prescribed manipulated variable). Prescribing the manipulated variable $u(0)$ to relatively small values, leads compared with 3PC-3, to much more damped behaviour, somewhat larger S_e for much smaller S_u, smaller y_m and somewhat smaller k_3.

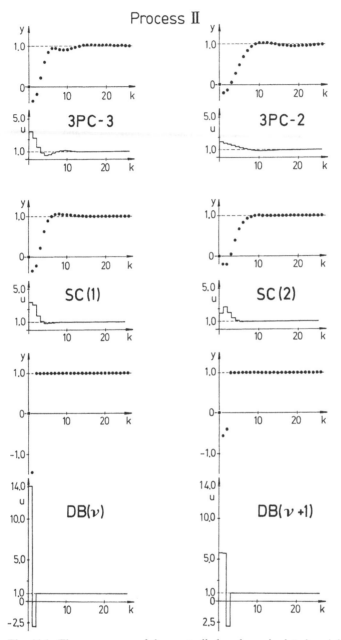

Fig. 11.1. Time responses of the controlled and manipulated variable for different control algorithms and process II (nonminimum phase behaviour)

2PC–2 (PI-behaviour)

In comparison to 3PC-3 this controller gives a somewhat larger volume of S_e together with smaller S_u, much smaller $u(0)$, larger y_m and larger k_3, in addition a somewhat smaller computational effort ℓ_Σ.

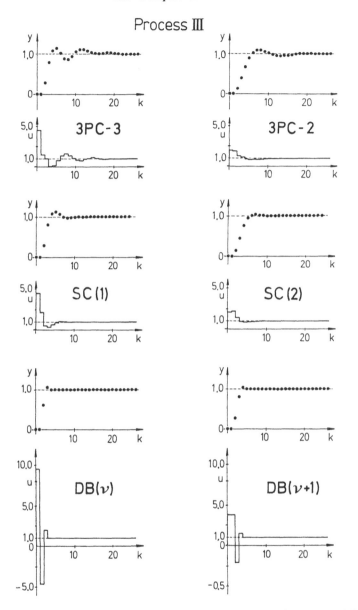

Fig. 11.2. Time responses of the controlled and manipulated variables of various control algorithms for process III (lowpass process)

2PC–1 (PI-behaviour, with prescribed manipulated variable)

Compared with 2PC-2 $u(0)$ was chosen to be larger, resulting in somewhat larger values of S_e and S_u, together with large values of y_m and k_m. This shows inferior performance compared with 2PC–2.

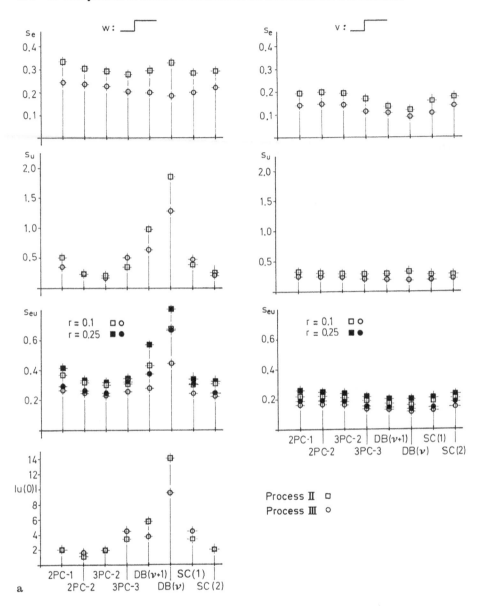

Fig. 11.3a. Characteristic values of the control behaviour of different control algorithms

SC − 1 (state controller for $r = 0.043$)

As $u(0)$ is about the same as for 3PC − 3 one can immediately compare with that control algorithm. From Fig. 11.2 results a better damped behaviour. S_e and S_u have changed a little but are now smaller, y_m is smaller and k_3 is much smaller. The computational effort ℓ_Σ, however, is six times larger.

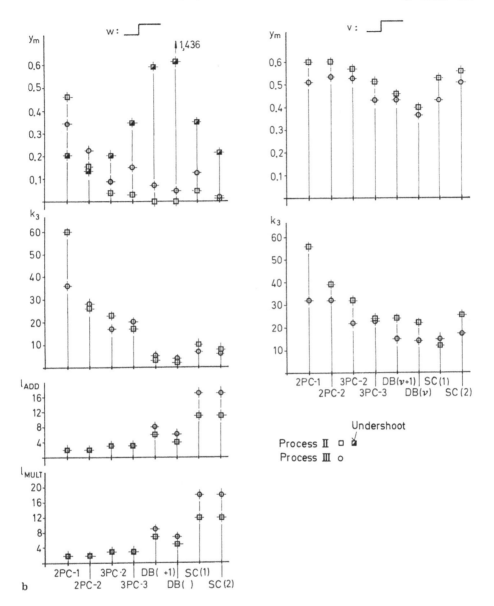

Fig. 11.3b. Characteristic values of the control behaviour of different control algorithms

SC – 2 (state controller for $r=0.18$)

$u(0)$ is the same as for 3PC – 2. Compared with that controller a more damped behaviour results. Furthermore S_e and S_u are about the same, whereas y_m and k_3 are essentially smaller. Compared with SC – 1: S_e is larger, S_u is smaller, y_m is smaller and k_3 is about the same.

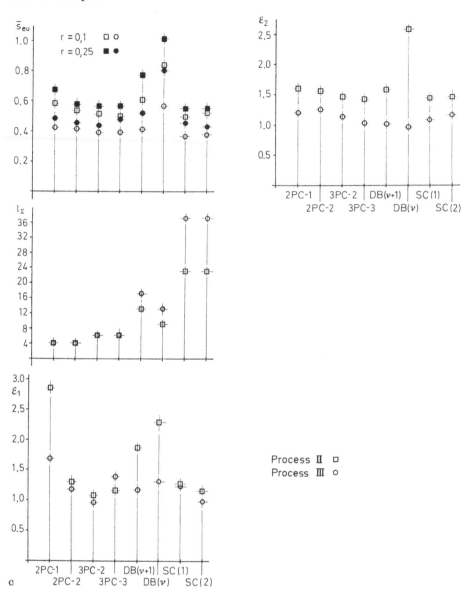

Fig. 11.3c. Characteristic values of the control behaviour of different control algorithms

DB(v) (deadbeat controller)

The steady state for $k=4$ is reached by taking into account a very large value $u(0)$ and large changes $\Delta u(k)$. When compared with all other control algorithms it gives the smallest S_e for the largest S_u, the largest $u(0)$ and a relatively small value of y_m together with the smallest k_3. The computational effort between samples is about twice that for 3PC − 3.

DB($v+1$) (deadbeat controller, with prescribed manipulated variable)

By increasing the settling time by one sampling unit to $k=5$ one can decrease the first manipulated variable $u(0)$ and the following $\Delta u(k)$ considerably compared with DB(v). S_e becomes somewhat larger for much smaller S_u; y_m and k_3 become larger. The computational effort ℓ_Σ increases because of the increase in order by one. Compared with 3PC − 3 it gives about the same S_e and larger S_u, but smaller $u(0)$, smaller y_m much smaller k_3 and three times of the computational effort ℓ_Σ.

A general evaluation of the control behaviour of all control algorithms is possible by using the quadratic performance criterion S_{eu}. This performance criterion expresses both the behaviour of the controlled and the manipulated variables, weighted by r. For smaller weighting of the manipulated variable $r=0.1$ the best results can be obtained with 3PC − 2, SC − 2 and SC − 1 and for larger weighting $r=0.25$ with 3PC − 2, SC − 2 and 2PC − 2. The parameter-optimized control algorithms 3PC − 3 and 3PC − 2 differ only little from the state control algorithms SC − 1 and SC − 2.

Process II (nonminimum phase behaviour)

From Fig. 11.1 it can be seen that the deadbeat algorithm DB(v) is unsuitable here, as the manipulated variable changes excessively. An increase of the manipulated effort S_u does not lead as it did for process III (and for all control algorithms) to a decrease of the mean squared control error S_e. Too large a manipulated effort or too large a value of $u(0)$ lead to an inferior control performance. The smallest values of S_e for still small S_u, i.e. relatively good S_{eu}, result for $r=0.1$ and 0.25 from controllers 2PC − 2, 3PC − 2, 3PC − 3, SC − 1 and SC − 2. Both a small undershoot and a small overshoot can be obtained using 3PC − 2.

Behaviour for the Disturbance v(k) at the Process Input

Step changes of the disturbance variable $v(k)$ lead to approximately the same characteristics with processes I and II.
Better behaviour: DB(v), DB($v+1$), SC − 1, 3PC − 3
                    (i.e. controllers with the highest gain)
Worse behaviour: 2PC − 2, 2PC − 1, 3PC − 2

The differences of the characteristic values are, however, smaller than for the setpoint changes as the step changes of the disturbance $v(k)$ excite the higher frequencies of the controlled variable less than the disturbance $w(k)$ for the design case. To evaluate the behaviour for both disturbances, $w(k)$ and $v(k)$, the following measure has been calculated

$$\bar{S}_{eu} = (S_{eu})_w + (S_{eu})_v.$$

A good control behaviour can, on average, be obtained using the following:

Process III: $r=0.1$:   SC − 1, SC − 2, 3PC − 2, 3PC − 3;
                   $r=0.25$: SC − 2, 3PC − 2, SC − 1, 2PC − 2.

Process II: $r=0.1$:   SC $-$ 1, 3PC $-$ 3, 3PC $-$ 2, SC $-$ 2;
$\qquad\qquad$ $r=0.25$: SC $-$ 1, SC $-$ 2, 3PC $-$ 2, 3PC $-$ 3, 2PC $-$ 2.

Sensitivity to Inexact Process Models

In most cases the process model is just an approximation to the real process, so control algorithms cannot be judged without considering their sensitivity to errors in the process model. In both theoretically and experimentally (identified) obtained process models, errors in individual parameters rarely occur independently, so that the sensitivity to single parameters can only lead to incomplete conclusions.

In the following, the sensitivity of the control algorithms under consideration is treated for imprecisely identified process models. The processes II and III were therefore identified several times with four different online parameter estimation methods for two different disturbance/signal ratios $\eta = 0.1$ and 0.2 and for three different identification times [3.13]. The synthesis of the control algorithms was then performed with these identified models, and the resulting control variable $\hat{y}(k)$ using the imprecise process models and the resulting controlled variable $y(k)$ using the exact process model (the real process) were calculated. The resulting identification or modelling error can be represented by

$$\Delta y(k) = \hat{y}(k) - y(k). \tag{11.3.1}$$

Hence, the rms control error

$$\delta_y = \left[\sum_{k=0}^{N} \Delta y^2(k) / \sum_{k=0}^{N} y_0^2(k) \right]^{1/2} \tag{11.3.2}$$

can be determined. $y_0(k)$ is the controlled variable for the exact model with its matched control algorithm. The error of the controlled variable δ_y is considered as a function of the error of the impulse response δ_g of the process model

$$\delta_g = \left[\overline{\Delta g^2(k)} / \overline{g^2(k)} \right]^{1/2} \tag{11.3.3}$$

$$\Delta g(k) = \hat{g}(k) \qquad - g(k). \tag{11.3.4}$$
$$\qquad\qquad \text{identi-} \qquad \text{exact}$$
$$\qquad\qquad \text{fied}$$

To reduce the influence of statistical fluctuations, the standard deviations σ_{δ_g} and σ_{δ_y} of these errors were calculated in each case for five identification runs. For the control algorithm 3PC $-$ 3 it was shown in [8.5] that for both investigated processes and for $0 \leq \sigma_{\delta_g} \leq 0.2$ there is an approximate linear dependence according to the relation $\sigma_{\delta_y} = f(\sigma_{\delta_g})$. This also occurred for all the other algorithms. A direct relationship between the single parameter errors of the models could not be seen. The errors of the weighting function in which the input/output behaviour of the process is expressed can therefore be used to show a relationship between the inexact process model and differences in the control performance of the loop. Then values of the model sensitivity of the loop given by

$$\varepsilon_1 = \sigma_{\delta_y}/\sigma_{\delta_g} \qquad\qquad (11.3.5)$$

can be determined. The smaller the value of ε_1 the smaller the influence of the inexact model on the behaviour of the closed loop. Fig. 11.3c shows that the sensitivity of process II is generally larger than for process III. The smallest sensitivity results for both processes when control algorithms $3PC-2$ and $SC-2$ are in use, the largest sensitivity being with $2PC-1$. A large sensitivity is shown by the deadbeat controller $DB(v)$ for process II. This can be explained by the nonminimum phase behaviour of this process. For process III, however, $DB(v)$ has about the same sensitivity as for $3PC-3$. The deadbeat controller $DB(v+1)$ is for both processes less sensitive than $DB(v)$.

Computational Effort Between Samples

To evaluate the computational effort during the operating phase of a control algorithm the following values are used:

$\ell_\Sigma = \ell_{add} + \ell_{mult}$: number of calculations
ℓ_{add}: number of the additions and subtractions
ℓ_{mult}: number of the multiplications and divisions

Table 11.3 shows that the parameter-optimized control algorithms have the smallest computational effort between two samples, whereas the state control algorithms have the highest and the deadbeat control algorithms have an average computational effort between two samples.

Synthesis Effort Required by the Control Algorithms

Synthesis effort depends on the storage and computation time required for the design of the control algorithms. Both depend on the software system (including mathematical routines) of the digital computer used. The values given in Table 11.3 are for a process computer Hewlett-Packard HP 2100 A with 24 K core memory, an external disk storage and hardware floating point arithmetic. The synthesis computation time is particularly small with deadbeat controllers, larger for state controllers and the greatest for parameter-optimized controllers. Note that the

Table 11.3. Computational effort between samples and the computational effort for synthesizing various control algorithms, process III and process computer HP 2100 A

Control algorithm	3PC-2	3PC-3	SC	DB(v)	DB(v+1)
Computation time between samples ℓ_Σ	6	6	34	14	18
Synthesis computation time s	20...30	40...60	1	0.004	0.004
Synthesis storage [words]	1881	1881	1996	342	342

parameter optimization used the Hooke-Jeeves method, which requires relatively little storage; the stopping rule was $|\Delta q| = 0.01$. The storage required for synthesis is like that of the synthesis computation time – smallest for the deadbeat controller, larger for the state controller and greatest for the parameter-optimized controller.

Relationship Between Control Performance and Manipulation Effort

Fig. 11.4 shows the control performance S_e as a function of the required manipulation effort S_u for different control algorithms in the design case with step changes of the setpoint. If the first-order control algorithms 2PC – 1 and 2PC – 2 are excluded then a direct relationship between S_e and S_u can be observed for the other control algorithms of second and higher orders. For process III an increase of S_u leads to a decrease of S_e in the following way

$$\left.\begin{array}{l} 3\,PC-2 \\ SC-2 \end{array}\right\} \begin{array}{l} \text{1st group} \\ u(0) = 2,0 \end{array}$$

$$\left.\begin{array}{l} SC-1 \\ 3\,PC-3 \\ DB(v+1) \end{array}\right\} \begin{array}{l} \text{2nd group} \\ u(0) = 3,81 \dots 4,56 \end{array}$$

$$\left.\begin{array}{l} DB(v) \end{array}\right\} \begin{array}{l} \text{3rd group} \\ u(0) = 9,52\,. \end{array}$$

Therefore groups can be associated with particular values of the initial manipulated variable $u(0)$. Furthermore, it can be seen that, starting with the first group, small improvements of the control performance S_e are always obtained by increasing the manipulation effort S_u. For process II there is a small improvement in S_e given by

$$\left.\begin{array}{l} 3\,PC-2 \\ SC-2 \end{array}\right\} \begin{array}{l} \text{1st group} \\ u(0) = 2,0 \end{array}$$

$$\left.\begin{array}{l} 3\,PC-3 \\ SC-1 \end{array}\right\} \begin{array}{l} \text{2nd group} \\ u(0) = 3,44 \dots 3,49\,. \end{array}$$

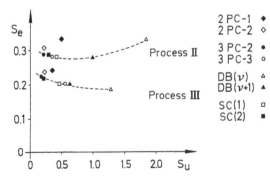

2 PC-1	◆
2 PC-2	◇
3 PC-2	●
3 PC-3	○
DB(v)	△
DB(v+1)	▲
SC(1)	□
SC(2)	■

Fig. 11.4. Relationship between the control performance S_e and the manipulation effort S_u for the investigated control algorithms and processes II and III

With an increasing manipulation effort S_u, however, the control performance becomes worse, first for $DB(v+1)$ with $u(0)=5.84$ and then for $DB(v)$ with $u(0)=14.09$. It can further be seen that for the same S_u the first-order control algorithm $2PC-1$ and $2PC-2$ lead to an inferior control performance S_e for both processes compared with control algorithms of second and higher order. For the same S_u the control performance for process II is worse than for process III. Fig. 11.4 therefore shows that for control algorithms of second and higher order and for other control algorithms there is a relationship between the reachable control performance S_e and the required manipulation effort S_u. However, this is true only for the design case in which there is a step change in the setpoint $w(k)$ [8.5].

11.4 Comparison of the Dynamic Control Factor

Section 11.3 compared the closed-loop performance of different control algorithms of step changes in the setpoint $w(k)$ and in the process input $v(k)$. For stochastic disturbances $n(k)$, chapter 13 shows the corresponding simulation results using parameter-optimized controllers.

To evaluate the control performance for different input signal spectra the *dynamic control factor* [5.14]

$$R(z) = \frac{1}{1 + G_R(z)G_P(z)} \qquad (11.4.1)$$

is useful, as the closed-loop response to an input signal is

$$\left. \begin{aligned} y(z) &= R(z)\ G_R(z)\ G_P(z)\ w(z) \\ y(z) &= R(z)\ n(z) \end{aligned} \right\} \qquad (11.4.2)$$

and a disturbance $v(k)$ passed through an arbitrary filter $G_{Pv}(z)=n(z)/v(z)$ gives

$$y(z) = R(z)\ G_{Pv}(z)\ v(z) = R(z)\ n(z). \qquad (11.4.3)$$

(11.4.3) includes (11.4.2) with $v(z)=n(z)$, $G_{Pv}(z)=1$ or $v(z)=w(z)$ and $G_{Pv}(z) = G_R(z)G_P(z)$. For deterministic disturbances the spectral densities are given by

$$|n(z)| = |G_{Pv}(z)|\ |v(z)| \qquad (11.4.4)$$

with $z = e^{T_0 i\omega}$ and $0 \leq \omega \leq \omega_s$, where ω_s is the Shannon frequency $\omega_s = \pi/T_0$ (see section 3.2). The amplitude spectral density of the controlled variable is therefore

$$|y(z)| = |R(z)|\ |n(z)| = |R(z)|\ |G_{Pv}(z)|\ |v(z)|. \qquad (11.4.5)$$

For stationary stochastic disturbances with power spectral density

$$S_{nn}(z) = \sum_{\tau=-\infty}^{\infty} R_{nn}(\tau)z^{-\tau} \qquad (11.4.6)$$

where

$$R_{nn}(\tau) = E\{[n(k) - \bar{n}] [n(k+\tau) - \bar{n}]\}$$

is the autocovariance function, it is

$$S_{nn}(z) = |G_{Pv}(z)|^2 S_{vv}(z) \tag{11.4.7}$$

and the power density of $y(k)$ becomes

$$S_{yy}(z) = |R(z)|^2 S_{nn}(z) = |R(z)|^2 |G_{Pv}(z)|^2 S_{vv}(z). \tag{11.4.8}$$

The magnitude of the dynamic control factor $|R(z)|$ or its squared value $|R(z)|^2$ indicate how much the amplitude or power spectra are reduced by the control loop. Therefore in the following the dependence of $|R(z)|$ on the frequency ω in the range $0 \leq \omega \leq \omega_s$ is shown for different controllers. The effect of different weighting of the manipulated variable is also shown.

To determine the dynamic control factor $R(z) = y(z)/n(z)$ also for state controllers with observers, the following procedure was applied: A low-pass process with several small time constants is described by the following transfer function

$$G_P(s) = \frac{y(s)}{u(s)} = \frac{1}{(1+4,2s)(1+1s)(1+0,9s)(1+0,6s)(1+0,55s)^2} \tag{11.4.9}$$

The process was simulated on an analog computer and identified by a digital computer after perturbation by a pseudorandom binary input signal. The method of "correlation and least-squares parameter estimation" and an order-search program [3.13, 30.1–30.3] led to the transfer function, for a sampling time of $T_0 = 2$ sec, of

$$G_P(z) = \frac{y(z)}{u(z)} = \frac{0,0600z^{-1} + 0,1617z^{-2} + 0,0328z^{-3}}{1 - 0,9470z^{-1} + 0,2164z^{-2} - 0,0005z^{-3}}. \tag{11.4.10}$$

Based on this model various control algorithms were then designed with the aid of the same digital computer for step changes in the setpoint (see chapter 30). $|R(z)|$ was then determined experimentally through measurement of the frequency response of the closed loop which comprised both the analogue computer and the

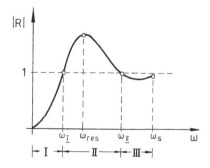

Fig. 11.5. Frequency regions of the dynamic control factor $|R(z)|$. $\omega_s = \pi/T_0$. ω_{res}: resonance frequency

process control computer, leading to the results described below. The dynamic control factor can, as is well-known, be divided into three main regions [5.14] (c.f. Fig. 11.5):

Region I: $0 \leq \omega < \omega_I \to 0 \leq |R| < 1$ (low frequencies)
            Disturbances $n(k)$ are reduced.
Region II: $\omega_I \leq \omega < \omega_{II} \to 1 < |R|$ (medium frequencies)
            Resonance effect. Disturbances $n(k)$ are amplified.
Region III: $\omega_{II} \leq \omega < \omega_s \to |R| \approx 1$ (high frequencies)
            Disturbances $n(k)$ are unaffected.

The effectiveness of the closed loop is therefore restricted to region I. Invariably, parameter changes of a controller are such that a decrease of $|R|$ in one region is followed by an increase in another region [5.14]. The graph of the magnitude of the dynamic control factor is shown for different controllers in Fig. 11.6. The controller

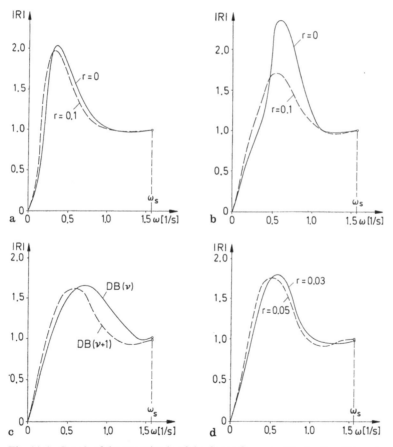

Fig. 11.6. Graph of the magnitude of the dynamic control factor for different controllers and different weightings on the manipulated variable or different $u(0)$. **a** Parameter optimized controller 2 PC-2 (PI); **b** Parameter optimized controller 3 PC-3 (PID); **c** Deadbeat-controller; **d** State-controller with observer

Table 11.4. Controller parameters for different dynamic control factors

Controller parameters	2 PC-2		3 PC-3	
	$r=0$	$r=0.1$	$r=0$	$r=0.1$
$q_0 = u(0)$	1.9336	1.5781	3.6072	2.4141
q_1	-1.5586	-1.2266	-4.8633	-2.9219
q_2	–	–	1.9219	1.0000
K	1.9336	1.5781	1.6957	1.4141
c_D	–	–	1.1475	0.7072
c_I	0.1939	0.2225	0.3992	0.3481
ω_{res}	0.35	0.33	0.55	0.60

Controller parameters	DB (v)	DB ($v+1$)
$q_0 = u(0)$	3.9292	2.2323
q_1	-3.7210	-0.4171
q_2	0.8502	-1.1240
q_3	-0.0020	0.3660
q_4	–	-0.0009
p_0	1.0000	1.0000
p_1	-0.2359	-0.1340
p_2	-0.6353	-0.4628
p_3	-0.1288	-0.3475
p_4	–	-0.0556
ω_{res}	0.73	0.58

Controller parameters	SC	
	$r=0.03$	$r=0.05$
k_1	2.6989	2.3466
k_2	3.1270	2.5798
k_3	2.3777	1.9358
k_4	1.0000	1.0000
$u(0)$	2.3777	1.9358
ω_{res}	0.57	0.50

Table 11.5. Change of $|R(z)|$ for different weights on the manipulated variable

		$|R(z)|$ becomes			
		region I	region II	region III	
controller	change at design	$0 \leqq \omega < \omega_I$	$\omega_I \leqq \omega < \omega_{res}$	$\omega_{res} \leqq \omega < \omega_{II}$	$\omega_{II} \leqq \omega \leqq \omega_s$
2 PC-2	$r=0 \rightarrow 0.1$	greater	greater	smaller	–
3 PC-3	$r=0 \rightarrow 0.1$	greater	greater/smaller	smaller	–
DB	$v \rightarrow v+1$	greater	greater	smaller	–
SC	$r=0.03 \rightarrow 0.05$	greater	greater	smaller	–

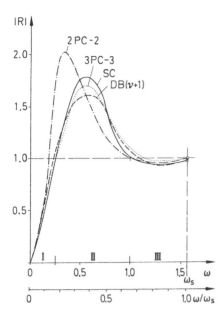

Fig. 11.7. Magnitude of the dynamic control factor for four different controllers 2PC-2: $u(0) = 1.93$, 3PC-3: $u(0) = 2.41$, DB($v + 1$): $u(0) = 2.23$, SC: $u(0) = 2.38$

parameters are summarized in Table 11.4 and the effect of a higher weighting on the manipulated variable is shown in Table 11.5. It can be seen that $|R|$ increases in region I, and therefore disturbances at low frequencies are less damped and the control performance becomes worse. The same happens in the ascending region at low frequencies in region II. However, in the descending part in region II, beyond the resonance peak ($\omega > \omega_{res}$) $|R|$ decreases for all controllers and the control performance is correspondingly improved. There are insignificant changes in region III. For all controllers it can be concluded that a higher weight on the manipulated variable or a smaller $u(0)$ decreases the resonance peak and moves it to a lower frequency. To appreciate the variation in the dynamic control factor for a state controller the reader is referred to (10.1.11) to (10.1.14) and the corresponding remarks and references.

This discussion again shows that evaluation of control behaviour depends significantly on the frequency spectrum of the exciting signals, especially from (11.4.4). Only if very low frequency signals act on the closed loop can a small value of r or a large value of $u(0)$ be chosen. Components near the resonance frequency require a large r or a small $u(0)$. If medium or high frequency signals are acting which are not specially filtered (see chapter 28), the deadbeat controller DB(v) should not be applied (Fig. 11.6c). For the other controllers r can be chosen to be larger or $u(0)$ smaller.

Fig. 11.7 shows the dynamic control for different controllers. The weight on the manipulated variable was chosen such that after a step change in the setpoint the manipulated variable $u(0)$ is about the same, i.e. $u(0) \approx 1.93 \ldots 2.41$. $|R(z)|$ does not differ very much for 3PC $-$ 3, DB($v + 1$) and SC. Only 2PC $-$ 2 shows a significantly higher resonance peak at lower frequencies. SC is the best in region I, DB($v + 1$) in region II, and in region III SC is once again the best.

The dynamic control factor is not only useful for evaluating control performance as a function of the disturbance signal spectrum. (10.1.10) shows that the dynamic control factor is identical to the *sensitivity function* $S(\mathbf{\Theta}_n, z)$ of the closed loop which determines the effect of changes in the process behaviour. Small $|R(z)|$ not only means a good control performance but also a small sensitivity (see chapter 10).

11.5 Conclusions for the Application of Control Algorithms

The most important properties of the studied control algorithms are summarized in Table 11.6 for the given test processes with proportional action, low-pass behaviour and nonminimum phase behaviour.

Parameter-optimized Control Algorithms

The three-parameter-control algorithms with PID-behaviour are better than the two-parameter-control algorithms with PI-behaviour, as they give better control for smaller manipulation effort, quicker settling with smaller overshoot and smaller sensitivity to inexact process models. Parameter-optimized control algorithms of low-order are characterized by an especially small computing time between samples, but the synthesis effort is relatively large in the numerical parameter optimization and performance criterion computation through simulation (section 5.4.1). However, a design with smaller computational effort is described in section 5.4 and 26.5. Unlike all other control algorithms, simple tuning rules can be applied to low-order parameter-optimized control algorithms. Parameter-optimized control algorithms are therefore recommended for the following cases:

— Easy controllable processes (low/medium order, small deadtime)
— Common requirements on the control performance
— Inexact or precise process models
— Parameter setting through tuning rules
— Many control loops
— Controller synthesis performed either only once or more often
— Feedforward adaptation of controller parameters as a function of the operating point
— easy to understand for operating personnel.

These controllers are therefore suitable for a wide range of processes.

State-control Algorithms

The control performance attainable with state control algorithms differs only a little from that of the three-parameter control algorithms for the considered test processes. For the same initial manipulated variable $u(0)$, state control algorithms result in somewhat more damping of the controlled variable and in a smaller settling time. The computational effort between samples is higher for processes beginning with second order. For computer-aided design the synthesis effort is

Table 11.6. Evaluation of the most important properties of the control algorithms. 1: "good" "small" 2: "medium" 'medium" 3: "bad" "large"

Control algorithm		Control behaviour		Sensitivity to inexact process modes		Computational effort between samples	Synthesis effort
		process III	process II	process III	process II		
PI	2PC-1	2	2	2	3	1	3,2
	2PC-2	2	2	2	2	1	3,2
PID	3PC-3	1	1	2	2	1	3,2
	3PC-2	1	1	1	1	1	3,2
Dead-	DB(v)	3	3	2	3	2	1
beat	DB($v+1$)	2	3	2	3	2	1
State-	SC-1	1	1	2	2	2	2
contr.	SC-2	1	1	1	1	2	2

medium-sized. The appropriate selection of the numerous weighting factors should be done in an interactive way. State-control algorithms are especially preferred in the following cases:

— Processes difficult to control (high order, large deadtime, allpass behaviour)
— High requirements on the control performance
— Precise process models
— Design by digital computers
— Small number of control loops
— Unstable processes which need a feedback of state variables for stabilization.

Deadbeat-control Algorithms

Because of large changes in the manipulated variable, the deadbeat control algorithm of order $v=m+d$ cannot be recommended for small sample times. If the sample time is large enough, however, deadbeat-control algorithms of order $v+1$ are more suited, as they result in smaller changes of the manipulated variable. Settling time and overshoot are smaller for deadbeat-control algorithms than for the three-parameter algorithms. The computation time between samples is about three times larger for the fourth-order process. A main advantage of the deadbeat-control algorithms is computational simplicity of their synthesis. Therefore the following applications can be recommended:

— Asymptotically stable processes (lowpass processes)
— Precise process models
— Design by digital computers
— Controller synthesis to be repeated many times (adaptive control).

These statements are valid for the test processes investigated here, but they can be generalized for similar linear processes without excessive error.

Since higher-order control algorithms only furnish good control performance if precise mathematical process models are known, modelling becomes very significant. If process identification- and controller design methods are combined in the computer, self-tuning or adaptive control algorithms are obtained. Chapter 26 shows how the control behaviour can be improved with deadbeat- and state controllers compared with (adaptive) PID-controllers.

Chapter 9 has already discussed briefly proportional action processes with large deadtime, and further results comparing various control algorithms for stochastic disturbances are given in chapter 13 and 14. It is difficult to choose control algorithms for a multivariable process, as multivariable processes can differ considerably. The advantages and disadvantages of parameter-optimized and state-control algorithms must be investigated in each special case (see part E).

Appendix A

A1 Table of z-Transforms and Laplace-Transforms

The following table contains some frequently used time functions $x(t)$, their Laplace-transforms $x(s)$ and z-transforms $x(z)$. *The sample time is T_0.* More functions can be found in [2.15], [2.11] and [2.25].

$x(t)$	$x(s)$	$x(z)$
1	$\dfrac{1}{s}$	$\dfrac{z}{z-1}$
t	$\dfrac{1}{s^2}$	$\dfrac{T_0 z}{(z-1)^2}$
t^2	$\dfrac{2}{s^3}$	$\dfrac{T_0^2 z(z+1)}{(z-1)^3}$
t^3	$\dfrac{6}{s^4}$	$\dfrac{T_0^3 z(z^2+4z+1)}{(z-1)^4}$
e^{-at}	$\dfrac{1}{s+a}$	$\dfrac{z}{z-e^{-aT_0}}$
$t \cdot e^{-at}$	$\dfrac{1}{(s+a)^2}$	$\dfrac{T_0 z e^{-aT_0}}{(z-e^{-aT_0})^2}$
$t^2 \cdot e^{-at}$	$\dfrac{2}{(s+a)^3}$	$\dfrac{T_0^2 z e^{-aT_0}(z+e^{-aT_0})}{(z-e^{-aT_0})^3}$
$1-e^{-at}$	$\dfrac{a}{s(s+a)}$	$\dfrac{(1-e^{-aT_0})z}{(z-1)(z-e^{-aT_0})}$
$at-1+e^{-at}$	$\dfrac{a^2}{s^2(s+a)}$	$\dfrac{(aT_0-1+e^{-aT_0})z^2+(1-aT_0 e^{-aT_0}-e^{-aT_0})z}{(z-1)^2(z-e^{-aT_0})}$
$e^{-at}-e^{-bt}$	$\dfrac{b-a}{(s+a)(s+b)}$	$\dfrac{z(e^{-aT_0}-e^{-bT_0})}{(z-e^{-aT_0})(z-e^{-bT_0})}$
$1-(1+at)e^{-at}$	$\dfrac{a^2}{s(s+a)^2}$	$\dfrac{z}{z-1}-\dfrac{z}{z-e^{-aT_0}}-\dfrac{aT_0 e^{-aT_0}z}{(z-e^{-aT_0})^2}$

Table (continued)

$x(t)$	$x(s)$	$x(z)$
$\sin \omega_1 t$	$\dfrac{\omega_1}{s^2+\omega_1^2}$	$\dfrac{z \sin \omega_1 T_0}{z^2 - 2z \cos \omega_1 T_0 + 1}$
$\cos \omega_1 t$	$\dfrac{s}{s^2+\omega_1^2}$	$\dfrac{z(z - \cos \omega_1 T_0)}{z^2 - 2z \cos \omega_1 T_0 + 1}$
$e^{-at} \sin \omega_1 T$	$\dfrac{\omega_1}{(s+a)^2+\omega_1^2}$	$\dfrac{z \cdot e^{-aT_0} \sin \omega_1 T_0}{z^2 - 2z \cdot e^{-aT_0} \cos \omega_1 T_0 + e^{-2aT_0}}$
$e^{-at} \cos \omega_1 t$	$\dfrac{s+a}{(s+a)^2+\omega_1^2}$	$\dfrac{z^2 - z \cdot e^{-aT_0} \cos \omega_1 T_0}{z^2 - 2z \cdot e^{-aT_0} \cos \omega_1 T_0 + e^{-2aT_0}}$

A2 Table of Some Transfer Elements with Continuous and Sampled Systems
see p. 293 and 294

A3 Test Processes for Simulation

Various "test processes" have been used in this book to simulate the typical dynamical behaviour of processes in order to test control systems with various control algorithms, identification and parameter estimation methods and adaptive control algorithms. These test processes are models of processes with various pole-zero configurations and dead times. They were chosen with regard to several viewpoints and summarized in the following. The discrete-time transfer functions $G(z)$ were determined by z-transformation from the continuous-time transfer function $G(s)$ with a zero-order hold, c.f. (3.4.10) or section 3.7.3 (except process I).

Process I: second-order, oscillating behaviour

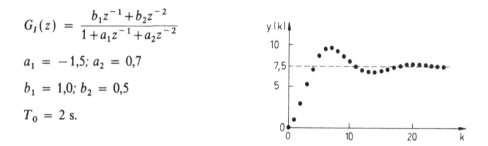

$$G_I(z) = \frac{b_1 z^{-1} + b_2 z^{-2}}{1 + a_1 z^{-1} + a_2 z^{-2}}$$

$a_1 = -1{,}5;\; a_2 = 0{,}7$

$b_1 = 1{,}0;\; b_2 = 0{,}5$

$T_0 = 2$ s.

There is no corresponding transfer function $G(s)$ in this case. From Åström, K. J. and Bohlin, T. (1966), c.f. [3.13].

A2 Table of some dynamic elements with continuous-time and discrete-time signals

Process Identification	Continuous-time signals s-transfer function	transient function	s-domain (o zero, x pole) (no pole no zero)	discrete-time signals, sample time T_0 z-transfer function (with zero-order hold)	b_0	transient function	z-domain (o zero, x pole) T_0 small / T_0 large (no pole no zero)
P	K		no pole no zero	b_0	$b_0 = K$		no pole no zero
PT_1	$\dfrac{K}{1+T_1 s}$		$x\ {-}\frac{1}{T_1}$	$\dfrac{b_1 z^{-1}}{1 + a_1 z^{-1}}$	$b_1 = K[1 - \exp(-\frac{T_0}{T_1})]$ $a_1 = -\exp(-\frac{T_0}{T_1})$		
PT_2 damping $D>1$	$\dfrac{K}{(1+T_1 s)(1+T_2 s)}$		$x\ {-}\frac{1}{T_2}\quad x\ {-}\frac{1}{T_1}$	$\dfrac{b_1 z^{-1} + b_2 z^{-2}}{1 + a_1 z^{-1} + a_2 z^{-2}}$	$z_1 = \exp(-\frac{T_0}{T_1})\quad z_2 = \exp(-\frac{T_0}{T_2})$ $b_1 = \frac{K}{T_1 - T_2}[T_1(1-z_1) - T_2(1-z_2)]$ $b_2 = \frac{K}{T_1 - T_2}[T_2 z_1(1-z_2) - T_1 z_2(1-z_1)]$ $a_1 = -(z_1 + z_2)\quad a_2 = z_1 z_2$	$T_{O2} < \frac{T_1 + T_2}{2}$ $T_{O2} > \frac{T_1 + T_2}{2}$	
$D=1$	$\dfrac{K}{(1+Ts)^2}$		$x\ {-}\frac{1}{T}$	$\dfrac{b_1 z^{-1} + b_2 z^{-2}}{1 + a_1 z^{-1} + a_2 z^{-2}}$	$z_0 = \exp(-\frac{T_0}{T})$ $b_1 = K[1 - z_0(\frac{T_0}{T} + 1)]$ $b_2 = K z_0[z_0 + \frac{T_0}{T} - 1]$ $a_1 = -2 z_0\quad a_2 = z_0^2$	$T_0 < T$ $T_0 > T$	
$D<1$	$\dfrac{K}{1 + 2DTs + T^2 s^2}$		$\overset{\times}{\underset{\times}{\ }}\ \frac{\sqrt{1-D^2}}{T}$ ${-}\frac{D}{T}$	$\dfrac{b_1 z^{-1} + b_2 z^{-2}}{1 + a_1 z^{-1} + a_2 z^{-2}}$	$\delta = \frac{D}{T}\quad \omega = \frac{\sqrt{1-D^2}}{T}\quad z_0 = \exp(-D \frac{T_0}{T})$ $b_1 = K[1 - z_0(\cos\omega T_0 + \frac{\delta}{\omega}\sin\omega T_0)]$ $b_2 = K z_0[z_0 - \cos\omega T_0 + \frac{\delta}{\omega}\sin\omega T_0]$ $a_1 = -2 z_0 \cos\omega T_0\quad a_2 = z_0^2$	$T_0 < \frac{2\pi}{\omega}$ $\frac{3\pi}{2\omega} > T_0 > \frac{\pi}{2\omega}$	$T_0 < \frac{\pi}{2\omega}$ $\frac{3\pi}{2\omega} > T_0 > \frac{\pi}{2\omega}$
I	$\dfrac{1}{T_I s}$		$\overset{\times}{}$	$\dfrac{b_1 z^{-1}}{1 - z^{-1}}$	$b_1 = \frac{T_0}{T_I}$		
$I T_1$	$\dfrac{1}{T_I s(1 + T_1 s)}$		$x\ {-}\frac{1}{T_1}$	$\dfrac{b_1 z^{-1} + b_2 z^{-2}}{1 + a_1 z^{-1} + a_2 z^{-2}}$	$z_1 = \exp(-\frac{T_0}{T_1})$ $b_1 = [T_0 - T_1(1-z_1)]/T_I$ $b_2 = [T_1(1-z_1) - T_0 z_1]/T_I$ $a_1 = -(1 + z_1)\quad a_2 = z_1$	$T_0 < T_1$ $T_0 > T_1$	
PI	$K(1 + \frac{1}{T_I s})$		$o\ {-}\frac{1}{T_I}$	$\dfrac{b_0 + b_1 z^{-1}}{1 - z^{-1}}$	$b_0 = K$ $b_1 = K(\frac{T_0}{T_I} - 1)$	$T_0 > T_I$	$T_0 < T_I$ $T_0 > 2 T_I$

Process	Continuous-time signals			discrete-time signals, sample time T_0			
Identifi-cation	s-transfer function	transfer function	s-domain o zero; x pole	z-transfer function (with zero-order hold)		transfer function	z-domain o zero; x pole T_0 small, T_0 large

PIT_1

s-transfer function: $(1+\frac{1}{T_I s})\frac{K}{1+T_1 s}$

$$\frac{b_2 z^{-1}+b_2 z^{-2}}{1+a_1 z^{-1}+a_2 z^{-2}}$$

$z:=\exp(-\frac{T_0}{T_1})$

$b_1:=\frac{K}{T_I}[T_0+(\frac{T_1}{T_I}-T_1)(1-z_1)]$

$b_2:=-\frac{K}{T_I}[T_0 z_1+(\frac{T_1}{T_I}-T_1)(1-z_1)]$

$a_1:=-(1+z_1)\qquad a_2:=z_1$

$T_0<T_1$; $T_0>T_1$

DT_1

s-transfer function: $\frac{T_D s}{1+T_1 s}$

$$\frac{b_0+b_1 z^{-1}}{1+a_1 z^{-1}}$$

$b_0:=\frac{T_D}{T_1}\qquad b_1:=-\frac{T_D}{T_1}$

$a_1:=-\exp(-\frac{T_0}{T_1})$

$\frac{T_D}{T_1}$

PDT_1

s-transfer function: $K\frac{1+T_D s}{1+T_1 s}$; $K\frac{1-T_s}{1+T s}$

$$\frac{b_0+b_1 z^{-1}}{1+a_1 z^{-1}}$$

$b_0:=K\frac{T_D}{T_1}$

$b_1:=K(1-\frac{T_D}{T_1}\cdot\exp(-\frac{T_0}{T_1}))\qquad a_1:=-\exp(-\frac{T_0}{T_1})$

$b_0:=-K$

$b_1:=K(2-\exp(-\frac{T_0}{T_1}))$

$a_1:=-\exp(-\frac{T_0}{T_1})$

$T_0<T_1$; $T_0>T$

ALLPASS 1st ORDER

o bei $|s|=+\infty$
x bei $|s|=-\infty$

PDT_2

s-transfer function: $K\frac{1+T_D s}{(1+T_1 s)(1+T_2 s)}$

$$\frac{b_1 z^{-1}+b_2 z^{-2}}{1+a_1 z^{-1}+a_2 z^{-2}}$$

$z_1:=\exp(-\frac{T_0}{T_1})\qquad z_2:=\exp(-\frac{T_0}{T_2})$

$b_1:=\frac{K}{T_1-T_2}[z_1(T_0-T_1)-z_2(T_0-T_2)+T_1-T_2]$

$b_2:=\frac{K}{T_1-T_2}[z_1(T_1 z_2+T_2-T_D)-z_2(T_2 z_1+T_1-T_D)]$

$a_1:=-(z_1+z_2)\qquad a_2:=z_1 z_2$

$0<T_2<T_D<T_1$

$T_0<\frac{T_1+T_2}{2}$; $T_0>\frac{T_1+T_2}{2}$

T_t

s-transfer function: $K\exp(-sT_1)$

$$Kz^{-d}$$

$T_1=dT_0$

d-times

K

Controller

PID

s-transfer function: $K(1+T_D s+\frac{1}{T_I s})$

$$\frac{q_0+q_1 z^{-1}+q_2 z^{-2}}{1-z^{-1}}$$

$q_0=K(1+\frac{T_0}{2T_I}+\frac{T_D}{T_0})$

$q_1=-K(1+\frac{T_0}{2T_I}+2\frac{T_D}{T_0})$

$q_2=K\frac{T_D}{T_0}$

(Discretization with trapezoidal integration)

q_0

Process II: second order, nonminimum phase behaviour

$$G_{II}(s) = \frac{K(1-T_1s)}{(1+T_1s)(1+T_2s)}$$

$K = 1; \; T_1 = 4 \text{ s}; \; T_2 = 10 \text{ s}$

$$G_{II}(z) = \frac{b_1 z^{-1} + b_2 z^{-2}}{1 + a_1 z^{-1} + a_2 z^{-2}}$$

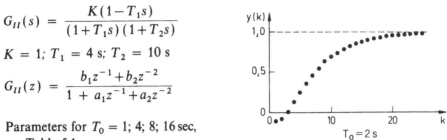

Parameters for $T_0 = 1; \; 4; \; 8; \; 16$ sec,
see Table 5.1.

Process III: third-order with deadtime, low-pass behaviour

$$G_{III}(s) = \frac{K(1+T_4s)}{(1+T_1s)(1+T_2s)(1+T_3s)} e^{-T_t s}$$

$K = 1; \; T_1 = 10 \text{ s}; \; T_2 = 7 \text{ s};$

$T_t = 4 \text{ s}; \; T_3 = 3 \text{ s}; \; T_4 = 2 \text{ s}.$

$$G_{III}(z) = \frac{b_1 z^{-1} + b_2 z^{-2} + b_3 z^{-3}}{1 + a_1 z^{-1} + a_2 z^{-2} + a_3 z^{-3}} z^{-d}$$

Parameters for $T_0 = 1; \; 4; \; 8; \; 16$ sec;
see Table 5.2
Process I, II and III: c.f. [24.9, 3.13].

Process IV: fifth-order, low-pass behaviour. Model of a steam superheater

$$G_{IV}(s) = -\frac{(1+13{,}81\ s)^2(1+18{,}4s)}{(1+59s)^5} \left[\frac{K}{\%}\right]$$

$$G_{IV}(z) = \frac{b_2 z^{-1} + b_2 z^{-2} + b_3 z^{-3} + b_4 z^{-4} + b_5 z^{-5}}{1 + a_1 z^{-1} + a_3 z^{-2} + a_3 z^{-3} + a_4 z^{-4} + a_5 z^{-5}}$$

$T_0 = 20 \text{ s}:$
$a_1 = -3{,}562473 \quad b_1 = -1{,}73 \cdot 10^{-3}$
$a_2 = 5{,}076484 \quad b_2 = -1{,}831 \cdot 10^{-3}$
$a_3 = -3{,}616967 \quad b_3 = 2{,}143 \cdot 10^{-3}$
$a_4 = 1{,}288535 \quad b_4 = -5{,}95 \cdot 10^{-4}$
$a_5 = -0{,}183615 \quad b_5 = 4{,}9 \cdot 10^{-5}$

$T_0 = 40 \text{ s}:$
$a_1 = -2{,}538242 \quad b_1 = -9{,}725 \cdot 10^{-3}$
$a_2 = 2{,}577069 \quad b_2 = -2{,}1679 \cdot 10^{-2}$
$a_3 = -1{,}308245 \quad b_3 = 2{,}18 \cdot 10^{-3}$
$a_4 = 0{,}332064 \quad b_4 = 3{,}28 \cdot 10^{-4}$
$a_5 = -0{,}033714 \quad b_5 = -3{,}6 \cdot 10^{-5}$

Process V: twovariable process "evaporator and superheater of a drum steam generator" due to Fig. 18.1.1

$$G_{11}(s) = G_{IV}(s)$$

$$G_{21}(s) = \frac{1,771}{(1+153,5s)(1+24s)(1+15s)} \left[\frac{K}{\%}\right]$$

$$G_{22}(s) = \frac{0,96}{695s(1+15s)} \left[\frac{bar}{\%}\right]$$

$$G_{12}(s) = \frac{0,0605}{695s} \left[\frac{bar}{\%}\right]$$

$T_0 = 20$ s:
$G_{11}(z) = G_{IV}(z)$ see process IV

$$G_{21}(z) = \frac{2,476 \cdot 10^{-2} z^{-1} + 5,744 \cdot 10^{-2} z^{-2} + 7,859 \cdot 10^{-3} z^{-3}}{1 - 1,576 z^{-1} + 0,7274 z^{-2} - 0,1006 z^{-3}}$$

$$G_{22}(z) = \frac{0,01237 z^{-1} + 0,00798 z^{-2}}{1 - 1,264 z^{-1} + 0,264 z^{-2}}$$

$$G_{12}(z) = \frac{0,001741 z^{-1}}{1 - z^{-1}}$$

The derivations of these models of processes IV and V can be looked up in: Isermann, R.; Baur, U. and Blessing, P.: Test-case C for comparison of different identification methods.
Proc. 5th IFAC-Congress, Boston 1975.

Process VI: third-order, low-pass behaviour

$$G_{VI}(s) = \frac{K}{(1+T_1 s)(1+T_2 s)(1+T_3 s)}$$

$K = 1; T_1 = 10$ s; $T_2 = 7,5$ s; $T_3 = 5$ s

$T_0 = 4$ s:

$$G_{VI}(z) = \frac{b_1 z^{-1} + b_2 z^{-2} + b_3 z^{-3}}{1 + a_1 z^{-1} + a_2 z^{-2} + a_3 z^{-3}}$$

$a_1 = -1,7063; a_2 = +0,9580; a_3 = -0,1767$
$b_1 = 0,0186; b_2 = 0,0486; b_3 = 0,0078$

For $T_0 = 2; 6; 8; 10; 12$ sec, see Table 3.4.

Process VII: second-order, low-pass behaviour

$$G_{VII}(s) = \frac{K}{(1+T_2s)(1+T_3s)}$$

$K = 1;\ T_2 = 7,5\ \text{s};\ T_3 = 5\ \text{s}$

$$G_{VII}(z) = \frac{b_1z^{-1}+b_2z^{-2}}{1+a_1z^{-1}+a_2z^{-2}}$$

$T_0 = 4$ s:
$a_1 = -1,036;\ a_2 = 0,2636$
$b_1 = 0,1387;\ b_2 = 0,0889$

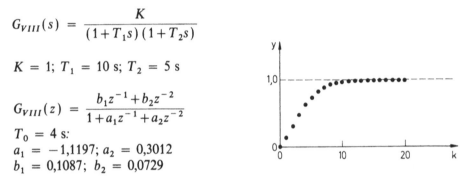

Process VIII: second-order, low-pass behaviour

$$G_{VIII}(s) = \frac{K}{(1+T_1s)(1+T_2s)}$$

$K = 1;\ T_1 = 10\ \text{s};\ T_2 = 5\ \text{s}$

$$G_{VIII}(z) = \frac{b_1z^{-1}+b_2z^{-2}}{1+a_1z^{-1}+a_2z^{-2}}$$

$T_0 = 4$ s:
$a_1 = -1,1197;\ a_2 = 0,3012$
$b_1 = 0,1087;\ b_2 = 0,0729$

For $T_0 = 1; 2; 6; 8; 12$ sec see Table 3.5.

Process IX: second-order, oscillating behaviour (with corresponding process for continuous-time signals (distinct from process I))

$$G_{IX}(s) = \frac{K}{1+2DTs+T^2s^2}$$

$K = 1;\ D = 0,5;\quad T = 5\ \text{s}$

$$G_{IX}(z) = \frac{b_1z^{-1}+b_2z^{-2}}{1+a_1z^{-1}+a_2z^{-2}}$$

$T_0 = 1$ s:
$a_1 = -1,7826;\ a_2 = 0,8187$
$b_1 = 0,01867;\ b_2 = 0,01746$

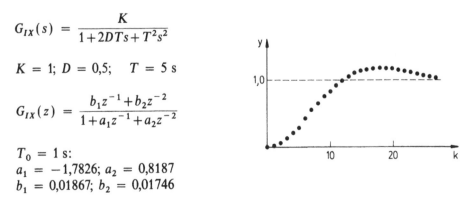

Process X: integral acting second-order process (see Table 26.5)

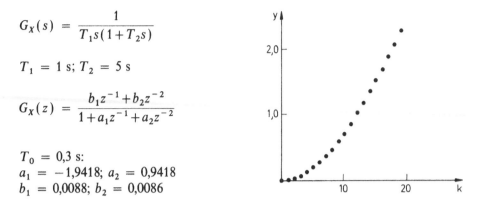

$$G_X(s) = \frac{1}{T_1 s(1+T_2 s)}$$

$T_1 = 1 \text{ s}; T_2 = 5 \text{ s}$

$$G_X(z) = \frac{b_1 z^{-1} + b_2 z^{-2}}{1 + a_1 z^{-1} + a_2 z^{-2}}$$

$T_0 = 0,3$ s:
$a_1 = -1,9418; a_2 = 0,9418$
$b_1 = 0,0088; b_2 = 0,0086$

Process XI: instable second-order process (see Table 26.5)

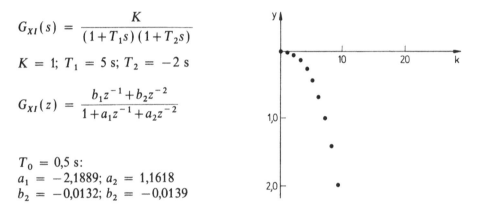

$$G_{XI}(s) = \frac{K}{(1+T_1 s)(1+T_2 s)}$$

$K = 1; T_1 = 5 \text{ s}; T_2 = -2 \text{ s}$

$$G_{XI}(z) = \frac{b_1 z^{-1} + b_2 z^{-2}}{1 + a_1 z^{-1} + a_2 z^{-2}}$$

$T_0 = 0,5$ s:
$a_1 = -2,1889; a_2 = 1,1618$
$b_2 = -0,0132; b_2 = -0,0139$

A4 On the Differentiation of Vectors and Matrices

Let the vector x be a function of the parameters $a_1, a_2, \ldots, a_n$. The partial derivative of this vector with respect to the single parameters is sought. Here a partial differentiation operator is defined as the vector:

$$\frac{\partial}{\partial a} = \begin{bmatrix} \dfrac{\partial}{\partial a_i} \\ \vdots \\ \dfrac{\partial}{\partial a_n} \end{bmatrix}$$

As this is defined as a column vector, the operator cannot be applied to:

$$x = \begin{bmatrix} x_1 \\ \vdots \\ x_p \end{bmatrix}$$

but only to x^T. This results in:

$$\frac{\partial x^T}{\partial a} = \begin{bmatrix} \dfrac{\partial x_1}{\partial a_1} & \dfrac{\partial x_2}{\partial a_1} & \cdots & \dfrac{\partial x_p}{\partial a_1} \\ \vdots & \vdots & & \vdots \\ \dfrac{\partial x_1}{\partial a_n} & \dfrac{\partial x_2}{\partial a_n} & \cdots & \dfrac{\partial x_p}{\partial a_n} \end{bmatrix}.$$

If x is the inner product of two other vectors:

$$x = v^T w = [v_1 \ldots v_p] \begin{bmatrix} w_1 \\ \vdots \\ w_p \end{bmatrix} = v_1 w_1 + \ldots + v_p w_p$$

i.e. a scalar, this then gives

$$\frac{\partial}{\partial a} [v^T w] = \frac{\partial v^T}{\partial a} w + \frac{\partial w^T}{\partial a} v$$

$$= \begin{bmatrix} \dfrac{\partial v_1}{\partial a_1} w_1 + \ldots + \dfrac{\partial v_p}{\partial a_1} w_p \\ \vdots \\ \dfrac{\partial v_1}{\partial a_n} w_1 + \ldots + \dfrac{\partial v_p}{\partial a_n} w_p \end{bmatrix} + \begin{bmatrix} \dfrac{\partial w_1}{\partial a_1} v_1 + \ldots + \dfrac{\partial w_p}{\partial a_1} v_p \\ \vdots \\ \dfrac{\partial w_1}{\partial a_n} v_1 + \ldots + \dfrac{\partial w_p}{\partial a_n} v_p \end{bmatrix}.$$

If the elements of the vector v can be considered independent of the parameters a_i and if $w = a$, then it follows that:

$$\frac{\partial}{\partial a} [v^T a] = v.$$

If, on the other hand, the elements of w are independent of the parameters a_i and $v = a$, then

$$\frac{\partial}{\partial a} [a^T w] = w.$$

The above pair of equations is also valid for the matrices V and W instead of the vectors v and w

$$\frac{\partial}{\partial a}[V\,a]^T = V^T$$

$$\frac{\partial}{\partial a}[a^T W] = W.$$

Let A be a quadratic matrix, then

$$\frac{\partial}{\partial x}[x^T A\,y] = A\,y$$

$$\frac{\partial}{\partial y}[x^T A\,y] = A^T x$$

$$\frac{\partial}{\partial x}[x^T A\,x] = 2\,A\,x \quad A \text{ symmetrical.}$$

Appendix B

Problems

In the following several problems are given for chapters 3, 5, 6, 7, 8 and 9. They are marked according to the individual sections, e.g. B 3.3.5 indicates the fifth problem for section 3.3. The solutions can be looked up in Appendix C.

B 3.1.1 Discretization of a differential equation
A spring mass damper system with characteristic loop frequency $\omega_0 = 0.51/s$ and damping $D = 0.7$ is described by the following differential equation

$$a_2 \ddot{x}(t) + a_1 \dot{x}(t) + x(t) = b_0 w(t)$$

with $a_2 = 1/\omega_0^2 = 4\,s^2$, $a_1 = 2D/\omega_0 = 2.8\,s$ and $b_0 = 2$. Assume a sampling time $T_0 = 1\,s$ and determine the difference equation through discretization.

B 3.2.1 Shannon's sampling theorem
Determine the sampling time T_0 with which harmonic oscillations with frequencies $f = 1$ and 50 Hz have to be sampled in order to regain the continuous signals through low pass filtering of the sampled signal.

B 3.3.1 Computation of the z-transform
Determine the z-transform for $x(t) = ct$.

B 3.3.2 Computation of the z-transform
Determine the z-transform of the following ramp function for sampling time $T_0 = 4\,s$

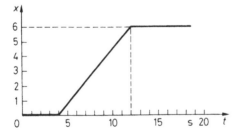

B 3.3.3 Initial and final values

Determine the initial and final values of the discrete-time function $x(k)$ with the z-transforms

a) $x(z) = \dfrac{0,6z}{z^2 - 1,7z + 0,7}$

b) $x(z) = \dfrac{1,5z}{z^2 - 1,732z + 1}$

c) $x(z) = \dfrac{z^2 - 0,8z + 1}{z^2 - 1,4z + 0,4}$.

B 3.3.4 Calculation of z-inverse transform

Given the function

$$x(z) = \frac{1,6z^2 - 0,8}{z^3 - 2,6z^2 + 2,2z - 0,6}$$

with sampling time $T_0 = 1$ s. Determine $x(k)$ explicitly and as numerical values for $k = 0.1, \ldots, 4$. Check the result by division.

B 3.3.5 Calculation of the z-transform

Given the integral acting continuous transfer element $G(s) = y(s)/u(s) = 1/s$. Determine the z-transform of the sampled output signal, when the input signal is a unit step and is one of the following:

a) continuous-time
b) sampled without holding element
c) sampled with zero-order hold.

Calculate the resulting transfer functions.

B 3.4.1 Calculation of an output variable with the convolution sum

Given a process with transfer function

$$G(s) = \frac{K}{1 + T_1 s} e^{-T_t s}$$

with $K = 1$; $T_1 = 1$ s; $T_t = 3$ s and sampling time $T_0 = 1$ s. Determine the behaviour of the output variable for a rectangular pulse with height 2 and duration $3 T_0$. Assumed is a preconnected zero-order hold.

B 3.4.2 Calculation of $HG(z)$ from $G(s)$

Determine the z-transfer function for the following s-transfer functions:

a) $G(s) = \dfrac{1}{T_2 s (1 - T_1 s)}$

   $T_1 = 1$ s; $T_2 = 5$ s; $T_0 = 1$ s.

b) $G(s) = \dfrac{2}{(s+1)(2s+1)^2}$

   $T_0 = 2$ s

c) $G(s) = \dfrac{1}{1 + a_1 s + a_2 s^2} e^{-T_t s}$

   $a_1 = 10$ 1/s; $a_2 = 21$ 1/s^2; $T_t = 8$ s; $T_0 = 4$ s

B 3.4.3 Properties of z-transfer functions
Given the following z-transfer functions

$$G_1(z) = \frac{0,02z^{-2} - 0,07z^{-3}}{z^{-1} - 2z^{-2} + 0,75z^{-3}}$$

$$G_2(z) = \frac{3,12z + 1,56}{z^2 + z + 0,34}$$

$$G_3(z) = \frac{2,55z^{-1}}{z^{-2} + 0,7z^{-3}}$$

$$G_4(z) = \frac{12z^3 - 6z^2}{z^4 + 5z^3 + 6z^2}$$

a) Which $G(z)$ are realizable?
b) Which are stable or unstable?
c) How large is the gain factor to the realizable and stable $G(z)$?

B 3.4.4 Calculation of the impulse response
For a process with zero-order hold

$$HG(z) = \frac{0,2z^{-1}}{1 - 0,8z^{-1}}$$

is assumed. The impulse response $g(k)$ and the step response $h(k)$ for $k = 0, 1, 2, 3, 4$ is to be determined.

B 3.4.5 Connection of linear sampled-data systems
Given

$$G_1(s) = \frac{1}{T_1 s}; \; G_2(s) = \frac{K_2}{1 + T_2 s}.$$

with $T_1 = 10$ s; $K_2 = 2$; $T_2 = 5$ s.

Determine $G(z)$ for a sampling time $T_0 = 2$ s
a) when connected in series according Fig. 3.11a
b) when connected in series according Fig. 3.11b

B 3.4.6 Sampled-data control loop

The process according problem B 3.4.2a is to be controlled with a proportionally acting sampled-data controller with $G_R(z) = K_R = 2$. Determine the command transfer function $G_w(z)$ for a structure presented in Fig. 3.11c.

B 3.4.7 Parameters from impulse response
Given the impulse response

a) How large is the dead time d if the process has no jump discontinuity ($b_0 = 0$)?
b) Determine the parameters of $G(z)$ for $m = 2$.
c) Calculate the step response from $g(k)$.

B 3.5.1 Poles and zeros
Determine for the following z-transfer functions:

$$G_1(z) = \frac{z^2+0,4z+0,04}{z^3+0,4z^2-0,05z}$$

$$G_2(z) = \frac{z^{-2}}{2+z^{-1}}$$

$$G_3(z) = \frac{2z^{-1}+z^{-2}}{z^{-1}+0,6z^{-2}+0,09z^{-3}}$$

$$G_4(z) = \frac{z^2+0,5z}{z^2+1,6z+1,28}$$

$$G_5(z) = \frac{-0,21z^{-1}+0,25z^{-2}}{1-1,65z^{-1}+0,67z^{-2}}$$

a) Location of the poles.
b) Location of the zeros.
c) Which ones are stable?
d) Which ones have zeros inside the unit circle
e) Which ones can be realized?
f) Does a corresponding $G(s)$ exist?

B 3.5.2 Location of the poles
Given the process

$$G(s) = \frac{1}{(1+T_1 s)(1+T_2 s)}; \ T_1 = 10 \text{ min}; \ T_2 = 5 \text{ min}$$

Determine with and without zero-order hold the z-transfer function for sampling time $T_0 = 4$ min and check stability. How do locations of poles and zeros change, if the sampling time is doubled and bisected?

B 3.5.3 Stability limit of a control loop
A sampled-data process

$$HG_p(z) = \frac{0,2z^{-1}+0,1z^{-2}}{1-1,5z^{-1}+0,7z^{-2}}$$

is controlled by a discrete P-controller. How làrge is the critical controller gain K_{Rcrit}?

B 3.5.4 Location of poles. Stability.
Which of the following transfer pulse functions

a) $G(z) = \dfrac{10}{(z-1)}$

b) $G(z) = \dfrac{2(z+1)}{(z-1)^2}$

c) $G(z) = \dfrac{3}{(z-0,8)}$

are asymptotically stable, critically stable or unstable?

B 3.5.5 Stability
Are the discrete-time systems with the characteristic equations

a) $A(z) = z^3-3z^2+2,25z-0,5 = 0$
b) $A(z) = z^3+5z+3z +2 = 0$
c) $A(z) = z^4+9z^3+3z^2+9z+1 = 0$

stable?

B 3.5.6 Stability limit
Determine the stable range of the loop gain K_0 for a discrete P-controller with process VI (Appendix A).

B 3.5.7 Stable range of controller parameters
Determine the stable areas of the parameters of a PD-controller

$$u(k) = q_0 e(k) + q_1 e(k-1)$$

which is operated with a second-order process in closed-loop control loop. For the numerical values use process VIII (Appendix A). In which area may q_0 be chosen, if $q_1 = -5$?

B 3.5.8 Location of zeros
Given the allpass process

$$G(s) = \frac{(s-1)}{s(s+2)}$$

Determine with a zero-order hold location of poles and zeros for sampling time $T_0 = 0.1$ s and 2 s.

B 3.5.9 Location of zeros
Given the allpass process

$$G(s) = \frac{(s-1)}{(s+1)(s+2)}$$

Determine with a zero-order hold location of poles and zeros for sampling time $T_0 = 0.1$ s and 2 s.

B 3.6.1 State representation
A process is described by the difference equation

$$y(k) + 1.2y(k-1) + 0.72y(k-2) + 0.32y(k-3)$$
$$= 0.4u(k-1) - 0.4u(k-2) + 0.84u(k-3)$$

a) Determine the state representation in controllable canonical form.
b) Determine the characteristic equation.
c) Is the process stable? (One root is $z_1 = -0.8$)
d) Is the process controllable and observable?

B 3.6.2 State representation
Given a state model with

$$A = \begin{bmatrix} 0 & 0 & 0.1 \\ 1 & 0 & -0.7 \\ 0 & 1 & 1.5 \end{bmatrix} \quad b = \begin{bmatrix} 0.1 \\ 0.06 \\ 0.04 \end{bmatrix} \quad c = \begin{bmatrix} 0 \\ 0 \\ 1 \end{bmatrix}$$

a) Which canonical form does it represent?
b) How large is the gain factor?
c) How large are the stationary state variables $x(\infty)$ for $u(k) = 1$?

B 3.6.3 Properties of the state model
Given the discrete process:

$$G(z) = \frac{0{,}2z^{-2}}{1-0{,}8z^{-1}}$$

a) State the controllable canonical form
b) Where are the poles located?
c) Is the process controllable and observable?
d) Determine the difference equation from the state representation

B 5.1.1 Discretization of a PID-controller
Given a continuous PID-controller with $K=3$; $T_1=10$ min; $T_D=1$ min. Determine the parameters q_0, q_1, q_2 of the discrete-time control algorithm with rectangular integration and sampling time $T_0=30$ s.

B 5.2.1 PID-controller with prescribed manipulated variable.
Design a discrete-time PID-controller for the process:

$$HG_P(z) = \frac{0{,}1z^{-1}+0{,}1z^{-2}}{1-1{,}1z^{-1}+0{,}3z^{-2}}$$

in such a way that after a step change of the setpoint from 0 to 1, the manipulated variable $u(0)=3$ and $u(1)=1.2$ can be generated. For the transient response of the controller is valid $\lim_{k\to\infty} (u(k)-u(k-1))=0.5$ for $e(k)=1(k)$. Determine the controller parameters q_0, q_1, q_2.

B 5.2.2 PID-controller with prescribed manipulated variable.
For process VI (Appendix) the parameters of a PID-control algorithm are to be determined in such a way that after a step change of the command variable $w(k)$ the manipulated variables are $u(0)=u(1)$. How large is q_1, if $q_0=2.5$? How large is q_2, if the controller gain is $K=1.5$?

B 5.2.3 Stability range of a control loop
Process $HG_P(z)=\dfrac{b_1z^{-1}}{1+a_1z^{-1}}$ with $a_1 = -0.3276$ and $b_1 = 0.6321$ is to be controlled with the following PI-control algorithm:

$$u(k) = u(k-1) + q_0e(k) +q_1e(k-1)$$

Determine the stability conditions for the controller parameters. Draw the stability area into a diagram $q_0=f(q_1)$, (for further reference see Fig. 5.3).

B 6.2.1 Cancellation controller
Calculate for test process VIII (Appendix) with $T_0=4$ s a cancellation controller for the prescribed behaviour.
a) $G_w(z)=\beta_1z^{-1}/(1+\alpha_1z^{-1})$ with $\alpha_1 = -0.6$ and $\beta_1 =0.4$. Determine the behaviour of $u(k)$ and $y(k)$ for $w(k)=1(k)$ (unit step).
b) $G_w(z)=z^{-1}$. Show the behaviour of $u(k)$ and $y(k)$ for $w(k) =1(k)$.

B 7.1.1 Deadbeat controller
Calculate for test process VIII (Annex) for $T_0=4$ s a deadbeat controller without prescribed manipulated variable and determine the behaviour of $u(k)$ and $y(k)$.

B 7.1.2 Deadbeat controller

Determine which one of the following processes cannot be controlled in practice with a deadbeat controller

$$G_1(z) = \frac{0.1z^{-1}+0.14z^{-2}}{1-z^{-1}+0.24z^{-2}} \quad G_2(z) = \frac{0.02z^{-1}+0.02z^{-2}}{1-2z^{-1}+0.96z^{-2}}$$

B 7.2.1 Deadbeat controller with prescribed manipulated variable

A deadbeat controller with prescribed variable is to be determined for the process

$$HG_P(z) = \frac{0.2z^{-1}+0.2z^{-2}}{1-1.5z^{-1}+0.7z^{-2}}$$

in such a way that the manipulated variable $u(0)$ has just half the size of the manipulated variable of a deadbeat controller without prescribed manipulated variable. Calculate $u(k)$ and $y(k)$ for $w(k)=1(k)$.

B 8.1.1 Optimal state feedback

Given a first-order process:

$$x(k+1) = -a_1 x(k) + u(k)$$
$$y(k) = b_1 x(k)$$

with $a_1 = -0.6$ and $b_1 = 0.4$.

a) Determine the gain factor $k_{(N-j)}$ of an optimal state feedback through recursive solution of the (scalar) Riccati difference equation according to (8.1.31) for $j=1, 2, \ldots, 5$ with weighting factors $q=1$ and $r=1$.

b) Determine the value of the performance criterion (8.1.32) if the process was in steady state for $u(k)=1$ before closing the feedback.

c) Determine the characteristic equation and the pole location.

B 8.3.1 State feedback through pole assignment

For a process with

$$A = \begin{bmatrix} 0 & 1 & 0 \\ 0 & 0 & 1 \\ 0 & -0.6 & 1.5 \end{bmatrix} \quad b = \begin{bmatrix} 0 \\ 0 \\ 1 \end{bmatrix} \quad c = \begin{bmatrix} -0.01 \\ 0.11 \\ 0 \end{bmatrix}$$

the following are to be determined:

a) gain factor k

b) $G(z)$ and poles

c) state feedback in such a way that all poles become zero

d) behaviour of $u(k)$ and $y(k)$ for $k=0, 1, \ldots, 4$ with $x(0)$ from steady state for $u(k)=1(k)$.

B 8.3.2 State feedback through pole assignment

Given the process

$$A = \begin{bmatrix} 0 & 1 \\ -0.35 & 1.2 \end{bmatrix} \quad b = \begin{bmatrix} 0 \\ 1 \end{bmatrix} \quad c = \begin{bmatrix} 0.06 \\ 0.1 \end{bmatrix}$$

a) Transform the state feedback into diagonal form

b) Determine the transfer function

c) Determine a state feedback in such a way that the poles of the sampled closed-loop system are $\lambda_{1,2}=0.5\pm0.3\ i$.

B 8.3.3 State feedback

Given the state model of a process in observable canonical form with

$$A = \begin{bmatrix} 0 & 0 & 0{,}1 \\ 1 & 0 & -0{,}7 \\ 0 & 1 & 1{,}5 \end{bmatrix} \quad b = \begin{bmatrix} 0{,}1 \\ 0{,}06 \\ 0{,}04 \end{bmatrix} \quad c = \begin{bmatrix} 0 \\ 0 \\ 1 \end{bmatrix}$$

Use

$$k^T = [5 \ \ 4{,}5 \ \ 3{,}77]$$

as state feedback

a) Determine the characteristic equation and the closed-loop poles
b) Calculate the steady states $x(\infty)$ and $y(\infty)$ if the process (without state feedback) for the constant input signal $u(k)=1$
c) Determine the first manipulated variable $u(0)$ if, originating from steady state b), the above indicated state feedback is switched on at time instant $k=0$.

B 8.3.4 State feedback with feedforward control

Given the process

$$A = \begin{bmatrix} 0{,}5 & 0 \\ -2 & 0{,}8 \end{bmatrix} \quad b = \begin{bmatrix} -0{,}5 \\ 0 \end{bmatrix} \quad c = \begin{bmatrix} -0{,}2 \\ 0{,}18 \end{bmatrix}$$

a) Determine the eigenvalues of the process
b) Determine a state feedback control law such that the poles become $\lambda_1 = 0.1$ and $\lambda_2 = 0.2$
c) A constant disturbance variable $v(k)=1(k)=\text{const.}$ acts upon a process with state feedback $k^T = [-2 \ 0.42]$ so that

$$x(k+1) = A \, x(k) + bu(k) + dv(k)$$

with $d^T = [1.5 \ 2]$
$v(k)$ is assumed to be measurable and value f of a feedforward control is to be determined

$$u(k) = -k_1^T x(k) + f \, v(k)$$

so that in steady state $y(\infty)$ is valid.

B 8.6.1 State observers

Given the processes with transfer function

$$G_{P1}(z) = \frac{0{,}6z^{-1}}{1-0{,}1z^{-1}+0{,}2z^{-2}}; \ G_{P2}(z): \text{ testprocess VIII (Appendix)}$$

a) Write the state representation in controllable and observable canonical form
b) What are the observer equations in controllable and observable canonical form?
c) The observer feedback h is to be determined in such a way that the observer obtains

α) deadbeat behaviour
β) the poles G_{P1}: $z_1 = 0.8$ $z_2 = 0.5$
G_{P2}: $z_1 = 0.5$ $z_2 = 0.5$

d) Calculate the state controllers for the poles of the characteristic equation

G_{P1}: $z_1 = 0{,}2$ $z_2 = -0{,}4$
G_{P2}: $z_1 = 0{,}1$ $z_2 = 0{,}5$

e) Determine the to-be-programmed algorithms of the resulting observers and state feedback using controllable canonical form representation for the deadbeat observer.

B 8.6.2 State variable observer

Write down the equations of an observer in observable canonical form for the process of problem B 8.3.3. It may be assumed that the characteristic equation of the observer is given by:

$$z^3 - 0,5z^2 + 0,2z - 0,05 = 0$$

B 9.1.1 Model of a process with deadtime

Determine for process VII with additional deadtime $T_1 = 12$ s

a) z-transfer function

b) state representation in controllable canonical form (according to (9.1.6) and (9.1.7))

B 9.2.1 Predictor controller

Given the continuous linear process:

$$G(s) = \frac{K}{1 + T_1 s} e^{-T_t s}$$

with $K = 2$; $T_1 = 10$ s; $T_t = 16$ s; $T_0 = 4$ s.

Design a predictor controller and determine the behaviour of the manipulated variable $u(k)$ for a step change of $w(k) = 1(k)$.

Appendix C

Results of the Problems

Results of the problems presented in Appendix B are given in the following as well as some hints for their solution.

C 3.1.1 $\quad 7,8x(k) - 10,8x(k-1) + 4x(k-2) = 2w(k)$

C 3.2.1 $\quad \omega = 2\pi f; T_0 \leq 1/2f; T_0 \leq 0,5$ s und $0,01$ s

C 3.3.1 $\quad x(z) = \dfrac{cT_0 z^{-1}}{(1-z^{-1})^2}$.

Use power series expansion for $(1-z)^{-m}$

C 3.3.2 $\quad x(z) = 3\,z^{-2}(1+z^{-1})/(1-z^{-1})$.
$x_1(t) = 3(t-3T_0)/4 \quad t \geq 3T_0$
$x_2(t) = -3(t-T_0)/4 \quad t \geq 3T_0$

C 3.3.3
  a) $x(0) = 0; x(\infty) = 2$
  $\quad N(z) = (z-1)(z-0,7)$
  b) $x(0)=0; x(\infty)$ does not exist, since for the poles
  $\quad$ of $(z-1)x(z)$ $|z| = 1$ is valid.
  c) $x(0) = 1; x(\infty) = 2$
  $\quad N(z) = (z-1)(z-0,4)$

C 3.3.4 $\quad x(k) = 3 + 2(u-1) - 1,4e^{\ln 0,6(u-1)}$
$\quad x(0) = 0; x(1) = 1,6; x(2) = 4,16; x(3) = 6,496$
$\quad x(4) = 8,6976.$

C 3.3.5
  a) $y(z) = \mathscr{L}\{G(s)u(s)\} = \mathscr{L}\{1/s^2\} = \dfrac{T_0 z}{(z-1)^2}$

  b) $y(z) = \mathscr{L}\{G(s)\}u(z) = \dfrac{z^2}{(z-1)^2}$

  c) $y(z) = \dfrac{z-1}{z}\mathscr{L}\left\{\dfrac{G(s)}{s}\right\}u(z) = \dfrac{T_0 z}{(z-1)^2}$

Step responses: Division of numerator and denominator. According to (3.4.33) it follows that:

k	0	1	2	3
a)	0	1	2	3
b)	1	2	3	4
c)	0	1	2	3

C 3.4.1 $HG(z) = \dfrac{b_1 z}{z + a_1} z^{-4}$; $a_1 = -e^{-\frac{T_0}{T_1}} = -e^{-0.5} = -0.6065$

$$b_1 = \left(1 - e^{-\frac{T_0}{T_1}}\right) = 1 - e^{-0.5} = 0.3935$$

$$g(k) = 3^{-1}\{HG(z)\} = b_1 e^{-\frac{T_0}{T_1} k}$$

$$y(k) = \sum_{v=0}^{k} u(v) g(k-v)$$

k	0	1	2	3	4	5	6	7	8
$g(k)$	0	0	0	0	0,3935	0,2387	0,1447	0,0878	0,0533
$y(k)$	0	0	0	0	0,7869	1,2642	1,5537	0,9424	0,5716

C 3.4.2

a) $HG(z) = \dfrac{0,0736z^{-1} + 0,0528z^{-2}}{1 - 1,3679z^{-1} + 0,3679z^{-2}}$

b) $HG(z) = \dfrac{0,2578z^{-1} + 0,3983z^{-2} + 0,03489z^{-3}}{1 - 0,8711z^{-1} + 0,2349z^{-2} - 0,01832z^{-3}}$

(perform partial fraction expansion)

c) $HG(z) = \dfrac{0,21z^{-1} + 0,11z^{-2}}{1 - 0,83z^{-1} + 0,15z^{-2}} z^{-2}$

C 3.4.3

$G_1(z)$: realizable; unstable
$G_2(z)$: realizable; stable; $K = 2$
$G_3(z)$: not realizable
$G_4(z)$: realizable, unstable

C 3.4.4

k	0	1	2	3	4
$g(k)$	0	0,2	0,16	0,128	0,1024
$h(k)$	0	0,2	0,36	0,488	0,5904

C 3.4.5   a) $G(z) = \dfrac{0,066z}{z^2 - 1,67z + 0,67}$

b) $G(z) = \dfrac{0,04z^2}{z^2 - 1,67z + 0,67}$

C 3.4.6 $G_w(z) = \dfrac{0,4z^{-1} + 0,1057z^{-2}}{1 - 1,2207z^{-1} + 0,1057z^{-2}}$

C 3.4.7

a) $d = 2$

b) $a_1 = -1,7 \quad a_2 = 1,04 \quad b_1 = 0 \quad b_2 = 0,5$

c) $h(0) = 0; \; h(1) = 0; \; h(2) = 0,5; \; h(3) = 1,35; \; h(4) = 2,275;$
$h(5) = 2,9635$

C 3.5.1

$G_1(z)$:   a) $z_1 = 0; \; z_2 = 0.1; \; z_3 = -0.5$ b) $z_{01.02} = -0.2$
c) stable d) yes e) yes f) no

$G_2(z)$:   a) $z_1 = -0.5; \; z_2 = 0$ b) none
c) stable d) – e) yes f) no

$G_3(z)$:   a) $z_{1,2} = -0.3$ b) $z_{01} = -0.5$
c) yes d) yes e) yes f) yes

$G_4(z)$:   a) $z_{1,2} = -0.8 \pm i0.8$ b) $z_{01} = -0.5;$
$z_{02} = 0$ c) no d) yes e) yes f) yes

$G_5(z)$:   a) $z_1 = 0.928 \; z_2 = 0.722$ b) $z_{01} = 1.19$
c) stable d) no e) yes f) yes.

C 3.5.2

T_0	without holding element poles	zeros	with holding element poles	zeros
2 min	0,8187; 0,6703	0	as without	−0,8187
4 min	0,6703; 0,4493	0	holding	−0,6703
8 min	0,4493; 0,2019	0	element	−0,4493

With increasing sampling time poles and zeros move toward the origin.

C 3.5.6 Application of bilinear transformation and stability conditions of example 3.5.1 yields $K_{Rcrit} = 3$.

C 3.5.4
a) critically stable, since simple pole on the unit circle
b) unstable, since double poles on the unit circle (state the homogeneous difference equation)
c) asymptotically stable, since pole inside the unit circle.

C 3.5.5

a) $A(1) = -0,25 < 0 \rightarrow$ unstable

b) $A(1) = 11 > 0; A(-1) = -3 < 0 \rightarrow$ unstable

c) $A(1) = 23 > 0; A(-1) = -13 < 0 \rightarrow$ unstable

C 3.5.6 $A(1) > 0: K_0 > -1; |b_0'| > |b_2'|: K_0 < 5.$
$$-1 < K_0 < 5.$$

C 3.5.7

$A(1) > 0 \rightarrow q_1 > -0,9994 - q_0$

$-A(-1) > 0 \rightarrow q_1 > -67,62 + q_0$

$|a_0'| < 1 \rightarrow |q_1| < 13,71$

$$|b_0'| > |b_2'| \rightarrow q_0 < \frac{131,85}{13,75 - 1,49 q_1} - \frac{q_1^2 + 21,57 q_1}{13,75 - 1,49 q_1}$$

$q_1 = -5 \rightarrow q_0 < 4,0006$

C 3.5.8

$T_0 = 0.1$ s: poles: $z_1 = 1; z_2 = 0,8187$

zero: $z_{01} = 1,1054$

$T_0 = 2$ s: poles: $z_1 = 1; z_2 = 0,01831$

zero: $z_{01} = -2,7222$

(zeros are located outside the unit circle).

C 3.5.9

$$HG(z) = \frac{1}{2} \cdot \frac{(4e^{-T_0} - 3e^{-2T_0} - 1)z + (4e^{-2T_0} - e^{-3T_0} - 3e^{-T_0})}{z^2 - (e^{-T_0} + e^{-2T_0})z + e^{-3T_0}}$$

$T_0 = 0.1$ s: poles: $z_1 = 0,9048; z_2 = 0,8187$

zero: $z_{01} = 1,1057$

$T_0 = 1$ s: poles: $z_1 = 0,3679; z_2 = 0,1353$

zero: $z_{01} = 9,3431$

$T_0 = 2$ s: poles: $z_1 = 0,1353; z_2 = 0,0183$

zero: $z_{01} = -0,6527$

The zero for $T_0 = 0.1$ and 1 s is located outside the unit circle, for $T_0 = 2$ s, however, inside the unit circle.

C 3.6.1

a) $A = \begin{bmatrix} 0 & 1 & 0 \\ 0 & 0 & 1 \\ -0,32 & -0,72 & -1,2 \end{bmatrix}$ $b = \begin{bmatrix} 0 \\ 0 \\ 1 \end{bmatrix}$ $c = \begin{bmatrix} 0,84 \\ -0,4 \\ 0,4 \end{bmatrix}$

b) $\det[zI - A] = \det \begin{bmatrix} z & -1 & 0 \\ 0 & z & -1 \\ 0,32 & 0,72 & (z+1,2) \end{bmatrix}$

$$= z^3 + 1,2z^2 + 0,72z + 0,32 = 0$$

c) $z_{2,3} = -0,2 \pm 0,6i \rightarrow$ stable

d) $Q_s = \begin{bmatrix} 0 & 0 & 1 \\ 0 & 1 & -1,2 \\ 1 & -1,2 & 0,72 \end{bmatrix}$ $\det Q_s = -1 \rightarrow$ controllable
(precondition)

$Q_B = \begin{bmatrix} 0,84 & -0,4 & 0,4 \\ -0,128 & 0,552 & -0,88 \\ 0,2816 & 0,5056 & 1,608 \end{bmatrix}$ $\det Q_B = 1,098 \rightarrow$ observable.

C 3.6.2

a) Observable canonical form

b) $G(z) = \dfrac{0,04z^{-1} + 0,06z^{-2} + 0,1z^{-3}}{1 - 1,5z^{-1} + 0,7z^{-2} - 0,1z^{-3}}$

$K \quad = G(1) = \dfrac{0,2}{0,1} = 2$

c) $x(\infty) = [I - A]^{-1}b \cdot 1$

$= \begin{bmatrix} 1 & 0 & -0,1 \\ -1 & 1 & 0,7 \\ 0 & -1 & -0,5 \end{bmatrix}^{-1} \begin{bmatrix} 0,1 \\ 0,06 \\ 0,04 \end{bmatrix} = \begin{bmatrix} 0,3 \\ -1,04 \\ 2 \end{bmatrix}$

C 3.6.3

a) $G(z) = \dfrac{0,2z^{-1}}{1 - 0,8z^{-1}} z^{-1}; \; a_1 = -0,8; \; b_1 = 0,2$

$A = \begin{bmatrix} 0 & 1 \\ 0 & 0,8 \end{bmatrix} \quad b = \begin{bmatrix} 0 \\ 1 \end{bmatrix} \quad c = \begin{bmatrix} 0,2 \\ 0 \end{bmatrix}$

b) $\det[zI - A] = \det \begin{bmatrix} z & -1 \\ 0 & (z-0,8) \end{bmatrix} = z(z-0,8) = 0$

poles: $z_1 = 0; \; z_2 = 0,8$

c) $\det Q_s = \det \begin{bmatrix} 0 & 1 \\ 1 & 0,8 \end{bmatrix} = -1 \neq 0 \rightarrow$ controllable

$\det Q_B = \det \begin{bmatrix} 0,2 & 0 \\ 0 & 0,2 \end{bmatrix} = 0,04 \neq 0 \rightarrow$ observable

d) $x_1(k+1) = -a_1 x_1(k) + x_2(k); \; x_2(k) = u(k-1); \; x_1(k) = \dfrac{y(k)}{b_1}$

$y(k) + a_1 y(k-1) = b_1 u(k-2)$

C 5.1.1 $q_0 = 9; \; q_1 = -14,85; \; q_2 = 6$

C 5.2.1 $q_0 = 3; \; q_1 = -3,9; \; q_2 = 1,4$

C 5.2.2

(5.2.33) yields $q_1 = -q_0(1-q_0b_1) = -2,3838$
(5.2.15) yields $q_2 = q_0 - K = 1,0$

C 5.2.3

$N(z) = z^2 + (a_1 + q_0b_1 - 1)z + b_1q_1 - a_1$

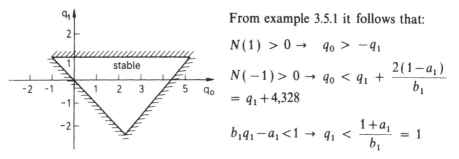

From example 3.5.1 it follows that:

$N(1) > 0 \rightarrow q_0 > -q_1$

$N(-1) > 0 \rightarrow q_0 < q_1 + \dfrac{2(1-a_1)}{b_1}$

$= q_1 + 4,328$

$b_1q_1 - a_1 < 1 \rightarrow q_1 < \dfrac{1+a_1}{b_1} = 1$

C 6.2.1

a) $G_R(z) = \dfrac{3,6798 - 4,1203z^{-1} + 1,1084z^{-2}}{1 - 0,3293z^{-1} - 0,6706z^{-2}}$

k	0	1	2	3	4	5	6	7	8	9
u(k)	3,68	−0,70	2,19	0,23	1,53	0,65	1,24	0,84	1,10	0,93
y(k)	0	0,40	0,64	0,78	0,87	0,92	0,95	0,97	0,98	0,99

b) $G_R(z) = \dfrac{9,1995 - 10,3007z^{-1} + 2,7707z^{-2}}{1 - 0,3293z^{-1} - 0,6706z^{-2}}$

k	0	1	2	3	4	5	6	7	8	9
u(k)	9,20	−7,27	6,55	−2,73	3,49	−0,67	2,12	0,25	1,5	0,66
y(k)	0	1	1	1	1	1	1	1	1	1

C 7.1.1 $G_R(z) = \dfrac{5,5081 - 6,1672z^{-1} + 1,6590z^{-2}}{1 - 0,5987z^{-2} - 0,4013z^{-2}}$

k	0	1	2	3
u(k)	5,5081	−0,6590	1	1
y(k)	0	0,5987	1	1

C 7.1.2

Process $G_2(z)$ is unstable. Therefore no deadbeat control.

C 7.2.1

$u(0) = 2.5$ for DB(v)
$u(0) = 1.25$ for DB($v+1$) prescribed

$$G_R(z) = \frac{1,25 - 0,625z^{-1} - 1z^{-2} + 0,875z^{-3}}{1 - 0,25z^{-1} - 0,5z^{-2} - 0,25z^{-3}}$$

k	0	1	2	3	4
$u(k)$	1,25	0,625	$-0,375$	0,5	0,5
$y(k)$	0	0,25	0,75	1,0	1,0

C 8.1.1

a) From (8.1.30) and (8.1.31) it follows that with initial value

$$k_{N-j} = \frac{a_1\, p_{N-j+1}}{r + p_{N-j+1}}$$

$$p_{N-j} = q + a_1^2 p_{N-j+1} - k_{N-j}^2 (r + p_{N-j+1})$$

with the starting value $p_N = q = 1$ result

$$k_{N-1} = 0,3 \qquad p_{N-1} = 1,18$$
$$k_{N-2} = 0,32477 \quad p_{N-2} = 1,19486$$
$$k_{N-3} = 0,32663 \quad p_{N-3} = 1,19598$$
$$k_{N-4} = 0,32677 \quad p_{N-4} = 1,19606$$
$$k_{N-5} = 0,32678 \quad p_{N-5} = 1,19607$$

b) In steady state is

$$x = a_1 x + 1 \rightarrow x(0) = 2,5$$

Herewith $I = x(0) p_0 x(0) \approx 2,5^2 \cdot 1,1961 = 7,4756$

c) characteristic equation $z + a_1 + k = 0$

$$\text{pole:} z_1 = -a_1 - k = 0,6 - 0,3268 = 0,2732$$

C 8.3.1

a) $K = 1$

b) $G(z) = \dfrac{-0,01 + 0,11z}{z(z^2 - 1,5z + 0,6)}$

pole: $z_1 = 0$; $z_2 = 0,75 \pm 0,1963i$

c) $k_1 = 0$; $k_2 = -0,6$; $k_3 = 1,5$

d) $x^T(0) = [10\ 10\ 10]$

k	0	1	2	3	4
$u(k)$	-9	6	0	0	0
$y(k)$	1	1	$-0,1$	0	0

→ deadbeat behaviour!

C 8.3.2

a) $A = \begin{bmatrix} 0,5 & 0 \\ 0 & 0,7 \end{bmatrix}$ $b = \begin{bmatrix} -3 \\ 5 \end{bmatrix}$ $c = \begin{bmatrix} 0,11 \\ 0,13 \end{bmatrix}$

b) $G(z) = \dfrac{0,1z^{-1} + 0,06z^{-2}}{1 - 1,2z^{-1} + 0,35z^{-2}}$

c) $k_1 = -0,01;\ k_2 = 0,2$

C 8.3.3

a) $z^3 - 0,5792z^2 + 0,1152z = 0$

$z_1 = 0;\ z_{2,3} = 0,2896 \pm 0,1770i$

b) $x_1(\infty) = 0,3;\ x_2(\infty) = -1,04;\ x_3(\infty) = 2;\ y(\infty) = 2$

c) $u(0) = -k^T x(0) = -4,36$

C 8.3.4

a) $z_1 = 0,5$ $z_2 = 0,8$

b) $k_1 = -2$ $k_2 = 0,42$

c) $x(\infty) = [A - b\,k^T]x(\infty) + [bf + d]v(k)$

$x(\infty) = [I - A + b\,k^T]^{-1}[bf + d]v(k)$

$0 = c^T[I - A + bk^T]^{-1}[bf + d] \cdot 1$

$f = 0,72$

C 8.6.1

Process $G_{P1}(z)$:

a) $A = \begin{bmatrix} 0 & 1 \\ -0,2 & 0,1 \end{bmatrix}$ $b = \begin{bmatrix} 0 \\ 1 \end{bmatrix}$ $c = \begin{bmatrix} 0 \\ 0,6 \end{bmatrix}$

$A = \begin{bmatrix} 0 & -0,2 \\ 1 & 0,1 \end{bmatrix}$ $b = \begin{bmatrix} 0 \\ 0,6 \end{bmatrix}$ $c = \begin{bmatrix} 0 \\ 1 \end{bmatrix}$

b) $\hat{x}(k+1) = \begin{bmatrix} 0 & 1-0,6h_2 \\ -0,2 & 0,1-0,6h_1 \end{bmatrix}\hat{x}(k) + \begin{bmatrix} 0 \\ 1 \end{bmatrix}u(k) + \begin{bmatrix} h_2 \\ h_1 \end{bmatrix}y(k)$

$\hat{x}(k+1) = \begin{bmatrix} 0 & -0,2-h_2 \\ 1 & 0,1-h_1 \end{bmatrix}\hat{x}(k) + \begin{bmatrix} 0 \\ 0,6 \end{bmatrix}u(k) + \begin{bmatrix} h_2 \\ h_1 \end{bmatrix}y(k)$

c) α) $h^T = [-0,2\ \ 0,1]$

β) $h^T = [0,2\ \ -1,2]$

d) $k^T = [-0,28\ \ 0,3]$

e) $\hat{x}_1(k+1) = -0,2y(k);\ \hat{x}_2(k+1) = \hat{x}_1(k) + 0,6u(k) + 0,1y(k)$

$u(k) = -\hat{x}_1(k) + 0,8333\hat{x}_2(k)$

Process $G_{P2}(z)$:

a) $A = \begin{bmatrix} 0 & 1 \\ -0,3012 & 1,1197 \end{bmatrix}$ $b = \begin{bmatrix} 0 \\ 1 \end{bmatrix}$ $c = \begin{bmatrix} 0,0729 \\ 0,1087 \end{bmatrix}$

$A = \begin{bmatrix} 0 & -0,3012 \\ 1 & 1,1197 \end{bmatrix}$ $b = \begin{bmatrix} 0,0729 \\ 0,1087 \end{bmatrix}$ $c = \begin{bmatrix} 0 \\ 1 \end{bmatrix}$

b) $\hat{x}(k+1) = \begin{bmatrix} -0,0729h_2 & 1-0,1087h_2 \\ -0,3012-0,0729h_1 & 1,120-0,0109h_1 \end{bmatrix} \hat{x}(k)$

$+ \begin{bmatrix} 0 \\ 1 \end{bmatrix} u(k) + \begin{bmatrix} h_2 \\ h_1 \end{bmatrix} y(k)$

$\hat{x}(k+1) = \begin{bmatrix} 0 & -0,3012-h_2 \\ 1 & 1,1197-h_1 \end{bmatrix} \hat{x}(k) + \begin{bmatrix} 0,0729 \\ 0,1087 \end{bmatrix} u(k) + \begin{bmatrix} h_2 \\ h_1 \end{bmatrix} y(k)$

c) $\alpha)$ $h^T = [-0,3012 \quad 1,1197]$

$\beta)$ $h^T = [-0,0512 \quad 0,1197]$

d) $k^T = [-0,2512 \quad 0,5197]$

e) $\hat{x}_1(k+1) = 0,0729u(k) - 0,3012y(k)$
$\hat{x}_2(k+1) = \hat{x}_1(k) + 0,1087u(k) + 1,1197y(k)$
$u(k) = -3,6728\hat{x}_1(k) - 2,3179\hat{x}_2(k)$

C 8.6.2

Characteristic equation

$z^3 + (h_1-1,5)z^2 + (h_2+0,7)z + (h_3-0,1) = 0$

$h_1 = 1; h_2 = -0,5; h_3 = 0,05$

$\hat{x}(k+1) = \begin{bmatrix} 0 & 0 & 0,05 \\ 1 & 0 & -1,2 \\ 0 & 1 & 0,5 \end{bmatrix} x(k) + \begin{bmatrix} 0,1 \\ 0,06 \\ 0,04 \end{bmatrix} u(k) + \begin{bmatrix} 0,05 \\ -0,5 \\ 1 \end{bmatrix} y(k)$

C 9.1.1

a) $G(z) = \dfrac{0,1387z^{-1} + 0,0889z^{-2}}{1 - 1,036z^{-1} + 0,2636z^{-2}} z^{-3}$

b) $x(k+1) = A\,x(k) + b\,u(k-3)$

$A = \begin{bmatrix} 0 & 1 \\ -0,2636 & 1,036 \end{bmatrix}$ $b = \begin{bmatrix} 0 \\ 1 \end{bmatrix}$ $c = \begin{bmatrix} 0,0889 \\ 0,1387 \end{bmatrix}$

c) $x(k+1) = A\,x(k) + b\,u(k)$

$$A = \begin{bmatrix} 0 & 1 & 0 & 0 & 0 \\ -0{,}2636 & 1{,}036 & 1 & 0 & 0 \\ 0 & 0 & 0 & 1 & 0 \\ 0 & 0 & 0 & 0 & 1 \\ 0 & 0 & 0 & 0 & 0 \end{bmatrix}; \quad b = \begin{bmatrix} 0 \\ 0 \\ 0 \\ 0 \\ 1 \end{bmatrix}; \quad c = \begin{bmatrix} 0{,}0889 \\ 0{,}1387 \\ 0 \\ 0 \\ 0 \end{bmatrix}$$

C 9.2.1

$$G_P(z) = \frac{0{,}6594 z^{-5}}{1-0{,}6703 z^{-1}}$$

$$G_R(z) = \frac{0{,}5 - 0{,}3352 z^{-1}}{1 - 0{,}6703 z^{-1} - 0{,}3297 z^{-5}}$$

$$u(k) = 0{,}5w(k)$$

$$u(0) = u(1) = u(2) = \ldots = 0{,}5$$

References

Chapter 1

1.1 Thompson, A: Operating experience with direct digital control. IFAC/IFIP Conference on Application of Digital Computers for Process Control, Stockholm 1964, New York: Pergamon Press

1.2 Giusti, A.L.; Otto, R.E.; Williams, T.J.: Direct digital computer control. Control Eng 9 (1962) 104−108

1.3 Evans, C.S.; Gossling, T.H.: Digital computer control of a chemical plant. 2. IFAC-Congress, Basel 1963

1.4 Ankel, Th.: Prozeßrechner in der Verfahrenstechnik, gegenwärtiger Stand der Anwendungen. Regelungstechnik 16 (1968) 386−395

1.5 Ernst, D.: Digital control in power systems. 4. IFAC/IFIP Symp. on Digital Computer Applications to Process Control, Zürich 1974. Lecture Notes „Control Theory" 93/94 Berlin: Springer 1974

1.6 Amrehn, H.: Digital computer applications in chemical and oil industries. 4 IFAC/IFIP Symp. on Digital Computer Applications to Process Control, Zürich 1974. Lecture Notes „Control Theory" 93/94 Berlin: Springer 1974

1.7 Savas, E.S.: Computer control of industrial processes. London: McGraw-Hill 1965

1.8 Miller, W.E. (Ed.): Digital computer applications to process control. New York: Plenum Press 1965

1.9 Lee, T.H.; Adams, G.E.; Gaines, W.M.: Computer process control: modeling and optimization. New York: Wiley 1968

1.10 Schöne, A.: Prozeßrechensysteme der Verfahrensindustrie. München: Hanser 1969

1.11 Anke, K.; Kaltenecker, H.; Oetker, R.: Prozeßrechner. Wirkungsweise und Einsatz. München: Oldenbourg 1971

1.12 Smith, C.L.: Digital computer process control. Scranton: Intext Educ. Publish. 1972

1.13 Harrison, T.J. (Ed.): Handbook of industrial control computers. New York: Wiley-Interscience 1972

1.14 Syrbe, M.: Messen, Steuern, Regeln mit Prozeßrechnern. Frankfurt: Akad. Verlagsges. 1972

1.15 Kaltenecker, H.: Funktionelle und strukturelle Entwicklung der Prozeßautomatisierung. Regelungstech. Prax. 23 (1981) 348−355

1.16 Ernst, D.: New trends in the application of process computers. 7th IFAC-Congress, Helsinki 1978. Proc. Oxford: Pergamon Press

1.17 Schreiber, J.: Present state and development of microelectronics. In: Mikroelektronik in der Antriebstechnik, ETG-Fachbericht 11, Offenbach: VDE-Verlag, 1982

1.18 Larson, R.E.; Hall, W.E.: Control technology development during the first 25 years of IFAC. 25th Aniversary of IFAC, Heidelberg 1982. Düsseldorf: Preprints VDI/VDE-GMR

1.19 Prince, B.: Entwicklungen und Trends bei MOS-Speicherbausteinen. Elektronik 10 (1983) 47−50

1.20 Isermann, R.: The role of digital control in engineering. Trans. of the South African Institute of Electrical Engineers 75 (1984) 3−21

1.21 Isermann, R.: Bedeutung der Mikroelektronik für die Prozeßautomatisierung. ETZ 106 (1985) 330−337, 474−478, 602−606

Chapter 2

2.1 Oldenbourg, R.C.; Sartorius, H.: Dynamik selbsttätiger Regelungen. München: Oldenbourg 1944 und 1951
2.2 Zypkin, J.S.: Differenzengleichungen der Impuls- und Regeltechnik. Berlin: VEB-Verlag Technik 1956
2.3 Jury, E.I.: Sampled-data control systems. New York: Wiley 1958
2.4 Ragazzini, J.R.; Franklin, G.F.: Sampled-data control systems. New York: McGraw-Hill 1958
2.5 Smith, O.J.M.: Feedback control systems. New York: McGraw-Hill 1958
2.6 Zypkin, J.S.: Theorie der Impulssysteme. Moskau: Staatl. Verl. für physikalisch-mathematische Lit. 1958
2.7 Tou, J.T.: Digital and sampled-data control systems. New York: McGraw-Hill 1959
2.8 Tschauner, J.: Einführung in die Theorie der Abtastsysteme. München: Oldenbourg 1960
2.9 Monroe, A.J.: Digital processes for sampled-data systems. New York: Wiley 1962
2.10 Kuo, B.C.: Analysis and synthesis of sampled-data control systems. Englewood-Cliffs, N.J.: Prentice Hall 1963
2.11 Jury, E.I.: Theory and application of the z-transform method. New York: Wiley 1964
2.12 Zypkin, J.S.: Sampling systems theory. New York: Pergamon Press 1964
2.13 Freeman, H.: Discrete-time systems. New York: Wiley 1965
2.14 Lindorff, D.P.: Theory of sampled-data control systems. New York: Wiley 1965
2.15 Strejc, V.: Synthese von Regelungssystemen mit Prozeßrechnern. Berlin: Adademie-Verlag 1967
2.16 Zypkin, J.S.: Theorie der linearen Impulssysteme. München: Oldenbourg 1967
2.17 Kuo, B.C.: Discrete-data control systems. Englewood-Cliffs, N.J.: Prentice Hall 1970
2.18 Cadzow, J.A.; Martens, H.R.: Discrete-time and computer control systems. Englewood-Cliffs, N.J.: Prentice Hall 1970
2.19 Ackermann, J.: Abtastregelung. Berlin: Springer 1972
2.20 Leonhard, W.: Diskrete Regelsysteme. Mannheim: Bibl. Inst. 1972
2.21 Föllinger, O.: Lineare Abtastsysteme. München: Oldenbourg 1974
2.22 Isermann, R.: Digitale Regelsysteme. 1. Aufl. Berlin: Springer-Verlag 1977
2.23 Isermann, R.: Digital control systems. Berlin: Springer-Verlag 1981
2.24 R.依扎尔曼 著 数字调节系统 机械工业出版社 chin. Übersetzung [2.22]. Verlag für mechan. Technik 1983
2.25 Изерман Р. Цифровые системы управления: Пер. с англ. — М.: Мир, 1984. russ. Übersetzung von [2.23]. Moskau: MIR 1984
2.25 Kuo, B.C.: Digital control systems. Tokyo: Holt-Saunders 1980
2.26 Franklin, G.F.; Powell, J.D.: Digital control of dynamic systems. Reading, Mass.: Addison-Wesley 1980
2.27 Strejc, V.: State space theory of discrete linear control. Prag: Acedemia 1981
2.28 Ackermann, J.: Abtastregelung. 2 Aufl. Bd I und II. Berlin: Springer-Verlag 1983
2.29 Åström, K.J.; Wittenmark, B.: Computer controlled systems. Englewood-Cliffs, N.J.: Prentice Hall 1984

Chapter 3

3.1 Kurzweil, F.: The control of multivariable processes in the presence of pure transport delays. IEEE Trans. Autom. Control (1963) 27–34
3.2 Koepcke, R.W.: On the control of linear systems with pure time delays. Trans. ASME (1965) 74–80
3.3 Tustin, A.: A method of analyzing the behaviour of linear systems in terms of time series. JIEE (London) 94 pt. IIA (1947) 130–142
3.4 Isermann, R.: Theoretische Analyse der Dynamik industrieller Prozesse. Mannheim: Bibliographisches Inst. 1971 Nr. 764/764a

3.5   Isermann, R.: Results on the simplification of dynamic process models. Int. J. Control (1973) 149−159

3.6   Campbell, D.P.: Process dynamics. New York: Wiley 1958

3.7   Profos, P.: Die Regelung von Dampfanlagen. Berlin: Springer-Verlag 1962

3.8   Gould, L.A.: Chemical process and control. Reading, Mass: Addison-Wesley 1969

3.9   Mac Farlane, A.G.J.: Dynamical system models. London: G.G. Harrap 1970

3.10 Gilles, E.D.: Systeme mit verteilten Parametern. München: Oldenbourg 1973

3.11 Isermann, R.: Experimentelle Analyse der Dynamik von Regelsystemen. Mannheim: Bibliographisches Inst. 1971 Nr. 515/515a

3.12 Eykhoff, P.: System identification. London: Wiley 1974

3.13 Isermann, R.: Prozeßidentifikation. Berlin: Springer-Verlag 1974

3.14 Wilson, R.G.; Fisher, D.G.; Seborg, D.E.: Model reduction for discrete-time dynamic systems. Int. J. Control (1972) 549−558

3.15 Gwinner, K.: Modellbildung technischer Prozesse unter besonderer Berücksichtigung der Modellvereinfachung. PDV-Entwicklungsnotiz PDV−E 51. Karlruhe: Ges. für Kernforschung 1975

3.16 Åström, K.J.; Hagander, P.; Sternby J.: Zeros of sampled systems. Automatica 20 (1984) 31−38

3.17 Tuschak, R.: Relation between transfer and pulse transfer functions of continuous processes. 8. IFAC-Kongreß, Kyoto. Oxford: Pergamon Press 1981

3.18 Isermann, R.: Practical aspects of process identification Automatica 16 (1980) 575−587

3.19 Litz, L.: Praktische Ergebnisse mit einem neuen modalen Verfahren der Ordnungsreduktion. Regelungstechnik 27 (1979) 273−280

3.20 Bonvin, D.; Mellichamp, D.A.: A unified derivation and critical review of model approaches to model reduction. Int. J. Control 35 (1982) 829−848

Chapter 5

5.1   Bernard, J.W.; Cashen, J.F.: Direct digital control. Instrum. Control Sys. 38 (1965) 151−158

5.2   Cox, J.B.; Williams, L. J.; Banks, R.S.; Kirk, G.J.: A practical spectrum of DDC chemical process control algorithms. ISA J. 13 (1966) 65−72

5.3   Davies, W.T.D.: Control algorithms for DDC. Instrum. Prac. 21 (1967) 70−77

5.4   Lauber, R.: Einsatz von Digitalrechnern in Regelungssystemen. ETZ-A 88 (1967) 159−164

5.5   Amrehn, H.: Direkte digitale Regelung. Regelungstech. Prax. 10 (1968) 24−31, 55−57

5.6   Hoffmann, M.; Hofmann, H.: Einführung in die Optimierung. Weinheim: Verlag Chemie 1973

5.7   Isermann, R.; Bux, D.; Blessing, P.; Kneppo, P.: Regel- und Steueralgorithmen für die digitale Regelung mit Prozeßrechnern − Synthese, Simulation, Vergleich −. PDV-Bericht Nr. 54 KFK-PDV. Karlsruhe: Ges. für Kernforschung 1975

5.8   Rovira, A.A.; Murrill, P.W.; Smith, C.L.: Modified PI algorithm for digital control. Instrum. Control Syst. Aug. (1970) 101−102

5.9   Isermann, R.; Bamberger, W.; Baur, W.; Kneppo, P.; Siebert, H.: Comparison and evaluation of six on-line identification methods with three simulated processes. IFAC-Symp. on Identification, Den Haag 1973. IFAC-Automatica 10 (1974) 81−103

5.10 Lee, W.T.: Plant regulation by on-line digital computers. S.I.T. Symp. on Direct Digital Control

5.11 Goff, K.W.: Dynamics in direct digital control, I and II. ISA J. Nov. (1966) 45−49, Dec. (1966) 44−54

5.12 Beck, M.S.; Wainwright, N.: Direct digital control of chemical processes. Control (1968) 53−56

5.13 Bakke, R.M.: Theoretical aspects of direct digital control. ISA Trans. 8 (1969) 235 – 250

5.14 Oppelt, W.: Kleines Handbuch technischer Regelvorgänge. Weinheim: Verlag Chemie 1960

5.15 Lopez, A.M.; Murrill, P.W.; Smith, C.L.: Tuning PI- and PID-digital controllers. Instrum. Control Syst. 42 (1969) 89 – 95

5.16 Takahashi, Y.; Chan, C.S.; Auslander, D.M.: Parametereinstellung bei linearen DDC-Algorithmen. Regelungstech. Prozeßdatenverarb. 19 (1971) 237 – 244

5.17 Takahashi, Y.; Rabins, M.; Auslander, D.: Control and dynamic systems. Reading. Mass.: Addison-Wesley 1969

5.18 Schwarz, Th.: Einstellregeln für diskrete parameteroptimierte Regelalgorithmen. Studienarbeit Nr. 72/74, Abt. für Regelungstech. und Prozeßdynamik (IVD), Univ. Stuttgart 1975

5.19 Unbehauen, H.; Böttiger, F.: Regelalgorithmen für Prozeßrechner. Bericht KFK-PDV 26. Karlsruhe: Ges. für Kernforschung 1974

5.20 Smith, C.L.: Digital computer process control. Scranton: Intext Educational Publ. 1972

5.21 Chiv, K.C.; Corripio, A.B.; Smith, C.L.: Digital control algorithms, Part III: Tuning PI and PID controllers. Instrum. Control Syst. 46 (1973) 41 – 43

5.22 Wittenmark, B.; Åström, K.J.: Simple self-tuning controllers. Symp. on Methods and Applications in Adaptive Control. Bochum 1980

5.23 Kofahl, R.; Isermann, R.: A simple method for automatic tuning of PID-controllers based on process parameter estimation. Boston: American Control Conference 1985

5.24 Kofahl, R.: Selbsteinstellende digitale PID-Regler-Grundlagen und neue Entwicklungen. VDI-Bericht Nr. 550. Düsseldorf: VDI-Verlag 1985, 115 – 130

5.25 Eckelmann, W.; Hofmann, W.: Vergleich von Regelalgorithmen in Automatisierungssystemen. Regelungstech. Praxis 25(1983) 423 – 426

5.26 Hensel, H.: Methoden des rechnergestützten Entwurfs und Echtzeiteinsatzes zeitdiskreter Mehrgrößenregelsysteme und ihre Realisierung in einem CAD-System. Interner Bericht. Inst. für Regelungstechnik, TH Darmstadt 1986

5.27 Radke, F.; Isermann R.: A parameteradaptive PID-controller with stepwise parameter optimization. Proc. 9th IFAC-Congress, Budapest 1984. Oxford: Pergamon Press

5.28 Tolle, H.: Optimierungsverfahren. Berlin: Springer-Verlag 1971

5.29 Wilde, D.J.: Optimum seeking methods. Englewood Cliffs, Mass.: Prentice Hall 1964

5.30 Horst, R.: Nichtlineare Optimierung. München: Hanser 1979

5.31 Hofer, E.; Lunderstädt, R.: Numerische Methoden der Optimierung. München: Oldenbourg 1975

5.32 Hengstenberg, J.; Sturm, B.; Winkler, O.: Messen, Steuern, Regeln in der Chemischen Technik. 3. Aufl. Bd. III Berlin: Springer-Verlag 1981

5.33 Hengstenberg, J.; Sturm, B.; Winkler O.: Messen, Steuern, Regeln in der Chemischen Technik. 3. Aufl. Bd IV. Berlin: Springer-Verlag 1983

5.34 Moore, C.F.; Smith, C.L.; Murrill, P.W.: Improved algorithm for direct digital control. Instrum. Control Syst. Jan (1970) 70 – 74

5.35 Bányasz, Cs.; Keviczky, L.: Direct methods for self-tuning PID-regulators. Proc. 6th IFAC-Symp. on Identification, Washington 1982. Oxford: Pergamon Press

5.36 Schwefel, H.P.: Numerische Optimierung von Computer-Modellen mittels Evolutionsstrategie. Basel: Birkhäuser-Verlag 1977

Chapter 6

6.1 Bergen, A.R.; Ragazzini, J.R.: Sampled-data processing techniques for feedback control systems. AIEE Trans. 73 (1954) 236

6.2 Strejc, V.: Synthese von Regelkreisen mit Prozeßrechnern. Mess. Steuern Regeln (1967) 201 – 207

6.3 Dahlin, E.B.: Designing and tuning digital controllers. Instrum. Control Sys. 41 (1968) 77 – 83 und 87 – 92

Chapter 7

7.1   Jury, E.I.; Schroeder, W.: Discrete compensation of sampled data and continuous control systems. Trans. AIEE 75 (1956) Pt. II

7.2   Kalman, R.E.: Diskussionsbemerkung zu einer Arbeit von Bergen, A.R. und Ragazzini, J.R. Trans. AIEE (1954) 236–247

7.3   Lachmann, K.H.; Goedecke, W.: Ein parameteradaptiver Regler für nichtlineare Prozesse. Regelungstechnik 30 (1982) 197–206

Chapter 8

8.1   Bellman, R.E.: Dynamic programming. Princeton: Princeton University Press 1957

8.2   Kalman, R.; Koepcke, R.V.: Optimal synthesis of linear sampling control systems using generalized performance indexes. Trans. ASME (1958) 1820–1826

8.3   Athans, M. Falb, P.L.: Optimal control. New York: McGraw-Hill 1966

8.4   Kwakernaak, H.; Sivan, R.: Linear optimal control systems. New York: Wiley-Interscience 1972

8.5   Kneppo, P.: Vergleich von linearen Regelalgorithmen für Prozeßrechner. Diss. Univ. Stuttgart. PDV-Bericht KFK-PDV 96. Karlsruhe: Ges. für Kernforschung 1976

8.6   Johnson, C.D.: Accomodation of external disturbances in linear regulators and servomechanical problems. IEEE Trans. Autom. Control AC 16 (1971)

8.7   Kreindler, E.: On servo problems reducible to regulator problems. IEEE Trans. Autom. Control AC 14 (1969)

8.8   Bux, D.: Anwendung und Entwurf konstanter, linearer Zustandsregler bei linearen Systemen mit langsam veränderlichen Parametern. Diss. Univ. Stuttgart. Fortschritt-Ber. VDI-Z Reihe 8, Nr. 21. Düsseldorf: VDI-Verlag 1975

8.9   Rosenbrock, H.H.: Distinctive problems of process control. Chem. Eng. Prog. 58 (1962) 43–50

8.10 Porter, B. Crossley, T.R.: Modal control. London: Taylor and Francis 1972

8.11 Gould, L.A.: Chemical process control. Reading Mass.: Addison-Wesley 1969

8.12 Föllinger, 0.: Einführung in die modale Regelung. Regelungstechnik 23 (1975) 1–10

8.13 Luenberger, D.G.: Observing the state of a linear system. IEEE Trans. Mil. Electron. (1964) 74–80

8.14 Luenberger, D.G.: Observers for multivariable systems. IEEE Trans. AC (1966) 190–197

8.15 Luenberger, D.G.: An introduction to observers. IEEE Trans. AC 16 (1971) 596–602

8.16 Levis, A.H.; Athans, M.; Schlueter, R.A.: On the behavior of optimal linear sampled data regulators. Preprints Joint Automatic Control Conf. Atlanta (1970), S. 695–669

8.17 Schumann, R: Digitale parameteradaptive Mehrgrößenregelung. Diss. T.H. Darmstadt. PDV-Bericht 217. Karsruhe: Ges. für Kernforschung 1982

8.18 Radke, F.: Ein Mikrorechnersystem zur Erprobung parameteradaptiver Regelverfahren. Diss. T.H. Darmstadt. Fortschritt Ber. VDI-Z. Reihe 8, Nr. 77, Düsseldorf: VDI-Verlag 1984

Chapter 9

9.1   Reswick, J.B.: Disturbance response feedback. A new control concept. Trans. ASME 78 (1956) 153

9.2   Smith, O.J.M.: Closer control of loops with dead time. Chem. Eng. Prog. 53 (1957) 217–219

9.3   Smith, O.J.M.: Feedback control systems. New York: McGraw-Hill 1958

9.4   Smith, O.J.M.: A controller to overcome dead time. ISA J. 6 (1958) 28–33

9.5 Giloi, W.: Zur Theorie und Verwirklichung einer Regelung für Laufzeitstrecken nach dem Prinzip der ergänzenden Rückführung. Diss. Univ. Stuttgart 1959

9.6 Schmidt, G.: Vergleich verschiedener Totzeitregelsysteme. Mess. Steuern Regeln 10 (1967) 71−75

9.7 Frank, P.M.: Vollständige Vorhersage im stetigen Regelkreis mit Totzeit. Regelungstechnik 16 (1968) 111−116 und 214−218

9.8 Mann, W.: Identifikation und digitale Regelung enes Trommeltrockners. Diss. T.H. Darmstadt. PDV-Bericht Nr. 189 Karlsruhe: Ges. für Kernforschung 1980

Chapter 10

10.1 Horowitz, I.M.: Synthesis of feedback systems. New York: Academic Press 1963

10.2 Kreindler, E.: Closed-loop sensitivity reduction of linear optimal control systems. IEEE Trans. AC 13 (1968) 254−262

10.3 Perkins, W.R.; Cruz, J.B.: Engineering of dynamic systems. New York: Wiley 1969

10.4 2nd IFAC-Symp on System Sensitivity and Adaptivity, Dubrovnik (1968). Preprints Yugoslav Committee for Electronics and Automation (ETAN), Belgrad/Jugoslawien

10.5 3rd IFAC-Symp. on Sensitivity, Adaptivity and Optimality, Ischia (1973). Proceedings instrument Soc. of America (ISA), Pittsburgh.

10.6 Tomovic, R.; Vucobratovic, M.: General sensitivity theory. New York: Elsevier 1972

10.7 Frank, P.M.: Empfindlichkeitsanalyse dynamischer Systeme. München: Oldenbourg 1976

10.8 Anderson, B.D.O.; Moore, J.B.: Linear optimal control. Englewood Cliffs, N.J.: Prentice Hall 1971

10.9 Cruz, J.B.: System sensitivity analysis. Stroudsburg: Dowen, Hutchinson and Ross 1973

10.10 Andreev, Y.N.: Algebraic methods of state space in linear object control theory. Autom. and Remote Control 39 (1978) 305−342

10.11 Frank, P.M.: The present state and trends using sensitivity analysis and synthesis in linear optimal control. Acta polytechnica, Práce CVUT v Praze, Vedeeká Konference 1982

10.12 Kreindler, E: On minimization of trajectory sensitivity. Int. J. Control 8 (1968) 89−96

10.13 Elmetwelly, M.M.; Rao, N.D.: Design of low sensitivity optimal regulators for synchroneous machines. Int. J. Control 19 (1974) 593−607

10.14 Byrne, P.C.; Burke, M.: Optimization with trajectory sensitivity considerations. IEEE Trans. Autom. Control 21 (1976) 282−283

10.15 Rillings, J.H.; Roy, R.J.: Analog sensitivity design of Saturn V launch vehicle. IEEE Trans. AC 15 (1970) 437−442

10.16 Graupe, D.: Optimal linear control subject to sensitivity constraints. IEEE Trans. AC 19 (1974) 593−594

10.17 Subbayyan, R.; Sarma, V.V.S.; Vaithiluigam, M.C.: An approach for sensitivity reduced design of linear regulators. Int. J. Control 9 (1978) 65−74

10.18 Krishnan, K.R.; Brzezowski, S.: Design of robust linear regulator with prescribed trajectory insensitivity to parameter variations. IEEE Trans. AC 23 (1978) 474−478

10.19 Verde, C.; Frank, P.M.: A design procedure for robust linear suboptimal regulators with preassigned trajectory insensitivity. CDC-Conference, Florida 1982

10.20 Verde, M.C.: Empfindlichkeitsreduktion bei linearen optimalen Regelungen. Diss. GH Duisburg 1983

10.21 Kalman, R.E.: When is a linear system optimal? Trans. ASME, J. Basic Eng. 86 (1964) 51−60

10.22 Safonov, M.G.; Athans, M.: Gain and phase margin for multivariable LQR Regulators. IEEE Trans AC 22 (1977) 173−179

10.23 Safonov, M.G.: Stability and robustness of multivariable feedback systems. Boston: MIT-Press, 1980

10.24 Frank, P.M.: Entwurf parameterunempfindlicher und robuster Regelkreise im Zeitbereich-Definitionen, Verfahren und ein Vergleich. Automatisierungstechnik 33 (1985) 233−240

10.25 Horowitz, I; Sidi, M.: Synthesis of cascaded multiple-loop feedback systems with large plantparameter ignorance. Automatica 9 (1973) 588−600

10.26 Ackermann, J.: Entwurfsverfahren für robuste Regelungen. Regelungstechnik 32 (1984) 143–150

10.27 Ackermann, J (Ed.): Uncertainty and control. Lecture Notes 76. Berlin: Springer-Verlag (1985)

10.28 Tolle, H.: Mehrgrößen-Regelkreissynthese, Bd. I und II, München: Oldenbourg 1983 und 1985

10.29 Kofahl, R.: Robustheitsanalyse zeitdiskreter, optimaler Zustandsregler. Interner Ber. Inst. f. Regelungstechnik, T.H. Darmstadt (1984)

10.30 Bux, D.: A new closed solution design of constant feedback control for systems with large parameter variations. 3rd IFAC-Symp. on Sensitivity, Adaptivity and Optimality, Ischia, Haly (1973), Inst. Soc. Am.

10.31 Bux, D.: Anwendung und Entwurf konstanter, linearer Zustandsregler bei linearen Systemen mit langsam veränderlichen Parameter. Diss. Univ. Stuttgart. Fortschritt Ber. VDI – Z. Reihe 8 Nr. 21. Düsseldorf: VDI-Verlag 1975

10.32 Isermann, R.; Eichner M.: Über die Lastabhängigkeit der Dampftemperatur-Regelung des Mehrgrößen-Regelsystems „Trommelkessel". Brennstoff, Wärme, Kraft 20 (1968) 453–459

Index

J. Ackermann

Sampled-Data Control Systems

Analysis and Synthesis, Robust System Design

1985. 152 figures. XIV, 596 pages.
(Communications and Control Engineering
Series).
ISBN 3-540-15610-0

Contents: Introduction. – Continuous Systems. –
Modelling and Analysis of Sampled-Data Systems. –
Controllability, Choice of Sampling Period and Pole
Assignment. – Observability and Observers. –
Control Loop Synthesis. – Geometric Stability
Investigation and Pole Region Assignment. – Design
of Robust Control Systems. – Multivariable
Systems. – Appendix A: Canonical Forms and
Further Results from Matrix Theory. – Appendix B:
The z-Transform. – Appendix C: Stability Criteria. –
Appendix D: Application Examples. – Literature. –
Index.

Springer-Verlag
Berlin Heidelberg New York
London Paris Tokyo Hong Kong

Lecture Notes in Control and Information Sciences

Editors: M. Thoma, A. Wyner

Volume 107: **Y. T. Tsay, L.-S. Shieh, S. Barnett**

Structural Analysis and Design of Multivariable Control Systems

An Algebraic Approach
1988. VI, 208 pages. ISBN 3-540-18916-5

Volume 108: **K. J. Reinschke**

Multivariable Control – A Graph-theoretic Approach

1988. 274 pages. In cooperation with
Akademie Verlag, Berlin (GDR).
ISBN 3-540-18899-1

Volume 109: **M. Vukobratović, R. Stojić**

Modern Aircraft Flight Control

1988. 86 figures. VIII, 288 pages. ISBN
3-540-19119-4

Volume 111: **A. Bensoussan, J. L. Lions** (Eds.)

Analysis and Optimization of Systems

1988. 171 figures. XIV, 1175 pages. (172 pages
in French). ISBN 3-540-19237-9

Volume 112: **V. Kecman**

State-Space Models of Lumped and Distributed Systems

1988. 94 figures. IX, 280 pages.
ISBN 3-540-50082-0

Volume 113: **M. Iri, K. Yajima** (Eds.)

System Modelling and Optimization

Proceedings of ther 13th IFIP Conference,
Tokyo, Japan, August 31–September 4, 1987
1988. 236 figures. X, 787 pages.
ISBN 3-540-19238-7

Springer

Mathematics of Control, Signals, and Systems

ISSN 0932-4194   Title No. 498

Managing Editors: B. W. Dickinson, Princeton University, and E. D. Sontag, Rutgers University

Associate Editors: K. J. Astrom, B. R. Barmish, R. W. Brockett, G. Cybenko, M. Fliess, B. A. Francis, Y. Genin, K. Glover, G. C. Goodwin, R. M. Gray, A. Isidori, B. Jakubczyk, T. Kailath, R. E. Kalman, P. Khargonekar, H. Kimura, P. R. Kumar, A. J. Laub, A. Manitius, S. I. Marcus, S. K. Mitter, H. V. Poor, W. J. Rugh, H. J. Sussman, J. C. Willems, W. M. Wonham, G. Zames

The main purpose of **Mathematics of Control, Signals, and Systems** is to provide a scholarly, international forum for the publication of original research in the areas of mathematical systems theory, control theory, and signal processing. In recent years there has been burgeoning interest in the study of mathematical problems in engineering – particularly in those areas often included under the broad heading of systems, signals, and control. This truly interdisciplinary field involves the combined efforts of mathematicians and engineers from many different backgrounds – from abstract algebra and differential geometry to signal processing and robotics. **MCSS** will feature full-length, original research papers, but will also publish applications of nontrivial theoretical results and expository papers of exceptional merit. The common thread of all papers will be systems, with emphasis on high-quality, mathematically rigorous contributions.

The scope of **MCSS** will encompass, e.g., algebraic and geometric aspects of linear and nonlinear control systems, stochastic control, stability, identification and estimation, robust and adaptive systems, numerical and computational techniques, large-scale systems, decentralized and hierarchical control, dynamic optimization, and optimal control.

Springer-Verlag
Berlin Heidelberg
New York London
Paris Tokyo
Hong Kong

Springer

CPSIA information can be obtained at www.ICGtesting.com
Printed in the USA
LVOW03s0702191014

409431LV00004B/54/P

9 783642 864193